IEE ELECTROMAGNETIC WAVES SERIES 45

Series Editors: Professor P. J. B. Clarricoats
Professor E. V. Jull

Parabolic equation methods for electromagnetic wave propagation

Other volumes in this series:

Volume 10 **Aperture antennas and diffraction theory** E. V. Jull
Volume 11 **Adaptive array principles** J. E. Hudson
Volume 12 **Microstrip antenna theory and design** J. R. James, P. S. Hall and C. Wood
Volume 13 **Energy in electromagnetism** H. G. Booker
Volume 14 **Leaky feeders and subsurface radio communications** P. Delogne
Volume 15 **The handbook of antenna design, Volume 1** A. W. Rudge, K. Milne, A. D. Olver, P. Knight (Editors)
Volume 16 **The handbook of antenna design, Volume 2** A. W. Rudge, K. Milne, A. D. Olver, P. Knight (Editors)
Volume 17 **Surveillance radar performance prediction** P. Rohan
Volume 18 **Corrugated horns for microwave antennas** P. J. B. Clarricoats and A. D. Olver
Volume 19 **Microwave antenna theory and design** S. Silver (Editor)
Volume 20 **Advances in radar techniques** J. Clarke (Editor)
Volume 21 **Waveguide handbook** N. Marcuvitz
Volume 22 **Target adaptive matched illumination radar** D. T. Gjessing
Volume 23 **Ferrites at microwave frequencies** A. J. Baden Fuller
Volume 24 **Propagation of short radio waves** D. E. Kerr (Editor)
Volume 25 **Principles of microwave circuits** C. G. Montgomery, R. H. Dicke, E. M. Purcell (Editors)
Volume 26 **Spherical near-field antenna measurements** J. E. Hansen (Editor)
Volume 27 **Electromagnetic radiation from cylindrical structures** J. R. Wait
Volume 28 **Handbook of microstrip antennas** J. R. James and P. S. Hall (Editors)
Volume 29 **Satellite-to-ground radiowave propagation** J. E. Allnutt
Volume 30 **Radiowave propagation** M. P. M. Hall and L. W. Barclay (Editors)
Volume 31 **Ionospheric radio** K. Davies
Volume 32 **Electromagnetic waveguides: theory and application** S. F. Mahmoud
Volume 33 **Radio direction finding and superresolution** P. J. D. Gething
Volume 34 **Electrodynamic theory of superconductors** S.-A. Zhou
Volume 35 **VHF and UHF antennas** R. A. Burberry
Volume 36 **Propagation, scattering and dissipation of electromagnetic waves** A. S. Ilyinski, G. Ya. Slepyan and A. Ya. Slepyan
Volume 37 **Geometrical theory of diffraction** V. A. Borovikov and B. Ye. Kinber
Volume 38 **Analysis of metallic antennas and scatterers** B. D. Popović and B. M. Kolundžija
Volume 39 **Microwave horns and feeds** A. D. Olver, P. J. B. Clarricoats, A. Kishk and L. Shafai
Volume 40 **Uniform stationary phase method** V. A. Borovikov
Volume 41 **Approximate boundary conditions in electromagnetics** T. B. A. Senior and J. L. Volakis
Volume 42 **Spectral theory and excitation of open structures** V. P. Shestopalov and Y. V. Shestopalov
Volume 43 **Open electromagnetic waveguides** T. Rozzi and M. Mongiardo
Volume 44 **Theory of nonuniform waveguides** B. Z. Katsenelenbaum, M. K. A. Thumm, L. Mercader Del Río, M. Pereyaslavets and M. Sorolla Ayza

A complete list of publications in this series may be obtained from the IEE

Parabolic equation methods for electromagnetic wave propagation

Mireille Levy

The Institution of Electrical Engineers

Published by: The Institution of Electrical Engineers, London, United Kingdom

The Institution of Electrical Engineers,
Michael Faraday House,
Six Hills Way, Stevenage,
Herts. SG1 2AY, United Kingdom

British Library Cataloguing in Publication Data

A CIP catalogue record for this book
is available from the British Library

ISBN 0 85296 764 0

Printed in England by TJ International, Padstow, Cornwall

Contents

Preface

Parabolic equation techniques have been used extensively in radiowave propagation modelling since the mid-1980s and have become the dominant tool for assessing clear-air effects on tropospheric links. They are also increasingly applied to smaller-scale scattering problems such as propagation in urban environments or radar-cross-section evaluation.

Although excellent texts on the application of parabolic equation methods to underwater acoustics are available, there is no systematic exposition in book form of their use for radiowave problems. In 1995, Jim Wait suggested that a monograph summarizing parabolic equation developments for electromagnetic applications would be useful. This book is the result of my attempt to carry out this task. Unfortunately Professor Wait died before completion of the book. I would like to express my deep gratitude for his support and his help with material on mode theory.

This book could not have been written without the crucial work of Jim Kuttler and Dan Dockery, who developed much of the formal framework for the use of parabolic equation techniques in radiowave propagation modelling.

I am grateful to the many people who helped with the preparation of the book. In particular I would like to thank Alexander Popov and his colleagues for providing most of the material on X-ray diffraction in Section 13.3.

Special thanks are due to Ken Craig and Andrew Zaporozhets. Many of the results in this book represent joint work with these two colleagues, whose insights and support have been invaluable.

M.F. Levy
Oxford
October 1999

Chapter 1

Introduction

The parabolic equation (PE) is an approximation of the wave equation which models energy propagating in a cone centred on a preferred direction, the paraxial direction. The parabolic approximation was introduced by Leontovich and Fock [83, 49] in the 1940s to treat the problem of diffraction of radiowaves around the Earth. They used the parabolic approximation to derive more simply the well-known Watson [163] and Van der Pol and Bremmer [157] results, and then extended the method to more complicated cases involving atmospheric refraction. Still in the 1940s, Malyuzhinets combined the parabolic approximation method with geometrical optics to develop a powerful theory of diffraction by obstacles (see [104] for an English language summary of this work). Russian workers pioneered the idea of simplifying the wave equation for certain types of radiowave propagation problems and solved a number of these problems in terms of special functions.

It took many years and the advent of digital computers for the parabolic approximation idea to be taken up again, this time with the aim of finding numerical solutions rather than closed-form expressions. Hardin and Tappert [56] introduced the very efficient split-step/Fourier solution of the parabolic equation for underwater acoustics problems, and Claerbout [28] developed finite-difference codes for geophysics applications. An excellent historical survey of early PE research can be found in [149]. Many workers in the field of underwater acoustics took up the PE method and huge advances were made [66]. Scalar PE techniques have become widely used in a variety of fields, like waterwave propagation [131, 42], optics [48, 168] or seismic wave propagation [32]. Vector PE methods have been derived to deal with more complicated problems such as acoustic wave propagation in poro-elastic media [36] or optics in anisotropic waveguides [169].

Most of the book is concerned with the application of PE methods to tropospheric propagation problems. Radiowaves in the VHF to millimetric range are affected by atmospheric refraction, and by terrain diffraction and reflection effects. Because of the very large size of the domains of interest relative to the wavelength, it is not possible to compute exact solutions of Maxwell's equations, and approximations must be sought. For many years, solutions were found in terms of geometrical optics and mode theory for refraction problems, and in terms of geometrical theory of diffraction for terrain problems. These have now largely been superseded by PE algorithms, which provide a fast and efficient numerical solution to most long-range radiowave propagation problems.

Ko *et al.* [73] introduced the use of the PE method to compute field strength in anomalous propagation environments. Quite quickly, the technique was taken up by a number of researchers, who developed efficient split-step/Fourier codes [40, 45, 41, 77, 46] and finite-difference algorithms [50]. At first PE methods were applied to propagation over the sea or flat terrain, but it was soon realized that they could also deal with propagation over irregular terrain [85, 107, 101, 11].

Initially the novel propagation tools based on PE models were extremely well received, since they could solve real-life problems which had remained intractable until then. However once the users became accustomed to the idea of computing radar coverage diagrams or path loss curves for complicated environments, the demand for faster algorithms began to grow. This led to the development of hybrid models, which combine ray-tracing and parabolic equation techniques to obtain fast solutions of very large radiowave propagation problems. This idea was pioneered by Hitney [58] with the Radio Physical Optics model, and is the basis for most current radar propagation assessment tools [93, 125].

In recent years, it was realized that PE methods may also be applied to object scattering problems, bridging the gap between rigorous methods like the method of moments or FDTD, which solve Maxwell's equations exactly, and approximate methods based on ray-tracing or physical optics. The last chapters of the book are devoted to object scattering applications of the PE.

This book does not attempt to present an exhaustive summary of parabolic equation developments relating to electromagnetic wave propagation. Rather, I have tried to describe the building blocks which are necessary to construct efficient PE models. For long-range radiowave propagation, PE methods have reached sufficient maturity to make such an organized description possible. Object scattering applications are more recent, and the exposition there is by necessity less systematic.

Most of the book is concerned with the two-dimensional scalar parabolic wave equation, to which many radiowave problems can be reduced. Chapter 2 presents the paraxial framework for the two-dimensional scalar wave equation and shows how PE solutions relate to the Fresnel-Kirchhoff approximation for classical diffraction problems. Chapter 3 describes the most commonly used PE algorithms, providing detailed derivations of the split-step/Fourier sine transform technique and of the Crank-Nicolson finite-difference scheme.

Chapter 4 is concerned with deriving parabolic equations suitable for tropospheric propagation from Maxwell's equations, using flattened Earth coordinates. In Chapter 5, we outline the basic principles of ray-tracing and mode theory. It is worth getting a flavour of how these techniques work, since for many years they were the only ones available. Beyond this, there is also a practical aspect. Although ray-tracing is no longer used on its own to compute the coverage of a transmitter, it is an essential part of the fast hybrid models which are presented later in the book. Also, models based on mode theory are crucial for validating PE methods in transhorizon propagation applications for which nothing else is available.

Chapter 6 is devoted to radiowave propagation over the sea, giving a number of examples of ducting propagation, together with comparisons with ray-tracing and mode theory results. Chapter 7 describes irregular terrain modelling with PE methods. Chapter 8 discusses domain truncation. Chapter 9 is concerned with the modelling of boundary conditions at the air/ground interface. Chapter 10 applies PE methods to the modelling of propagation over the rough sea. Chapter 11 presents hybrid models which provide considerable speed-up in coverage calculations.

The last three chapters are concerned with object scattering. Chapter 12 gives the building blocks for two-dimensional scattering applications. Scalar three-dimensional problems are treated in Chapter 13, which presents applications to X-ray diffraction and to building scatter. The vector PE method, which solves a paraxial version of Maxwell's equations, is discussed in Chapter 14 .

Some of the more technical mathematical derivations are given in the Appendices. Appendix A presents material on Airy functions. Appendix B gives far-field expressions, which are needed both for source modelling in long-range applications and for bistatic radar-cross-section calculations in scattering applications. Appendix C gives a mathematical framework for modal expansions. Finally, Appendix D proves the energy conservation results which are essential in the derivation of the vector PE.

Chapter 2

Parabolic equation framework

2.1 Introduction

In all that follows, we assume $\exp(-i\omega t)$ time-dependence of the fields, where ω is the angular frequency. Initially we work with Cartesian coordinates (x, y, z). In this chapter, we are concerned with two-dimensional electromagnetic problems where the fields are independent of the transverse coordinate y. There are then no depolarization effects, and all the fields can be decomposed into horizontally and vertically polarized components propagating independently.

For horizontal polarization, the electric field $\mathbf{E}$ only has one non-zero component E_y, while for vertical polarization, the magnetic field $\mathbf{H}$ only has one non-zero component H_y. We work with the appropriate field component ψ defined by

$$\psi(x, z) = E_y(x, z) \tag{2.1}$$

for horizontal polarization and

$$\psi(x, z) = H_y(x, z) \tag{2.2}$$

for vertical polarization.

We are going to solve the wave equation in a domain where the refractive index $n(x, z)$ has smooth variations, assuming that suitable boundary conditions can be defined at the domain boundaries. Typically the bottom boundary is the air/ground interface and the top boundary extends to infinity. We are interested in solving problems where energy propagates at small angles from a preferred direction, called the paraxial direction. Following the convention in radiowave propagation problems, we choose the positive x-direction as the paraxial direction. A schematic tropospheric propagation example is shown in

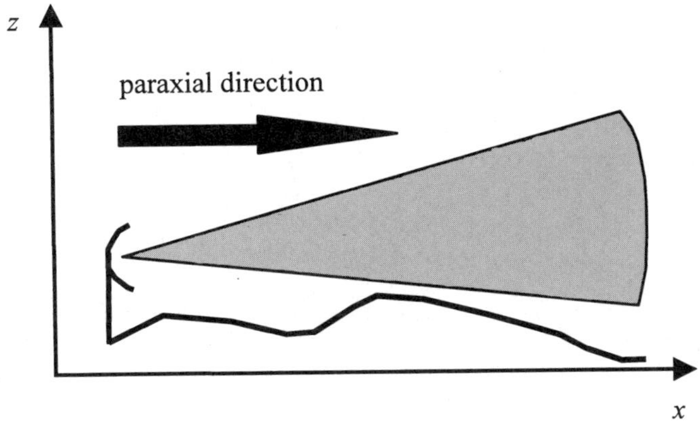

Figure 2.1 *Paraxial propagation in the troposphere, modelling energy propagating at angles close to the horizontal*

Figure 2.1. The propagation directions to be modelled constitute the paraxial cone.

If the propagation medium is homogeneous with refractive index n, the field component ψ satisfies the two-dimensional scalar wave equation

$$\frac{\partial^2\psi}{\partial x^2} + \frac{\partial^2\psi}{\partial z^2} + k^2 n^2 \psi = 0 \tag{2.3}$$

where k is the wave number in vacuum. In general, the refractive index varies with range x and height z, and eqn. 2.3 is not exact [25]. It is however a good approximation provided the variations of n remain slow on the scale of a wavelength.

2.2 Basic derivation

2.2.1 Paraxial wave equation

We introduce the reduced function associated with the paraxial direction x

$$u(x,z) = \mathrm{e}^{-ikx}\psi(x,z) \tag{2.4}$$

The point of using this reduced function is that it is slowly varying in range for energy propagating at angles close to the paraxial direction, which gives it convenient numerical properties.

The scalar wave equation in terms of u is

$$\frac{\partial^2 u}{\partial x^2} + 2ik\frac{\partial u}{\partial x} + \frac{\partial^2 u}{\partial z^2} + k^2\left(n^2 - 1\right)u = 0 \tag{2.5}$$

This can be formally factored as

$$\left\{\frac{\partial}{\partial x} + ik(1 - Q)\right\}\left\{\frac{\partial}{\partial x} + ik(1 + Q)\right\}u = 0 \tag{2.6}$$

where the pseudo-differential operator Q is defined by

$$Q = \sqrt{\frac{1}{k^2}\frac{\partial^2}{\partial z^2} + n^2(x, z)} \tag{2.7}$$

Pseudo-differential operators are constructed from partial derivatives and ordinary functions of the variables [150]. A formal mathematical framework is required to give a precise meaning to the square-root symbol in the expression of Q. The square-root corresponds to the composition of operators, in the sense that

$$Q\left(Q(u)\right) = \frac{1}{k^2}\frac{\partial^2 u}{\partial z^2} + n^2 u \tag{2.8}$$

must be satisfied for all functions u in a certain class. The construction of the appropriate square-root symbol is linked to the class of functions u on which it operates, and this in turn depends on the boundary conditions for the partial differential equation given by eqn. 2.5. In Section 2.4.3, we show how to construct Q for propagation in vacuum. More generally, we shall assume that Q can be defined unambiguously and that expansions for the ordinary square root function can be applied to Q.

There are errors inherent in the factorization given in eqn. 2.6: if the refractive index n varies with range x, the operator Q does not commute with the range derivative and the factorization is incorrect. Hence some care must be taken in the applications to make sure that the resulting error remains small.

The next step is to split the wave equation into the two terms defined by eqn. 2.6 and to look at functions satisfying one of the resulting pseudo-differential equations

$$\frac{\partial u}{\partial x} = -ik(1 - Q)u \tag{2.9}$$

$$\frac{\partial u}{\partial x} = -ik(1 + Q)u \tag{2.10}$$

Eqns. 2.9 and 2.10 correspond respectively to forward and back propagating waves. In a ray picture, forward propagation corresponds to rays propagating with increasing x, and backward propagation to rays propagating with decreasing x. Eqns. 2.9 and 2.10 are the outgoing and incoming parabolic wave equations.

In a range-independent medium where there are no commutator problems with the factorization of eqn. 2.6, a solution of either eqn. 2.9 or eqn. 2.10 will automatically satisfy the original reduced wave equation, eqn. 2.5. However such a solution does not in general correspond to the actual electromagnetic field. For example a solution of the outgoing equation 2.9 neglects the backscattered field. In order to get exact solutions of eqn. 2.5, eqns. 2.9 and 2.10 should be solved simultaneously in the coupled system

$$\begin{cases} u & = & u_+ + u_- \\ \dfrac{\partial u_+}{\partial x} & = & -ik(1-Q)u_+ \\ \dfrac{\partial u_-}{\partial x} & = & -ik(1+Q)u_- \end{cases} \tag{2.11}$$

The approximation we make by solving for each term separately is a paraxial approximation: for example, for the outgoing parabolic wave equation, we solve for energy propagating in a paraxial cone centred on the positive x-direction.

Eqns. 2.9 and 2.10 are pseudo-differential equations of first order in x (hence the "parabolic" terminology). They can be solved by marching techniques, given the field on the initial vertical and the boundary conditions at the top and bottom of the domain. The outgoing parabolic wave equation, eqn. 2.9, has the formal solution

$$u\,(x+\Delta x,.) = e^{ik\Delta x(-1+Q)}u\,(x,.) \tag{2.12}$$

The forward propagating field is obtained at a given range from the field at a previous range, and appropriate boundary conditions at the top and bottom of the domain, in other words the solution is marched in range. The computational gain is substantial compared to the elliptic wave equation, which is of second order in both x and z and must be solved simultaneously at all points of the integration domain. This is illustrated in Figure 2.2.

The splitting of the wave equation into two paraxial terms implies that only energy propagating inside a paraxial cone can be modelled. This limits the type of propagation which can be represented accurately. Consider for example a plane wave incident on a small perfectly conducting cylinder of diameter d, as shown in Figure 2.3. It will set off a surface wave travelling all around the

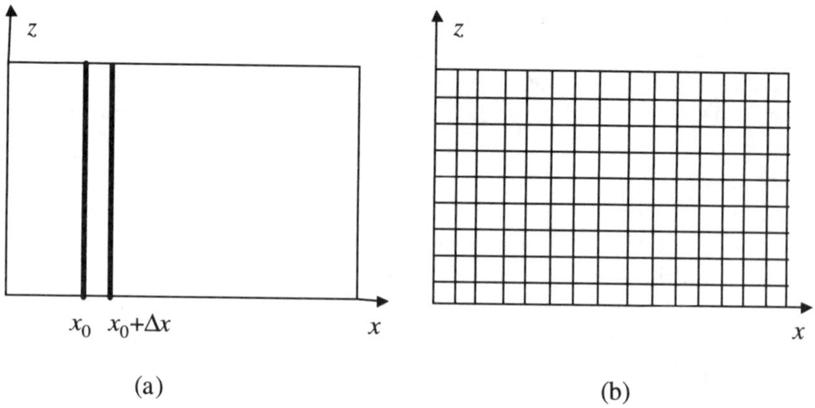

Figure 2.2 *Solution of (a) parabolic and (b) elliptic partial differential equations*

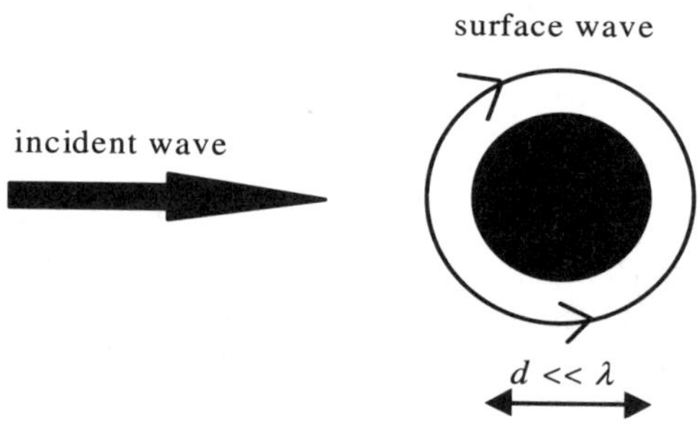

Figure 2.3 *Surface wave excited on perfectly conducting cylinder by incident electromagnetic field*

scatterer, and this surface wave is not correctly represented by the splitting. Correct representation involves solving the system of eqns. 2.11. However the paraxial representation is very accurate for a number of problems and also has substantial computational advantages.

The PE derived in this section provides a paraxial approximation of the two-dimensional scalar wave equation. Naturally the scalar framework is no longer adequate for general three-dimensional problems, where polarization aspects require a vector description. In that case scalar PEs have to be written for the electromagnetic field components and coupled through boundary conditions at interfaces and the divergence-free conditions. The vector PE framework is described in Chapter 14.

2.2.2 Cylindrical coordinates

For some radiowave propagation problems, it is convenient to work with cylindrical coordinates (x, θ, z), where x is now the distance from the cylindrical axis, and to assume azimuthal symmetry. If Ψ is an electromagnetic field component (transverse electric field E_θ for horizontal polarization, transverse magnetic field H_θ for vertical polarization), then Ψ satisfies the scalar wave equation in cylindrical coordinates

$$\frac{\partial^2 \Psi}{\partial x^2} + \frac{1}{x}\frac{\partial \Psi}{\partial x} + \frac{\partial^2 \Psi}{\partial z^2} + k^2 n^2 \psi = 0 \tag{2.13}$$

The appropriate reduced function u is given by

$$u(x, z) = \sqrt{kx}\,\mathrm{e}^{-ikx}\Psi(x, z) \tag{2.14}$$

The square-root factor takes cylindrical spreading into account. We rewrite eqn. 2.13 in terms of u to get

$$\frac{\partial^2 u}{\partial x^2} + 2ik\frac{\partial u}{\partial x} + \frac{\partial^2 u}{\partial z^2} + k^2(n^2 - 1)u = -\frac{1}{4x^2}u \tag{2.15}$$

or, in terms of the dimensionless coordinates $\xi = kx$ and $\zeta = kz$,

$$\frac{\partial^2 u}{\partial \xi^2} + 2i\frac{\partial u}{\partial \xi} + \frac{\partial^2 u}{\partial \zeta^2} + k^2(n^2 - 1)u = -\frac{1}{4\xi^2}u \tag{2.16}$$

We now make the far-field assumption $kx \gg 1$ and obtain the same reduced wave equation, eqn. 2.5, as in Cartesian coordinates. From then on, the derivation proceeds exactly as before. The far-field assumption is required to eliminate the singularity produced by the cylindrical source, which of course was not present in the Cartesian two-dimensional case. For radiowave propagation applications, near-field calculations are not normally required, so that the far-field assumption in the cylindrical framework is not a restriction to the applicability of the PE.

The underlying reason for using the reduced function 2.14 is that, in vacuum, the wave equation in cylindrical coordinates (eqn. 2.13) has the canonical outgoing solution

$$v(x, z) = H_0^{(1)}(kx) \tag{2.17}$$

where $H_0^{(1)}$ is the Hankel function of the first kind of order 0. In the far-field, the Hankel function is equivalent to the first term of its asymptotic expansion

$$H_0^{(1)}(kx) \sim \sqrt{\frac{2}{\pi kx}} e^{-\frac{i\pi}{4}} e^{ikx} \tag{2.18}$$

which corresponds up to a constant factor to the representation in eqn. 2.14. For solutions propagating at small angles from the horizontal, u is a slowly varying function of x. Putting

$$\psi(x, z) = \sqrt{kx}\, \Psi(x, z) \tag{2.19}$$

so the function ψ takes cylindrical spreading into account, but does not involve any phase modification, we have

$$\frac{\partial^2 \psi}{\partial x^2} + \frac{\partial^2 \psi}{\partial z^2} + k^2 n^2 \psi = -\frac{1}{4x^2}\psi \tag{2.20}$$

Hence, in the far-field, ψ satisfies approximately the usual two-dimensional wave equation, eqn. 2.3. We shall see in Chapter 4 that with a suitable choice of coordinates and a modified refractive index, the two-dimensional wave equation may also be used to model tropospheric radiowave propagation.

2.3 Approximations of the square-root operator

2.3.1 The standard parabolic equation

The simplest approximation of eqn 2.9 is obtained by using first-order Taylor expansions of the square-root and exponential functions. This yields the standard parabolic equation (SPE)

$$\frac{\partial^2 u}{\partial z^2}(x, z) + 2ik\frac{\partial u}{\partial x}(x, z) + k^2(n^2(x, z) - 1)u(x, z) = 0 \tag{2.21}$$

This simple form of the parabolic equation is extremely useful for solving long-range tropospheric radiowave propagation problems. The refractive index of the air is very close to unity and does not cause significant accuracy problems. The limitations of the standard PE approximation are due to its bad behaviour at large propagation angles: for a plane wave propagating at angle α from the horizontal, the first neglected term in the Taylor expansion is proportional to

$$\frac{1}{k^2}\left|\frac{\partial^2 u}{\partial z^2}\right| = \sin^2 \alpha \tag{2.22}$$

Hence the error is proportional to $\sin^4 \alpha$, going from 10^{-7} for an angle of 1° to 10^{-3} for an angle of 10° and over 10^{-2} for an angle of 20°. This shows that the SPE is a narrow-angle approximation of the parabolic wave equation. As propagation angles of interest in tropospheric problems are usually less than a few degrees, accuracy is very good for long-range calculations. We shall see in later chapters that the simplicity of the SPE makes it amenable to numerical solution by either split-step Fourier techniques or finite-difference methods, whereas the more sophisticated schemes described below cannot be implemented with Fourier techniques.

2.3.2 Overview of wide-angle schemes

For problems involving large propagation angles, more accurate expansions of the operator Q are required. A strategy that readily comes to mind is to use higher order polynomial expansions of the square-root; unfortunately this produces instability in numerical schemes [4, 5]. The right approach is to use rational approximations instead. This idea was introduced by Claerbout [28]. The Claerbout solution involves a Padé-(1,1) approximant of the form

$$\sqrt{1+Z} = \frac{1+aZ}{1+bZ} \tag{2.23}$$

where the operator Z is defined by

$$Z = \frac{1}{k^2}\frac{\partial^2}{\partial z^2} + n^2(x,z) - 1 \tag{2.24}$$

The choice of $a = 0.75, b = 0.25$ ensures that both sides of eqn. 2.23 have the same Taylor expansion of order 2. The error for a plane wave propagating at angle α from the horizontal is of the order of $(\sin\alpha)^6$, which makes it acceptable for propagation angles up to 45 degrees or so from the paraxial direction. Another way to look at the distortion introduced by this approximation of Z is to write the Claerbout equation as

$$\left(1+\frac{Z}{4}\right)\frac{\partial u}{\partial x} - ik\frac{Z}{2}u = 0 \tag{2.25}$$

and to look at the error made for plane wave solutions of the original equation in vacuum. If u is a reduced plane wave of the form $\exp\{ikx(1-\cos\theta) + ikz\sin\theta\}$, then

$$\left|\left(1+\frac{Z}{4}\right)\frac{\partial u}{\partial x} - ik\frac{Z}{2}u\right| = 2k\left(\sin\frac{\theta}{2}\right)^6 \tag{2.26}$$

The same analysis for the SPE gives

$$\left|\frac{\partial u}{\partial x} - ik\frac{Z}{2}u\right| = 2k\left(\sin\frac{\theta}{2}\right)^4 \tag{2.27}$$

which confirms the superior accuracy of the Claerbout approximation.

A very successful approach for modelling very wide propagation angles is the split-step Padé parabolic equation method pioneered by Collins [34]. The idea here is to approximate the exponential operator which appears in the marching solution given in eqn. 2.12 directly, rather than the square-root operator. This is done by using a Padé-(1,1) expansion in the form

$$e^{ik\Delta x\left(-1+\sqrt{1+Z}\right)} \sim 1 + \sum_{l=1}^{N}\frac{a_l Z}{1+b_l Z} \tag{2.28}$$

The $2N$ complex coefficients (a_l, b_l) are determined by Taylor expansion constraints and stability conditions. One important feature is their dependence on the range step Δx: the accuracy of the solution is improved by tayloring the Padé expansion to the discretization. The Padé expansion of eqn. 2.28 yields the split-step Padé solution

$$u\left(x+\Delta x, z\right) = u\left(x, z\right) + \sum_{l=1}^{N}\left(1+b_l Z\right)^{-1} a_l Z u\left(x, z\right) \tag{2.29}$$

which only involves second-order derivatives. This equation is marched in range by solving separately for each term in the sum. Since there is no interaction between the terms, the method lends itself quite well to parallel computing. As the number N of terms increases, so does the angular cone of validity. For example, with $N = 8$, propagation angles of up to 70 degrees can be modelled. The method is very stable and allows the use of large range steps.

This brief overview of wide-angle methods only gives a first idea of the possibilities for approximating the one-way wave equation, eqn. 2.9. Many other approximations are available, the optimal choice depending on the angular sector of interest and on the type of boundary condition to be modelled. We shall look at some of these approximations in more detail in later chapters.

2.4 Propagation in vacuum

2.4.1 Angular spectrum

We consider the special case of propagation in vacuum. This enables us to introduce spectral techniques for solving the parabolic equation, and to understand the method in terms of plane wave decomposition of the field [29, 115]. Fourier transform techniques provide a very powerful solution for certain types of partial differential equations. The idea is to transform the partial differential equation, solve the transformed equation (which is hopefully simpler than the original one if the transform is well chosen) in the spectral domain, and get back to the original domain with an inverse transform.

We introduce the Fourier transform $\mathcal{F}$ of a sufficiently well-behaved function $u(x, z)$ with respect to height. Putting $U = \mathcal{F}u$, we have

$$U(x,p) = \mathcal{F}u(x,p) = \int_{-\infty}^{+\infty} u(x,z)\mathrm{e}^{-2i\pi pz}dz \tag{2.30}$$

The inverse Fourier transform $\mathcal{F}^{-1}$ is given by

$$u(x,z) = \int_{-\infty}^{+\infty} U(x,p)\mathrm{e}^{2i\pi pz}dp \tag{2.31}$$

where $u = \mathcal{F}^{-1}U$. The appropriate mathematical framework for our use of Fourier transforms is that of generalized functions, or Schwartz distributions [146]. This allows formal manipulation of expressions which involve non-integrable functions. For example there is no difficulty dealing with the Fourier transform of $u(x,z) = \mathrm{e}^{ik\sin\alpha z}$ which is expressed in terms of Dirac delta functions.

We shall make use of the fact that if the function $u(x, z)$ is sufficiently regular, the Fourier transform of its second-order partial z-derivative is given by

$$\mathcal{F}\left(\frac{\partial^2 u}{\partial z^2}\right) = -4\pi p^2 \mathcal{F}u \tag{2.32}$$

whereas the Fourier transform of its partial x-derivative is given by

$$\mathcal{F}\left(\frac{\partial u}{\partial x}\right) = \frac{\partial\left(\mathcal{F}u\right)}{\partial x} \tag{2.33}$$

We note that the proof of eqn. 2.32, which uses integration by parts, requires $u(x, z)$ to be twice continuously differentiable in z on the whole real line. For problems involving an air/ground interface, the function $u(x, z)$ is only defined for z above ground level, and it is not possible in general to extend it to the whole real line in such a way that boundary conditions at the ground are automatically satisfied. We shall see in Chapter 3 that the Fourier transform can be replaced by a sine transform for problems involving a perfectly conducting surface. For more general problems, appropriate transforms must be defined which will take the boundary conditions at the ground into account (see Chapter 9).

We can look at the inverse Fourier transform expression given by eqn. 2.31 as the decomposition of the field $u(x, z)$ on a vertical into an angular spectrum $U(x, p)$ of plane waves depending on the spectral variable p. To get some insight into the physical meaning of p, consider a plane wave of the form

$$\psi(x, z) = \exp(ikx \cos\alpha + ikz \sin\alpha) \tag{2.34}$$

of amplitude 1 and propagating at angle α from the horizontal. The corresponding reduced function u is given by

$$u(x, z) = \exp\{ikx(\cos\alpha - 1) + ikz \sin\alpha\} \tag{2.35}$$

and its Fourier transform is

$$U(x, p) = \exp\{ikx(\cos\alpha - 1)\}\ \delta(p - \lambda \sin\alpha) \tag{2.36}$$

where δ is the Dirac function at the origin. In other words the spectral variable p corresponds to a plane wave propagating at an angle α from the horizontal given by

$$\sin\alpha = \frac{p}{\lambda} \tag{2.37}$$

For p greater than the wavelength λ, the angle α is complex and such that

$$\cos\alpha = i\sqrt{\sin^2\alpha - 1} \tag{2.38}$$

which ensures that the corresponding plane wave is evanescent. Finally the Fourier transform pair $\{\mathcal{F}, \mathcal{F}^{-1}\}$ allows us to switch between the field in the spatial domain and its angular spectrum representation in the spectral domain. We shall now solve the parabolic equation in vacuum by determining how the angular spectrum $U(x, p)$ evolves with range.

2.4.2 SPE in vacuum

If the background propagation medium is vacuum, the standard parabolic equation becomes

$$\frac{\partial^2 u}{\partial z^2}(x, z) + 2ik\frac{\partial u}{\partial x}(x, z) = 0 \tag{2.39}$$

Using the properties of the Fourier transform given by eqns. 2.32 and 2.33, we can write the Fourier transform of eqn. 2.39 as

$$-4\pi^2 p^2 U(x,p) + 2ik\frac{\partial U}{\partial x}(x,p) = 0 \tag{2.40}$$

This is now an ordinary differential equation which can be solved in closed form, giving

$$U(x,p) = \mathrm{e}^{-\frac{2i\pi^2 p^2 x}{k}} U(0,p) \tag{2.41}$$

Now it turns out that the inverse Fourier transform of the exponential propagator $\exp(-2i\pi^2 p^2 x/k)$ is given by

$$\mathcal{F}^{-1}\left\{\mathrm{e}^{-\frac{2i\pi^2 p^2 x}{k}}\right\}(x,z) = \sqrt{\frac{1}{\lambda x}}\mathrm{e}^{-\frac{i\pi}{4}}\mathrm{e}^{\frac{ikz^2}{2x}} \tag{2.42}$$

where λ is the wavelength. This is obtained by extending the well-known derivation for the integral of a Gaussian function $\mathrm{e}^{-\alpha t^2}$ where α is a positive real number to the case where α is complex with a non-negative real part. Getting back to the original domain, we obtain the convolution formula

$$u(x,z) = \sqrt{\frac{1}{\lambda x}}\mathrm{e}^{-\frac{i\pi}{4}} \int_{-\infty}^{+\infty} u(0,z')\mathrm{e}^{\frac{ik(z-z')^2}{2x}}\,dz' \tag{2.43}$$

This equation shows that the solution at any range is entirely determined by the initial reduced field $u(0,z)$. The only constraint on the initial field $u(0,z)$ is that it should not violate the SPE constraints, in other words it should only have narrow-angle spectral components.

The fact that we obtain a closed-form expression for the solution is specific to the case of propagation in a homogeneous medium. In general integration of the PE has to be numerical, and it is not possible to jump directly to an arbitrary range. Instead of using the closed-form convolution formula of eqn. 2.43, we can write the inverse of eqn. 2.41 as

$$u(x,z) = \mathcal{F}^{-1}\left\{\mathcal{F}\mathrm{e}^{-\frac{2i\pi^2 p^2 x}{k}}(u(0,z))\right\} \tag{2.44}$$

This equation provides a numerical recipe for calculating the field: transform the initial field $u(0,z)$, multiply by the spectral propagator and Fourier transform back. Fast Fourier Transforms (FFTs) provide an efficient numerical implementation, but have the drawback that suitable filters must be applied to avoid aliasing. This problem will be examined in detail in Chapter 8.

2.4.3 Square-root operator

For propagation in vacuum, it is actually straightforward to construct the square-root operator Q using Fourier transform techniques. To do this, we consider the multiplication operator M in spectral space defined by

$$MU(x,p) = \sqrt{1 - \frac{4\pi^2 p^2}{k^2}}\,U(x,p) \tag{2.45}$$

where the square-root is given by

$$\begin{cases} \sqrt{1 - \dfrac{4\pi^2 p^2}{k^2}} & \text{for} \quad |p| \leq \dfrac{k}{2\pi} \\ \\ i\sqrt{\dfrac{4\pi^2 p^2}{k^2} - 1} & \text{for} \quad |p| > \dfrac{k}{2\pi} \end{cases} \tag{2.46}$$

We now define the square-root operator Q by

$$Qu = \mathcal{F}^1 \{M\mathcal{F}u\} \tag{2.47}$$

It is easy to see that the operator Q^2 corresponds in spectral space to multiplication by M^2, which means that Q satisfies eqn. 2.8. With this definition of Q, the Fourier transform of eqn. 2.9 is the ordinary differential equation

$$\frac{\partial U}{\partial x} + ik(1 - M)U = 0 \tag{2.48}$$

which has the solution

$$U(x,p) = U(0,p)\mathrm{e}^{ik(M-1)x} \tag{2.49}$$

This shows that our choice of square-root is the only one possible for an outgoing solution. Finally the solution can be written as

$$u(x,z) = \mathcal{F}^{-1}\left\{\mathrm{e}^{ik(M-1)x}\mathcal{F}u_0\right\} \tag{2.50}$$

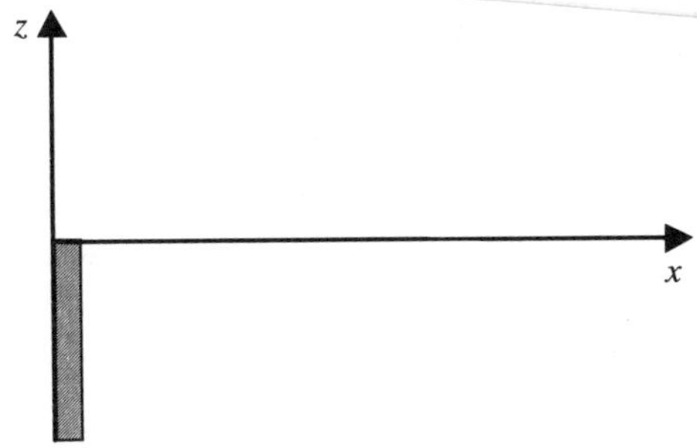

Figure 2.4 *Diffraction by a perfectly conducting half-plane screen*

where $u_0(z) = u(0, z)$ is the initial field. This time there are no restrictions on the angular spectrum of u_0, although u_0 should of course represent a forward propagating wave. The inverse Fourier transform of the exact paraxial propagator can be expressed in terms of a Hankel function

$$\mathcal{F}^{-1}\left\{e^{ik(M-1)x}\right\} = \frac{ikx}{2\sqrt{x^2+z^2}} e^{-ikx} H_1^{(1)}\left(k\sqrt{x^2+z^2}\right) \tag{2.51}$$

where $H_1^{(1)}$ is the Hankel function of the first kind of order 1. The proof can be found in Appendix B. This leads to a convolution expression similar to eqn. 2.43, which is used to derive near-field/far-field transformations in the paraxial framework (see Appendix B).

2.5 Analogy with Fresnel-Kirchhoff diffraction

In order to pinpoint the strengths and weaknesses of the paraxial framework, it is worth looking at the analogies with the Fresnel-Kirchhoff theory of diffraction [21, 29, 115]. Consider the case of an electromagnetic wave incident on a half-space diffraction screen located at zero range as shown in Figure 2.4.

The Fresnel-Kirchhoff formulation assumes that the field at some distance from the screen can be written as the integral of an angular spectrum of plane waves, weighted by a paraxial propagator. This is in fact equivalent to the parabolic equation formulation.

We assume for simplicity that the incident wave is a plane wave propagating horizontally. The reduced PE function u_i in the absence of the screen is then simply

$$u_i(x, z) = 1 \tag{2.52}$$

Neglecting edge current effects, we have at range zero

$$\begin{cases} u(0,z) & = & 1 \quad \text{for} \quad z > 0 \\ u(0,z) & = & 0 \quad \text{for} \quad z \leq 0 \end{cases} \tag{2.53}$$

Replacing with the reduced plane wave formulation in the convolution formula of eqn. 2.43, we get

$$u(x,z) = \sqrt{\frac{1}{\lambda x}} e^{-\frac{i\pi}{4}} \int_0^{+\infty} e^{\frac{ik(z-z')^2}{2x}} dz' \tag{2.54}$$

By shifting the integration variable, this becomes

$$u(x,z) = \sqrt{\frac{1}{\lambda x}} e^{-\frac{i\pi}{4}} \int_{-\infty}^{z} e^{\frac{ikz'^2}{2x}} dz' \tag{2.55}$$

In terms of the Fresnel integral

$$F(\nu) = \int_0^{\nu} e^{\frac{i\pi}{2}\tau^2} d\tau \tag{2.56}$$

the field can be written as

$$u(x,z) = \frac{1}{2} + \frac{\sqrt{2}}{2} e^{-\frac{i\pi}{4}} F(\nu) \tag{2.57}$$

where the dimensionless Fresnel parameter ν is given by

$$\nu = z\sqrt{\frac{1}{\lambda x}} \tag{2.58}$$

This is of course the well-known Fresnel formula for diffraction by a half-space screen [21]. Because the paraxial formulation neglects backscatter due to edge effects, it differs from the rigorous Sommerfeld expression [21]. In particular, it completely ignores polarization effects in the deep shadow zone. An interesting aspect of the Fourier transform formulation of eqn. 2.44 is that it provides an alternative way of computing Fresnel integrals through the use of FFTs.

Generalization of this analysis to an incident plane wave propagating at an angle α from the horizontal is straightforward, keeping α sufficiently small to satisfy the constraints of the SPE. The case of an arbitrary incident field can also be considered, provided the incident field does not violate the SPE constraints. A similar analysis could be conducted for an arbitrary screen, setting the initial field to zero at screened portions of the line $z = 0$. Extension of these ideas to three-dimensional screens leads to PE solution for optics problems [129].

Chapter 3

Parabolic equation algorithms

3.1 Introduction

In Chapter 2, we have seen how the parabolic equation could be interpreted in terms of Fourier transforms for the case of propagation in vacuum. The situation gets more complex when refractive index variations are present, since they cannot be dealt with directly with Fourier transform techniques. Section 3.2 introduces the powerful split-step/Fourier technique pioneered by Hardin and Tappert [56, 149], which replaces the original problem with propagation through a sequence of phase screens. At this stage the discussion is limited to the narrow-angle case, starting from the standard parabolic equation.

The numerical implementation of the split-step/Fourier solution depends on the boundary conditions to be satisfied at the top and bottom of the propagation domain. Section 3.3 treats the case where the field is zero on the bottom boundary, which for the moment is assumed to be flat. This simple set-up is actually crucial, since it covers the very important application of microwave propagation over a smooth sea. The solution is implemented with a discrete sine transform. More general split-step/Fourier methods will be presented in Chapter 9. Section 3.4 is devoted to a simple form of wide-angle PE approximation introduced by Thomson and Chapman [153], which is suitable for split-step Fourier implementation.

Split-step/Fourier methods are extremely attractive because of their numerical efficiency, but they lack flexibility for boundary modelling. For complicated boundaries, it is best to use finite-difference methods, which constitute the second main family of PE algorithms. The most commonly used finite-difference PE scheme is the Crank-Nicolson implementation of the SPE [79], which is presented in Section 3.5. This section also briefly describes the Crank-Nicolson implementation of the wider-angle Claerbout equation [28]. A much more detailed description of finite-difference PE methods is given in Chapter 12. The last section in this chapter is concerned with source modelling.

3.2 Split-step formulation of the SPE

We start with the standard parabolic equation given by eqn. 2.21. First we recast the SPE as

$$\frac{\partial u}{\partial x} = \frac{ik}{2}\left\{\frac{1}{k^2}\frac{\partial^2}{\partial z^2} + (n^2(x,z) - 1)\right\}u \tag{3.1}$$

This looks like an ordinary differential equation in range, except that the coefficient of u on the RHS is a differential operator in z. If we apply a Fourier transform in height z directly, we end up with a convolution term involving the Fourier transform of the refractive index profile, and there is no hope of a simple solution. Hence we must find a way to account separately for the action of the refractive index term.

3.2.1 The range-independent case

We start by assuming that the refractive index does not depend on range x. Then the solution can be formally expressed as

$$u(x + \Delta x, z) = e^{\delta(A+B)} \cdot u\,(x, z) \tag{3.2}$$

where we have used the notations

$$A = \frac{1}{k^2}\frac{\partial^2}{\partial z^2} \tag{3.3}$$

$$B = n^2(z) - 1 \tag{3.4}$$

$$\delta = \frac{ik\Delta x}{2} \tag{3.5}$$

Our goal will be achieved if we can split the two terms in the exponent and write the exponential as a product of factors containing A or B only. Indeed the action of factors in A can be expressed efficiently in terms of Fourier transforms, while factors in B act by simple multiplication. The simplest split is given by

$$S_1 = e^{\delta B}\, e^{\delta A} \tag{3.6}$$

This split is not exact for a non-homogeneous medium because the two operator terms in the exponent do not commute unless the refractive index is constant: if the refractive index n depends on height z, then

$$\frac{\partial^2 \{(n^2-1)u\}}{\partial z^2} \neq (n^2-1)\frac{\partial^2 u}{\partial z^2} \tag{3.7}$$

Now if the variations of the refractive index with height are relatively slow, the error incurred by splitting the exponential remains small. More precisely, we can define the commutator

$$[A, B] = AB - BA \tag{3.8}$$

From the definition of A and B, we have

$$[A, B]u = -\frac{1}{2k^2}\left(\frac{\partial^2 n^2}{\partial z^2}u + 2\frac{\partial n^2}{\partial z}\frac{\partial u}{\partial z}\right) \tag{3.9}$$

The error E caused by the split is

$$E = \mathrm{e}^{\delta B}\,\mathrm{e}^{\delta A} - \mathrm{e}^{\delta(A+B)} \tag{3.10}$$

If we expand the exponentials using Taylor series, we can express the error as a series in Δx. The dominant error term ϵ is given by

$$\epsilon = \frac{k^2}{8}(\Delta x)^2[A, B] \tag{3.11}$$

Hence the error is of second order in the range step. From eqn. 3.9, we see that it also depends on the height variations of the refractive index.

For this split, the split-step solution at the advanced range $x + \Delta x$ is given by

$$u(x+\Delta x, z) = \mathrm{e}^{\delta B} \cdot \left\{\mathrm{e}^{\delta A} \cdot u(x, z)\right\} \tag{3.12}$$

We can view the split-step solution as the field propagating through a series of phase screens, as shown in Figure 3.1. First we propagate the field through a slice of homogeneous medium represented by the A exponential. Then we apply a phase screen modulated by refractive index variations with the B exponential. The approximation made by replacing the original problem with the phase-screen problem decouples the diffractive effects (operator A) and the refractive effects (operator B), neglecting the effects of the commutator. The phase screens are orthogonal to the paraxial direction.

The split we have used here is the simplest possibility. An alternative split is given by

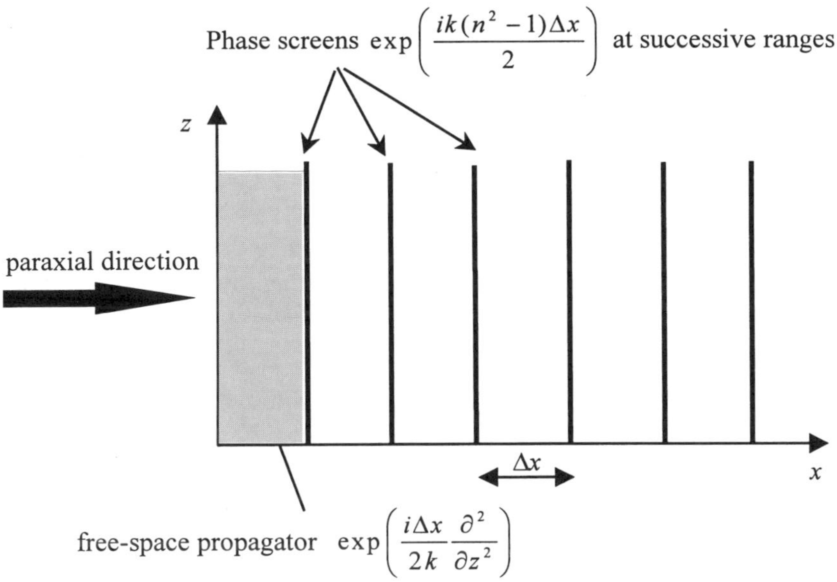

Figure 3.1 Phase-screen interpretation of split-step solution

$$S_2 = e^{\frac{1}{2}\delta B} e^{\delta A} e^{\frac{1}{2}\delta B} \tag{3.13}$$

The advantage of this second option is that it leads to an error of third order in Δx, while adding little computational overhead [66]. This improvement can be crucial in underwater acoustics, where sound speed gradients may be large. For radiowave propagation applications, where refractive index gradients usually remain small, the simple split of eqn. 3.12 is normally adequate.

3.2.2 Range dependence

If the refractive index also depends on range, the refractive index operator must be redefined as the integral

$$B_x = \frac{1}{\Delta x} \int_x^{x+\Delta x} n(\xi, z)\, d\xi \tag{3.14}$$

This is usually approximated by

$$B = n\left(x + \frac{1}{2}\Delta x, z\right) \tag{3.15}$$

which amounts to replacing the range-dependent medium by a sequence of slices where the refractive index depends linearly on range. If we use the simplest split S_1, the error resulting from this last approximation is given by

$$\begin{aligned} E &= e^{\delta A} e^{\delta B} - e^{\delta(A+B_x)} \\ &= e^{\delta A} e^{\delta B_x} - e^{\delta(A+B_x)} + e^{\delta A}\left(e^{\delta B} - e^{\delta B_x}\right) \end{aligned} \tag{3.16}$$

Since $B - B_x$ is of second order in the range step, the last term is of third order in the range step and can be neglected. The dominant error term is given by

$$\epsilon = \frac{k^2}{8}(\Delta x)^2 [A, B_x] \tag{3.17}$$

The expression for commutator $[A, B_x]$ is similar to that of eqn. 3.9, except that the height derivatives are replaced by their average over a slice.

3.2.3 Implementation of operators

The exponential operator in B is just straightforward multiplication by a scalar exponential. The main difficulty now is to construct the exponential pseudo-differential operator in A in a way that will make computations easy. We have seen in Section 2.4 how this can be done with Fourier transforms for propagation in a vacuum: then of course the refractive index is constant and no boundaries are present. By splitting the original exponential operator, we have already eliminated the problem of refractive index variations. However we still need to take the domain boundaries into account, and the Fourier transform idea needs to be adapted in order to do this correctly.

3.3 Split-step sine transform solution

One of the most important applications of PE modelling is the calculation of radar coverage for oversea propagation. At microwave frequencies and small grazing angles, it is a good approximation to assume that the impedance of the sea surface is infinite for both horizontal and vertical polarizations. For the moment we assume the sea is perfectly smooth, and treat the sea surface as a flat interface, located at $z = 0$. This means the reduced field u satisfies the homogeneous Dirichlet boundary condition at the sea surface

$$u(x, 0) = 0 \quad \text{for} \quad \text{all} \quad x \geq 0 \tag{3.18}$$

The usual boundary condition at infinity for wave propagation is the Sommerfeld outgoing radiation condition [37]. For three-dimensional problems, this can be stated as follows: for a field component ψ, we have

$$\lim_{r\to\infty} r\left\{\frac{\partial\psi(r\mathbf{e})}{\partial\mathbf{e}} - ik\psi(r\mathbf{e})\right\} = 0 \tag{3.19}$$

uniformly for all unit vectors **e**. Physically, this can be interpreted as the fact that all the sources are contained in a bounded volume. For two-dimensional problems, the Sommerfeld radiation condition is stated as

$$\lim_{r\to\infty} \sqrt{r}\left\{\frac{\partial\psi(r\mathbf{e})}{\partial\mathbf{e}} - ik\psi(r\mathbf{e})\right\} = 0 \tag{3.20}$$

uniformly for all unit vectors **e**. It follows from these definitions that the energy of the field remains bounded when averaged over spheres (in 3D) or circles (in 2D) of large radius r [37]. Here we shall ask for an additional slightly stronger condition, the finite energy condition on the reduced function u, given by

$$\int_0^{+\infty} |u(x,z)|^2\, dz < +\infty \tag{3.21}$$

for all positive ranges x.

In order to propagate solutions of the SPE satisfying the conditions given by eqns. 3.18 and 3.21, the appropriate transform is the Fourier sine transform $\mathcal{S}$ with respect to height. This is defined by

$$\mathcal{S}u(x,p) = U(x,p) = \int_0^{+\infty} u(x,z)\sin(2\pi pz)\, dz \tag{3.22}$$

With this definition of $\mathcal{S}$, the inverse transform $\mathcal{S}^{-1}$ is equal to $4\mathcal{S}$.

We wish to compute the effect of the diffractive operator $\exp(\delta B)$ discussed in the previous section. Treating the slice $x_0, x_0+\Delta x$ as a vacuum, we consider the field v which propagates in that slice. Then v is written as

$$v(x,z) = \mathrm{e}^{\frac{i(x-x_0)}{2k}\mathcal{B}} \cdot u(x_0,z) \tag{3.23}$$

with the boundary conditions that, at every range in the slice, v is of finite energy and vanishes on $z=0$. Equivalently, v satisfies the partial differential equation

$$\frac{\partial^2 v}{\partial z^2} + 2ik\frac{\partial v}{\partial x} = 0 \tag{3.24}$$

together with the conditions that v vanishes on the line $z = 0$, v is of finite energy and $v(x_0, z) = u(x_0, z)$. We are going to express v in terms of the sine transform $\mathcal{S}$.

Using integration by parts, we see that v satisfies

$$\begin{aligned}\mathcal{S}\left\{\frac{\partial^2 v}{\partial z^2}\right\}(x,p) &= \int_0^{+\infty} \frac{\partial^2 v}{\partial z^2}(x,z)\sin(2\pi pz)\,dz \\ &= -2\pi p\int_0^{+\infty} \frac{\partial v}{\partial z}(x,z)\cos(2\pi pz)\,dz \\ &= 2\pi p v(x,0) - 4\pi^2p^2\int_0^{+\infty} v(x,z)\sin(2\pi pz)\,dz \\ &= -4\pi^2p^2\mathcal{S}v(x,p)\end{aligned} \tag{3.25}$$

since $v(x,0) = 0$. In other words the sine transform of eqn. 3.24 for functions which satisfy the required conditions is

$$-4\pi^2p^2V(x,p) + 2ik\frac{\partial V}{\partial x}(x,p) = 0 \tag{3.26}$$

where V denotes the sine transform of v. The condition $v(x_0,z) = u(x_0,z)$ forces

$$V(x,p) = \mathrm{e}^{-\frac{i}{2k}\pi^2p^2(x-x_0)}\,U(x_0,p) \tag{3.27}$$

so that the only possible solution is

$$v(x,z) = 4\int_0^{+\infty} e^{-\frac{i}{2k}\pi^2p^2(x-x_0)}\,U(x_0,p)\sin(2\pi pz)\,dp \tag{3.28}$$

The definition of v shows that v vanishes on $z = 0$, and by Parseval's equality [146], we have for all ranges $x \geq x_0$

$$\int_0^{+\infty} |v(x,z)|^2dz = 2\int_0^{+\infty} |U(x_0,p)|^2dp = \int_0^{+\infty} |u(x_0,z)|^2dz < +\infty \tag{3.29}$$

so v is of finite energy. Two interesting properties come out of the derivation of eqn. 3.28. First, the solution is unique and given by eqn. 3.28. In Appendix B, we prove the unicity property for more general interfaces. Second, eqn. 3.29 shows that energy is conserved during the diffractive step. If the refractive index is real, the phase-screen modulation is of modulus 1, and hence the split-step solution has constant energy as it propagates. This corresponds to the fact that we have assumed perfect reflection at the air/ground interface.

We now have the complete split-step/sine transform solution of the SPE

$$u(x + \Delta x, z) = e^{\frac{ik(n^2-1)\Delta x}{2}} \mathcal{S}^{-1} \left\{ e^{\frac{-i\pi^2 p^2 \Delta x}{2k}} \mathcal{S}u(x, z) \right\} \tag{3.30}$$

We note that energy conservation holds for the complete split-step/sine transform provided the refractive index is real, since the refractive term is then of modulus 1. If the medium is lossy, energy decreases as the field propagates (see Section 6.6).

Clearly numerical implementation requires discretization. Discrete Fourier transforms immediately come to mind for implementation of the split-step/sine transform solution. The main difficulty is the truncation of the infinite height domain. We postpone the discussion of the truncation problem, looking first at a simpler waveguide problem which will give us a slightly different perspective.

3.3.1 Waveguides: discrete sine transform solution

Before tackling the discretization of eqn. 3.30, we consider the waveguide problem described in Figure 3.2. The integration domain is limited by the two horizontal lines $z = 0$ and $z = z_{\max}$, and we impose the boundary conditions

$$u(x, 0) = u\left(x, z_{\max}\right) = 0 \quad \text{for} \quad \text{all} \quad x \geq 0 \tag{3.31}$$

The refractive index may vary with range and height inside the domain. The continuous sine transform of the the previous section is not appropriate for this problem, because it does not match the boundary condition at the top of the domain. The best way ahead for the waveguide problem is discretize the PE in height z in order to apply an appropriate discrete sine transform, matching the boundary conditions at the top and bottom of the domain.

We choose a height increment Δz in such a way that $z_{\max} = L\Delta z$ for some integer L, and denote by $\tilde{u}(x) = (u_1(x), \ldots, u_{L-1}(x))$ the height-discretized version of u, where we have used the notation

$$u_j(x) = u(x, j\Delta z) \tag{3.32}$$

We can approximate the second order partial z-derivative by the usual finite-difference expression

$$\frac{\partial^2 u}{\partial z^2}(x, z) \sim \frac{u(x, z + \Delta z) + u(x, z - \Delta z) - 2u(x, z)}{\Delta z^2} = w(x, z) \tag{3.33}$$

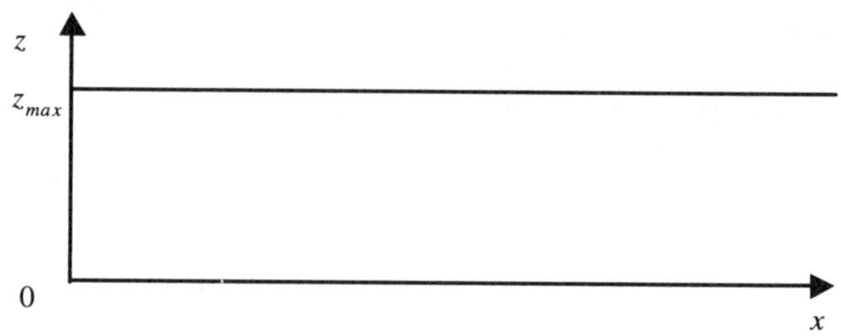

Figure 3.2 Waveguide problem

Let D^2 be the corresponding discrete operator. We write $D^2\tilde{u} = \tilde{w}$ where

$$\begin{aligned} w_1(x) &= \frac{u_2(x) - 2u_1(x)}{\Delta z^2} \\ w_j(x) &= \frac{u_{j+1}(x) + u_{j-1}(x) - 2u_j(x)}{\Delta z^2}, \quad j = 2, \ldots, L-2 \\ w_{L-1}(x) &= \frac{u_{L-2}(x) - 2u_{L-1}(x)}{\Delta z^2} \end{aligned} \tag{3.34}$$

This is indeed the correct discretization of the second order partial derivative at points Δz and $(L-1)\Delta z$ since u satisfies the waveguide boundary conditions

$$u(x,0) = u(x, L\Delta z) = 0 \quad \text{for} \quad \text{all} \quad x \tag{3.35}$$

We shall also need the discrete sine transform S of size L. For a vector $\tilde{u} = (u_1, \ldots, u_{L-1})$ of size $L-1$, $S\tilde{u} = \tilde{U}$ is given by

$$U_l = \sqrt{\frac{2}{L}} \sum_{j=1}^{L-1} u_j \sin\left(\frac{\pi j l}{L}\right), \quad j = 1, \ldots, L-1 \tag{3.36}$$

If L is a power of 2, the discrete sine transform can be implemented very efficiently through a Fast Fourier Transform of size $L/2$ [39, 130]. With our choice of normalization, the discrete sine transform is its own inverse, and it preserves the scalar product norm. We have

$$\sum_{l=1}^{L-1} |U_l|^2 = \sum_{j=1}^{L-1} |u_j|^2 \tag{3.37}$$

The discretized version of eqn. 3.24 is

$$D^2\tilde{u}(x) + 2ik\frac{\partial \tilde{u}}{\partial x} = 0 \tag{3.38}$$

Rearranging terms, we write the discrete sine transform $\widetilde{W}$ of $D^2\tilde{u}$ as

$$W_l = -4\sin^2\left(\frac{\pi l}{2L}\right) U_l, \quad l = 1, \ldots, L-1 \tag{3.39}$$

So we can write the discrete transform of eqn. 3.38 directly as

$$\frac{\partial U_l}{\partial x} = -\frac{2i}{k}\sin^2\left(\frac{\pi l}{2L}\right) U_l, \quad l = 1, \ldots, L-1 \tag{3.40}$$

The solution is given by

$$\widetilde{U}(x + \Delta x) = \widetilde{P}\widetilde{U}(x) \tag{3.41}$$

where the propagator $\widetilde{P}$ is defined by

$$\widetilde{P}_l = \exp\left(-\frac{2i\Delta x}{k}\sin^2\left(\frac{\pi l}{2L}\right)\right), \quad l = 1, \ldots, L-1 \tag{3.42}$$

Going back to the spatial domain, we obtain the complete discrete sine transform split-step PE solution as

$$\tilde{u}(x + \Delta x) = e^{\frac{ik(n^2-1)\Delta x}{2}} S\left\{\widetilde{P}S\{\tilde{u}(x)\}\right\} \tag{3.43}$$

We note that here too the solution is uniquely determined, and that a discrete version of the energy conservation property applies if the refractive index is real. The accuracy of the solution depends on that of the discretization of the second order derivative in height.

3.3.2 Discretization of the infinite problem

We wish to link the discrete and continuous transforms, hoping that $\mathcal{S}$ can be approximated in some sense by S. Assume that the energy of u is negligible above height $(L - 1/2)\Delta z$, so that we can truncate the domain at that height. Then writing $U = \mathcal{S}u$, we have since u vanishes at $z = 0$

$$
\begin{aligned}
U(x,p) &= \int_0^{\Delta z} u(x,z)\sin(2\pi pz)dz \\
&+ \sum_{j=1}^{L-1} \int_{(j-1/2)\Delta z}^{(j+1/2)\Delta z} u(x,z)\sin(2\pi pz)dz
\end{aligned} \tag{3.44}
$$

We now discretize the integral by writing

$$
U(x,p) \sim \Delta z \sum_{j=1}^{L-1} u(x,z)\sin(2\pi pj\Delta z) \tag{3.45}
$$

We put

$$
\Delta p = \frac{1}{2L\Delta z} \tag{3.46}
$$

and use the notation $\tilde{u}(x)$ for the height discretization of $u(x)$, as before. The discretization of the sine transform can now be written as

$$
U(x,l\Delta p) \sim \sqrt{\frac{L}{2}}\,\Delta z\, U_l(x), \quad l = 1,\ldots,L-1 \tag{3.47}
$$

where $\widetilde{U}(x) = (U_1(x),\ldots,U_{L-1}(x))$ is the discrete sine transform of $\tilde{u}(x)$. In other words, we have approximated the continuous sine transform with the discrete sine transform, up to a normalization factor.

We use the same approach in the spectral domain, assuming that the energy of $U(x,p)$ is negligible above $(L-1/2)\Delta p$. We then construct the angle-discretization $\widetilde{U}$ of U. The continuous and discrete versions of the inverse sine transforms are linked by

$$
u(x,j\Delta z) \sim 4\sqrt{\frac{L}{2}}\,\Delta p\, u_j(x), \quad j = 1,\ldots,L-1 \tag{3.48}
$$

Using the definition of Δp from eqn. 3.46 to simplify the normalization terms, we obtain the discrete version of the sine transform split-step PE solution as

$$
\tilde{u}(x+\Delta x) = e^{\frac{ik(n^2-1)\Delta x}{2}} S\left\{\widetilde{P'}S\{\tilde{u}(x)\}\right\} \tag{3.49}
$$

which is similar to the waveguide solution, except the propagator $\widetilde{P'}$ is given by

$$\widetilde{P'}_l = \exp\left(-\frac{i\pi^2 l^2 \Delta x}{2kL^2}\right), \quad l = 1, \ldots, L-1 \tag{3.50}$$

If we compare the discretization of the infinite problem with that of the waveguide problem, we see that for small angles, corresponding to small l/L ratios, the two propagators are equivalent, but that they differ substantially for larger angles. The reason for this is that the two approaches are really solving two very different problems. In the waveguide case, the zero boundary condition at the top of the domain is part of the specification of the problem, whereas for the infinite domain, we have to keep imposing an artificial boundary condition at each step to allow truncation at a finite height.

3.3.3 Sampling and domain truncation

We now look at the errors incurred in the discretization of the continuous problem. Clearly the error made by approximating the continuous sine transforms with discrete sine transforms depends on the choice of L and Δz (which together automatically determine Δp from eqn. 3.46). We have to make sure that the spatial z-domain and the spectral p-domain are both accurately represented. We first look at the spectral domain requirements. We recall that the spectral p-domain corresponds to the vertical angular spectrum of the field. The use of a Fourier sine transform rather than the full Fourier transform is simply a convenient way to represent the effects of a perfectly conducting boundary. Indeed we can write

$$u(x,z) = \frac{1}{2i}\int_0^{+\infty} U(x,p)\left(\mathrm{e}^{2i\pi pz} - \mathrm{e}^{-2i\pi pz}\right) dp \tag{3.51}$$

which expresses the fact that the vertical angular spectrum of u at any range is an odd function of p. From the derivation in Chapter 2, we have

$$\lambda p = \sin\alpha \tag{3.52}$$

where λ is the wavelength and α the propagation angle from the horizontal. We now see that the assumption $U(x,p) = 0$ for $p \geq L\Delta p$ can be interpreted as the fact that non-zero components of the vertical plane wave decomposition of u at each range should satisfy the Nyquist condition

$$\sin\alpha \leq \lambda L\,\Delta p = \frac{\lambda}{2\Delta z} \tag{3.53}$$

In practice it is not easy to impose this condition. In order to avoid aliasing, is is safest to apply a high-angle filter. In that case the inequality given by eqn. 3.53 is not strict enough, since angular components in the region of the filter will be modified. For simplicity we assume filtering takes place in the upper half of the domain. Then we must have

$$\sin 2\alpha_{\max} \leq \lambda L \, \Delta p = \frac{\lambda}{2\Delta z} \tag{3.54}$$

where $\alpha_{\max}$ is the maximum propagation angle from the horizontal to be represented without distortion. Assuming small angles, the criterion on height resolution is then

$$\Delta z \leq \frac{\lambda}{4\sin(\alpha_{\max})} \tag{3.55}$$

The transform size L is then determined by the truncation height. As for the spectral domain, it is not easy in practice to determine the height above which energy can be neglected. If no filtering is applied, this height could become very large and lead to prohibitively long computation times. It is more practical to consider instead the maximum height of interest H. Some precautions have to be taken in order to ensure no energy is reflected into the domain from features above H. Once H is determined, we must increase the domain above H and apply filtering in this upper layer to avoid spurious reflections from the top of the domain. If we assume this absorbing layer occupies the upper half of the domain, we must have

$$L \geq 2\frac{H}{\Delta z} \tag{3.56}$$

Clearly this condition can be relaxed somewhat with the use of more efficient filters. A commonly used filter is described in Section 8.2.

3.4 Wide-angle split-step algorithm

We shall see in Chapter 12 that it is possible to define finite-difference schemes which work for arbitrary large angles. This is not the case for split-step Fourier methods, because of the necessity to separate the refractive index effects from the diffractive part of the propagator. However the operator splitting proposed by Feit and Fleck [48] provides a good approximation for radiowave propagation applications. It is based on the approximation

$$\sqrt{1+a+b} \sim \sqrt{1+a} + \sqrt{1+b} - 1 \tag{3.57}$$

The error is of the order of $|ab|$. Going back to the operator Q, we write

$$Q = \sqrt{1+A+B} \tag{3.58}$$

where A and B are defined as in Section 3.2. The Feit and Fleck approximation gives

$$Q \sim \sqrt{1+A} + \sqrt{1+B} - 1 \tag{3.59}$$

The error is of the order of the coupled term AB. In a vacuum, expression 3.59 is exact. In order to evaluate the error, we follow [153] and consider the case of a homogeneous medium with $n = 1 + \delta n$. We can then compute the various pseudo-differential operators exactly by looking at Fourier transforms. We use the notations

$$Q_1 = 1 + \frac{A}{2} + \frac{B}{2} \tag{3.60}$$

$$Q_2 = \sqrt{1+A} + \sqrt{1+B} - 1 \tag{3.61}$$

for the narrow and wide-angle approximations of Q respectively. Under the Fourier transform in z defined in Chapter 2, the images of Q, Q_1 and Q_2 are respectively

$$\widehat{Q}(p) = \sqrt{1 + \delta n - \frac{4\pi^2 p^2}{k^2}} \tag{3.62}$$

$$\widehat{Q_1}(p) = 1 + \frac{\delta n}{2} - \frac{2\pi^2 p^2}{k^2} \tag{3.63}$$

$$\widehat{Q_2}(p) = 1 + \delta n + \sqrt{1 - \frac{4\pi^2 p^2}{k^2}} \tag{3.64}$$

Using Taylor expansions, we get the following error bounds:

$$\epsilon_1(p) = \left|\widehat{Q_1}(p) - \widehat{Q}(p)\right| \leq \frac{1}{8}\left[\frac{4\pi^2 p^2}{k^2} + \delta n\right]^2 \tag{3.65}$$

$$\epsilon_2(p) = \left|\widehat{Q_2}(p) - \widehat{Q}(p)\right| \leq \frac{\pi^2 p^2}{k^2}\delta n \tag{3.66}$$

Recalling that the Fourier variable p corresponds to a plane wave component of the form

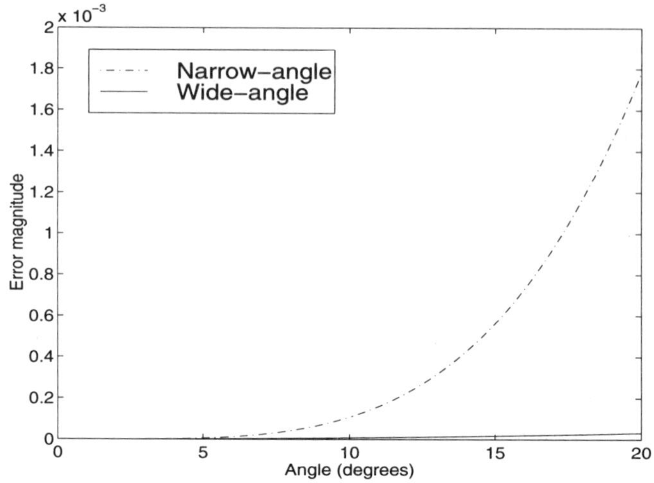

Figure 3.3 *Error in operator Q as a function of angle for $\delta n = 0.001$*

$$\psi(x, z) = \exp[ik(x\cos\alpha + z\sin\alpha)] \tag{3.67}$$

with

$$2\pi p = k\sin\alpha \tag{3.68}$$

we see that the bound on the narrow-angle error is of the order of $(\sin^2\alpha + \delta n)^2$ while the bound on the wide-angle error is of the order of $\delta n \sin^2\alpha$. Figure 3.3 shows numerical values of the errors as a function of angle for $\delta n = 0.001$. To put this in perspective, we note that extreme values of the surface atmospheric refractive index are of the order of 1.0004 [16], so that variations at low heights are well below this value of δn. At larger heights, variations are dominated by the exponential term in height introduced by conformal Earth flattening (see Chapter 6), and this remains less that 0.001 up to several kilometres altitude. By contrast, sound speed variations in underwater acoustics problems can be much larger, and this can lead to phase errors even with the wide-angle approximation [153].

The great advantage of the wide-angle PE resulting from eqn. 3.59 is that it can be solved with split-step Fourier methods just like the narrow-angle version. The wide-angle PE for the reduced PE function u is given by

$$\frac{\partial u}{\partial x} = ik\left(\sqrt{1+\frac{1}{k^2}\frac{\partial^2}{\partial z^2}}-1\right)u + ik(n-1)u \tag{3.69}$$

The split-step solution is obtained by solving successively for the refractive and diffractive terms. If we use the simplest split, we get the solution formally as

$$u(x+\Delta x, z) = \mathrm{e}^{ik\Delta x\left(\sqrt{1+\frac{1}{k^2}\frac{\partial^2}{\partial z^2}}-1\right)}\,\mathrm{e}^{ik(n-1)\Delta x}\cdot u(x,z) \tag{3.70}$$

We have already seen in the previous chapter how to interpret the wide-angle pseudo-differential operator in terms of Fourier transforms for propagation in a vacuum. For a flat perfectly conducting boundary, we can easily adapt the sine transform method of Section 3.3, simply replacing the narrow-angle propagator with the wide-angle propagator

$$P_w(p) = \exp\left\{ik\Delta x\left(\sqrt{1-\frac{\pi^2 p^2}{k^2}}-1\right)\right\} \tag{3.71}$$

where we take the square-root with non-negative imaginary part. We note that for this case the energy of the field decays slightly with range because of the presence of evanescent waves. In vacuum, the wide-angle PE is optimal, since no approximations at all are made beyond the original parabolic approximation of the wave equation, whereas the narrow-angle version distorts the beam at wider angles [76].

3.5 Finite-difference methods

In this section we introduce finite-difference implementations of the PE. The implementation we shall present here is based on an implicit finite difference scheme of the Crank-Nicolson type, which permits the modelling of arbitrary boundaries. It is an adaptation of techniques which have been implemented for underwater acoustic propagation. An excellent exposition can be found in [79].

We assume that the lower boundary is horizontal, located at $z = 0$. We start by defining the integration grid, which is fixed in the vertical direction, but not in range, so that it can adapt to the terrain shape. Let

$$z_j = j\Delta z, \quad j = 0, N \tag{3.72}$$

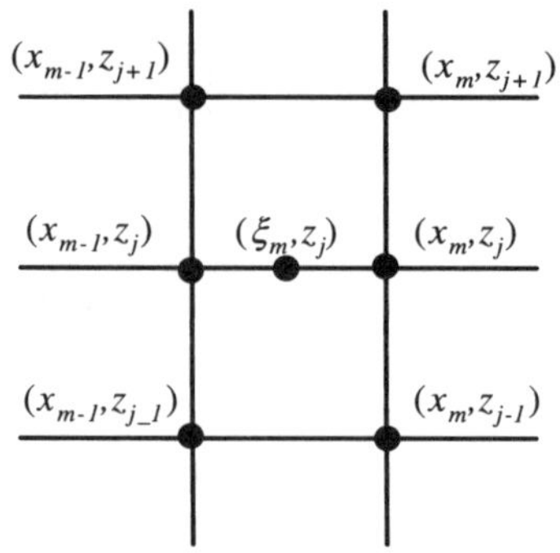

Figure 3.4 Finite-difference grid for Crank-Nicolson scheme

be the vertical grid points and let $x_0, \ldots, x_m, \ldots$ be the successive integration ranges. In order to advance the solution from range x_{m-1} to range x_m, we consider the midpoint

$$\xi_m = \frac{x_{m-1} + x_m}{2} \tag{3.73}$$

The basic idea is to write finite-difference expressions for the partial derivatives at point (ξ_m, z_j), involving only values of the function u at the corners of adjacent rectangles, as shown in Figure 3.4.

The central finite-difference approximation of the derivative in range is given by

$$\frac{\partial u}{\partial x}(\xi_m, z_j) \sim \frac{u(x_m, z_j) - u(x_{m-1}, z_j)}{\Delta x_m} \tag{3.74}$$

where

$$\Delta x_m = x_m - x_{m-1} \tag{3.75}$$

Using Taylor expansions, one can check that the error made with this approximation is of the order of $(\Delta x_m)^2$.

The central finite-difference approximation of the second order derivative in height is given by

$$\frac{\partial^2 u}{\partial z^2}(\xi_m, z_j) \sim \frac{u(\xi_m, z_{j+1}) + u(\xi_m, z_{j-1}) - 2u(\xi_m, z_j)}{\Delta z^2} \tag{3.76}$$

The error here is of the order of $(\Delta z)^4$. We note here that this expression only makes sense if all the points involved are inside the integration domain, which means that j cannot be 0 or N. Additional expressions incorporating the boundary conditions are needed in order to obtain a complete system of equations.

3.5.1 *Narrow-angle code*

We now combine eqns. 3.74 and 3.76 with the SPE, getting

$$\begin{aligned}\frac{u(\xi_m, z_{j+1}) + u(\xi_m, z_{j-1}) - 2u(\xi_m, z_j)}{\Delta z^2} & +2ik\frac{u(x_m, z_j) - u(x_{m-1}, z_j)}{\Delta x_m} \\ & +k^2(n^2(\xi_m, z_j) - 1)u(\xi_m, z_j) = 0\end{aligned} \tag{3.77}$$

The last stage is to approximate u at range ξ_m by averaging values at ranges x_{m-1} and x_m. We put

$$u_j^m = u(x_m, z_j) \tag{3.78}$$

$$b = 4ik\frac{\Delta z^2}{\Delta x} \tag{3.79}$$

$$a_j^m = k^2(n^2(\xi_m, z_j) - 1)\Delta z^2 \tag{3.80}$$

This gives

$$u_j^m(-2 + b + a_j^m) + u_{j+1}^m + u_{j-1}^m = u_j^{m-1}(2 + b - a_j^m) - u_{j+1}^{m-1} - u_{j-1}^{m-1} \tag{3.81}$$

for $j = 1, \ldots, N-1$. This provides $N-1$ equations. In order to complete the system, we need to include equations at the top and bottom of the domain. The finite-difference method is very flexible for the treatment of boundary conditions, as we shall see in later chapters. Here we treat the simplest possible case. Since we assume a perfectly conducting ground, u must be zero at the bottom of the domain. For the moment we will also impose $u(x, z_N) = 0$, using an absorbing layer as in the sine transform model to avoid parasitic reflections at the top of the domain.

We have now expressed values at range x_m as a function of values at range x_{m-1} in the form of a linear system. The solution is not explicitly given, but only implicitly, since we have to invert a matrix to obtain u at range x_m. This implicit scheme is of Crank-Nicolson type [145]. In matricial form, the scheme is given by

$$A_m U_m = V_m \tag{3.82}$$

where U_m is the vector giving the field at range x_m

$$U_m = \begin{pmatrix} u_0^m \\ \vdots \\ u_N^m \end{pmatrix} \tag{3.83}$$

and A_m is the tridiagonal matrix

$$A_m = \begin{pmatrix} 1 & 0 & 0 & 0 & 0 & \dots & 0 & 0 \\ 1 & \alpha_1^m & 1 & 0 & 0 & \dots & 0 & 0 \\ 0 & 1 & \alpha_2^m & 1 & 0 & \dots & 0 & 0 \\ \vdots & & & & & & & \vdots \\ 0 & \dots & & & 0 & 1 & \alpha_{N-1}^m & 1 \\ 0 & \dots & & & & 0 & 0 & 1 \end{pmatrix} \tag{3.84}$$

where

$$\alpha_j^m = -2 + b + a_j^m \tag{3.85}$$

The right hand side V_m is obtained from the field U_{m-1} at the previous range by the matrix multiplication

$$V_m = \begin{pmatrix} 1 & 0 & 0 & 0 & 0 & \dots & 0 & 0 \\ -1 & \beta_1^m & -1 & 0 & 0 & \dots & 0 & 0 \\ 0 & -1 & \beta_2^m & -1 & 0 & \dots & 0 & 0 \\ \vdots & & & & & & & \vdots \\ 0 & \dots & & & 0 & -1 & \beta_{N-1}^m & -1 \\ 0 & \dots & & & & 0 & 0 & 1 \end{pmatrix} U_{m-1} \tag{3.86}$$

where

$$\beta_j^m = 2 + b - a_j^m, \quad j = 1, \ldots, N-1 \tag{3.87}$$

Since the matrix A_m is tridiagonal, inversion with a Gaussian elimination scheme is straightforward provided A_m is non-singular. In what follows, we omit the m index for simplicity. In a first sweep, the lower subdiagonal is eliminated, with the forward loop

$$\begin{aligned} \Gamma_0 &= \frac{v_0}{\alpha_0} \\ \Gamma_j &= \frac{V_j - \Gamma_{j-1}}{\alpha_j}, \quad j = 1, \ldots, N \end{aligned} \tag{3.88}$$

The upper subdiagonal is then eliminated with a backsolving loop, giving the solution as

$$\begin{aligned} U_N &= \Gamma_N \\ U_j &= \Gamma_j - \frac{U_{j+1}}{\alpha_j}, \quad j = N-1, \ldots, 0 \end{aligned} \tag{3.89}$$

The number of operations is of order N. The Crank-Nicolson scheme has good numerical properties, which are described in Chapter 12.

3.5.2 Claerbout approximation

The Claerbout approximation mentioned in Chapter 2 can be implemented without difficulty with a finite-difference scheme similar to the one used for the SPE. The Claerbout equation can be written as

$$\left(1 + \frac{Z}{4}\right)\frac{\partial u}{\partial x} = ik\frac{Z}{2}u \tag{3.90}$$

As before we express this with finite differences at point (ξ_m, z_j), getting the system of equations

$$\begin{aligned} u_j^m(-2 + b' + a_j^m) + u_{j+1}^m + u_{j-1}^m &= \left\{u_j^{m-1}(-2 + \bar{b'} + a_j^m)\right. \\ &\quad \left. + u_{j+1}^{m-1} + u_{j-1}^{m-1}\right\} \end{aligned} \tag{3.91}$$

for $j = 1, \ldots, N-1$, where the a_j^ms are as before and

$$c = \frac{1 - ik\Delta x}{1 + ik\Delta x} \tag{3.92}$$

$$b' = \frac{(4k^2\Delta z)^2}{1 + ik\Delta x} \tag{3.93}$$

The resulting matrix is again tridiagonal and can be inverted as before.

3.6 Source modelling

Integration must start with the field on the initial vertical. In many propagation problems the source is assumed to be at zero range. Often the aperture field is not directly known, but instead the source is defined through its far-field beam pattern. The near-field/far-field transformation which relates aperture field and beam pattern then provides a convenient method to compute the initial field. For the SPE case, the aperture field u_{fs} in free space is given by

$$u_{\text{fs}}(0, z) = \sqrt{2\pi} \int_{-\infty}^{\infty} B(\lambda p) e^{2i\pi pz} dp \tag{3.94}$$

where $B(\theta)$ is the free space beam pattern as a function of elevation angle θ, where we assume that B is zero for angles of more than a few degrees (this narrow beam approximation corresponds to the narrow-angle SPE). The derivation can be found in Appendix B. Numerical implementation is most conveniently carried out with a Fast Fourier Transform.

The beam pattern is often given assuming a source at height zero, with zero elevation. To adapt this to arbitrary source height and elevation, we use the Fourier shift theorems. Let B_0 be the beam pattern for zero height and elevation. Shifting the source height to z_s corresponds to multiplying B_0 by $\exp(2i\pi z_s)$, while squinting the beam by angle θ_0 corresponds to multiplying the aperture field by $\exp(-2i\pi\theta_0)$.

Gaussian beam patterns are often used: besides having excellent numerical properties, they provide a good representation for paraboloid dish antennas. For zero height and elevation, the Gaussian beam pattern with half-power beamwidth β is given by

$$B(\theta) = A \exp\left(-2\log 2 \frac{\theta^2}{\beta^2}\right) \tag{3.95}$$

where A is a normalization constant. This beam pattern satisfies the half-power condition at angles $\pm\beta/2$ since

$$B^2\left(\frac{\beta}{2}\right) = \frac{1}{2} \tag{3.96}$$

For this case the Fourier transform is obtained explicitly. The aperture field corresponding to a source at height z_s for a Gaussian beam of half-power beamwidth β and elevation angle θ_0 is given by

$$u_{\text{fs}}(0,z) = A\frac{k\beta}{2\sqrt{2\pi\log 2}}\exp\left(-ik\theta_0 z\right)\exp\left(-\frac{\beta^2}{8\log 2}k^2(z-z_s)^2\right) \tag{3.97}$$

We shall see in Chapter 6 how to relate A to path loss, the quantity usually considered in radiowave propagation problems.

The previous discussion can be readily adapted to the case of a perfectly reflecting ground, using image theory. The expression for the aperture field is then

$$\begin{aligned} u(0,z) &= u_{\text{fs}}(0,z) - u_{\text{fs}}(0,-z) \\ &= 2i\sqrt{2\pi}\int_0^{+\infty}\{B(\lambda p) - B(-\lambda p)\}\sin(2\pi pz)dp \end{aligned} \tag{3.98}$$

For wide-angle versions of the parabolic equation, one must use a more accurate near-field/far-field transformation (see Appendix B).

Chapter 4

Tropospheric radiowave propagation

4.1 Introduction

One of the main applications of parabolic equation techniques is the calculation of radio coverage in the troposphere. Specific approximations need to be made in order to derive an appropriate scalar parabolic equation for tropospheric radiowave propagation from Maxwell's equations. We give a very brief overview of the basic notions of radio-meteorology in Section 4.2. More information can be found in [16, 55]. Scalar wave equations for horizontally and vertically polarized radiowaves are derived in Section 4.3. The next stage is to choose a coordinate system which simplifies the representation of structures following the Earth's surface. This is the purpose of the Earth flattening transformation, which is derived in Section 4.4.

We have now reduced several types of radiowave propagation problems to the two-dimensional scalar wave equation. The various frameworks are summarized in Section 4.5. We go back to tropospheric propagation in Section 4.6, where we set up the parabolic equations for tropospheric propagation of horizontally and vertically polarized radiowaves. Finally, we considered normalization of the output PE field in Section 4.7, relating the initial field to the far-field beam pattern.

4.2 Radio refractive index

At radiowave frequencies (below 100 GHz), propagation in the atmosphere is affected mainly by oxygen and water-vapour molecules. To a very good approximation, the air can be considered as a non-dispersive medium for radio frequencies. The real part of the radio-refractive index is then given by the Debye formula

$$n = 1 + 77.6 \times 10^{-6} \, \frac{P}{T} + 0.373 \, \frac{e}{T^2} \tag{4.1}$$

where P is atmospheric pressure in millibars, T temperature in Kelvin and e water vapour pressure in millibars.

The atmospheric refractive index remains very close to unity. Surface values of n, which are usually strongest, vary rarely exceed 1.0004 or so. It has become customary to use the refractivity N, defined by

$$N = 10^6 \times (n - 1) \tag{4.2}$$

Refractivity is expressed in N-units. In a well-mixed atmosphere, pressure, temperature and humidity decrease exponentially as a function of height z above the Earth's surface. This results in an exponential model for N, of the form

$$N(z) = N_0 \times \exp(-z/h_0) \tag{4.3}$$

where N_0 is the sea-level refractivity value and h_0 is the scale height. For the reference atmosphere defined by the International Telecommunications Union [63], we have

$$N(z) = 315 \times \exp(-0.136z) \tag{4.4}$$

where z is given in km. At low heights, less than a kilometre or so, the exponential refractivity profile is often approximated by a linear profile. For the reference atmosphere, the mean atmospheric gradient in the first kilometre is -40 N-units/km, corresponding to the linear approximation

$$N(z) = N_0 - 40z \tag{4.5}$$

where N_0 is the surface value of refractivity and z is given in km. The value of -40 N-units/km is an average gradient, representative of mid-latitude climates. Depending on local temperature and humidity, the low altitude gradient in a well-mixed atmosphere can vary widely [55].

Atmospheric mixing depends on meteorological conditions. It is not uncommon for layering to form, for example if a hot and dry air mass is advected from the land above a cool humid air mass over the sea, as shown in Figure 4.1. Then the refractive index can decrease extremely sharply and ducting can occur.

To get a first look at propagation effects, we assume a constant refractivity gradient α, given in N-units/km. Using geometrical optics, we consider the trajectory of a radio ray launched horizontally, as shown in Figure 4.2. It follows from Bouguer's law [27] that the trajectory is approximately a circular arc, with curvature κ given by

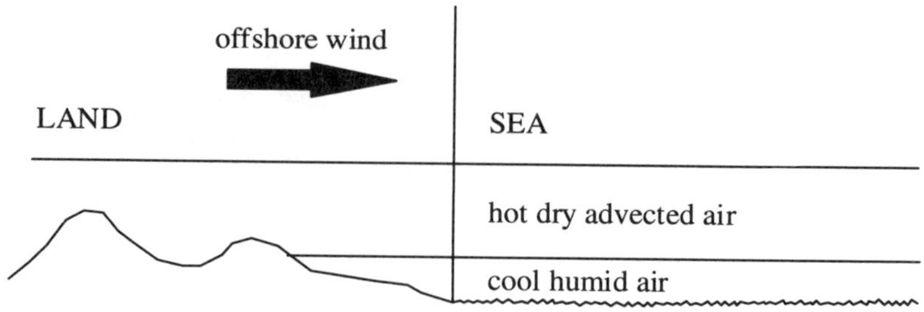

Figure 4.1 Atmospheric layering due to land/sea advection

$$\kappa = -\alpha\,10^{-6}\,\text{km}^{-1} \tag{4.6}$$

To see whether a ray will be refracted back towards the Earth or propagate away from the Earth, we compare the ray curvature to the curvature κ_0 of the Earth. Denoting the Earth's radius by a, given in km, we have

$$\kappa_0 = \frac{1}{a} \sim 157 \times 10^{-6}\,\text{km}^{-1} \tag{4.7}$$

The threshold value of refractivity gradient for rays to curve back towards the Earth is $\alpha_0 = -157$ N-units/km. If the surface of the Earth is a good reflector, ducting can occur when $\alpha < \alpha_0$, with rays bouncing up and down in the duct. For the standard well-mixed atmosphere, corresponding to $\alpha = -40$ N-units/km, rays curve downward, but not as fast as the Earth's surface. At optical frequencies, atmospheric refraction can be ignored. This corresponds to the case $\alpha = 0$, for which rays are straight lines. In standard well-mixed conditions the radio-horizon is always further than the optical horizon. As α decreases towards α_0, the radio-horizon gets further and further away. One speaks of superrefractive conditions when $\alpha_0 < \alpha < -80$ N-units/km. If α is positive, rays curve upwards, and the radio-horizon is closer than the optical horizon. One then speaks of subrefractive conditions.

One of the most important points to bear in mind is the fact that the real part of the atmospheric radio refractive index remains very close to 1 at all frequencies of interest, even in the presence of strong atmospheric layering. This removes normalization difficulties in the choice of a suitable parabolic approximation of the wave equation: the vacuum vavenumber is an obvious reference choice which works well. In other types of wave propagation problems, for example underwater acoustics, density, salinity and temperature variations in the ocean can result in very large variations of sound speed, and it is not

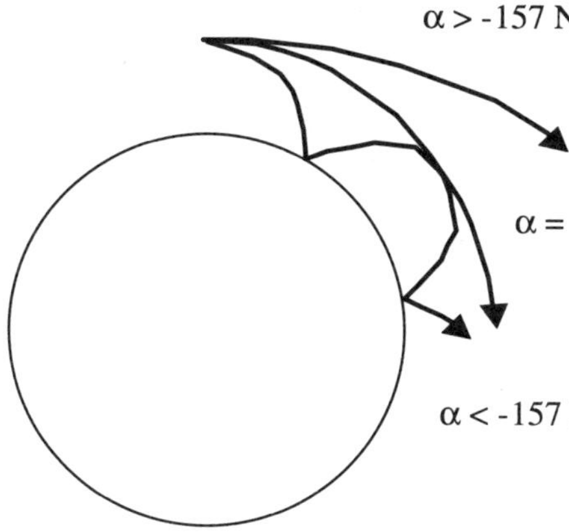

Figure 4.2 Ray curvature and refractivity gradient α

straighforward to determine an optimal reference wavenumber [66].

4.2.1 Atmospheric absorption

Atmospheric molecules can absorb radiowaves, particularly at certain frequencies. The main agents of atmospheric absorption are oxygen and water-vapour. Each of these has an absorption spectrum with many spectral lines. Atmospheric absorption is a complex quantum resonance phenomenon, which is further complicated by collisional effects at low altitudes. A given spectral line has effects not only at its centre frequency, but also on a whole interval of frequencies. This is usually described by an empirical shape function. Spectral models like MPM [97] can be used to compute the absorptive component of the refractive index, from pressure, temperature and humidity. MPM uses a line-by-line calculation based on spectral lines below 1000 GHz, supplemented by a correction term which takes into account the effect of the lines in the infra-red spectrum.

Figures 4.3 and 4.4 show oxygen and water-wapour absorption in dB/km for frequencies up to 100 GHz. The pressure is 1013.25 millibars and the temperature 15° Celsius. Two relative humidity levels are considered in Figure 4.4: 95% (nearly saturated) and 75% (average in temperate climates). For frequencies below 100 GHz, the spectral lines are near 22.3 GHz for water-vapour and cluster around 60 GHz for oxygen. The 22.3 GHz line is superimposed on an underlying increasing function which represents the combined effects of water-vapour lines at higher frequencies. We see that at frequencies above 20 GHz or so, water-vapour absorption can diminish the received signal by several tens of dB over a path of a few tens of kilometres.

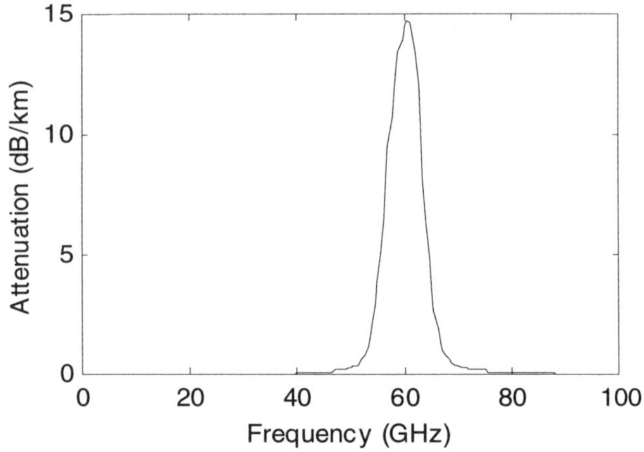

Figure 4.3 Oxygen absorption for P = 1013.25 mb, T = $15°C$

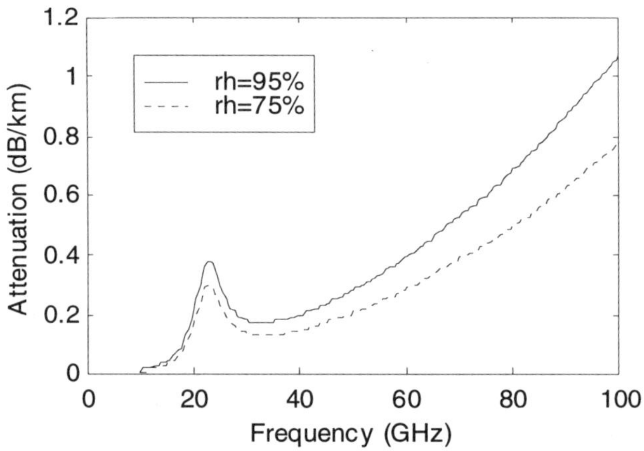

Figure 4.4 Water-vapour absorption for P = 1013.25 mb, T = $15°C$

4.3 Wave equations for radiowave propagation

In Chapter 2, we looked at electromagnetic problems which only depend on two Cartesian coordinates, or do not depend on azimuth in cylindrical coordinates, and exploited the fact that the solutions of Maxwell's equations can then be expressed in terms of vertically and horizontally polarized fields. These satisfy the scalar wave equation, which is of course much simpler to solve than the original Maxwell equations. In general the solution does not split, and

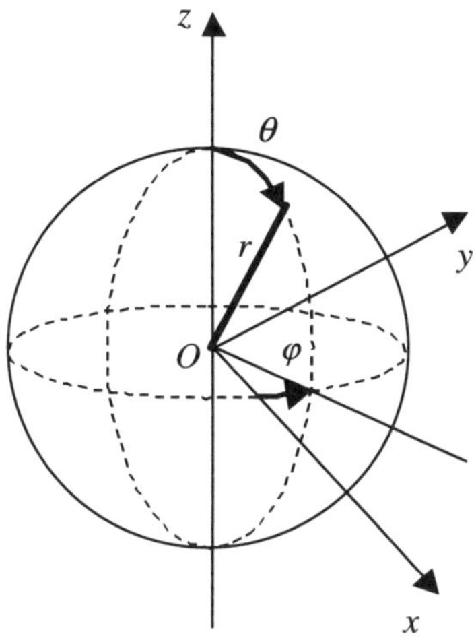

Figure 4.5 Spherical coordinates for tropospheric propagation, with origin O at the Earth's centre

the full set of Maxwell's equations must be considered, leading to the vector PE method presented in Chapter 14. Can we simplify Maxwell equations for radiowave propagation around the Earth? It is clear that Cartesian or cylindrical coordinates are not the right ones to use, since the presence of the Earth and of the atmosphere destroys the chances of a 2-dimensional framework.

The natural idea is to use spherical coordinates (r, θ, ϕ) with origin at the Earth's centre O, as shown in Figure 4.5. The source is located on the z-axis. We consider the atmosphere as a smoothly varying inhomogeneous propagation medium with the magnetic permeability of a vacuum. Assuming a time dependence in $\exp(-i\omega t)$ where ω is the angular frequency, we can write Maxwell's equations for the magnetic field $\mathbf{H}$ and the electric field $\mathbf{E}$ in the atmosphere as

$$\nabla \times \mathbf{E} = ikZ_0\mathbf{H} \tag{4.8}$$

$$Z_0 \nabla \times \mathbf{H} = -ikn^2\mathbf{E} \tag{4.9}$$

$$\nabla \cdot n^2\mathbf{E} = 0 \tag{4.10}$$

$$\nabla \cdot \mathbf{H} = 0 \tag{4.11}$$

where n is the atmospheric refractive index and $Z_0 = 120\pi$ Ohms is the impedance of vacuum [21]. Here we have used the SI units convention.

We write the curl eqns. 4.8 and 4.9 in spherical coordinates, getting the following system of six scalar equations:

$$\left\{\begin{array}{lcl}
\dfrac{1}{r^2\sin\theta}\left(\dfrac{\partial(rE_\phi\sin\theta)}{\partial\theta}-\dfrac{\partial(rE_\theta)}{\partial\phi}\right) & = & ikZ_0H_r \\
\dfrac{1}{r\sin\theta}\left(\dfrac{\partial E_r}{\partial\phi}-\dfrac{\partial(rE_\phi\sin\theta)}{\partial r}\right) & = & ikZ_0H_\theta \\
\dfrac{1}{r}\left(\dfrac{\partial(rE_\theta)}{\partial r}-\dfrac{\partial E_r}{\partial\theta}\right) & = & ikZ_0H_\phi \\
\dfrac{Z_0}{r^2\sin\theta}\left(\dfrac{\partial(rH_\phi\sin\theta)}{\partial\theta}-\dfrac{\partial(rH_\theta)}{\partial\phi}\right) & = & -ikn^2E_r \\
\dfrac{Z_0}{r\sin\theta}\left(\dfrac{\partial H_r}{\partial\phi}-\dfrac{\partial(rH_\phi\sin\theta)}{\partial r}\right) & = & -ikn^2E_\theta \\
\dfrac{Z_0}{r}\left(\dfrac{\partial(rH_\theta)}{\partial r}-\dfrac{\partial H_r}{\partial\theta}\right) & = & -ikn^2E_\phi
\end{array}\right. \tag{4.12}$$

Some symmetry property is needed in order to reduce the problem. If the environment depends only on the distance r from the centre of the Earth, the solutions can be expressed in terms of Debye potentials [21, 49]. The spherical symmetry holds of course if the atmosphere is concentrically layered and the Earth is treated as a perfect homogeneous sphere. However we wish to impose a slightly less constricting assumption on the environment since we want to treat the range-dependent case. We shall use the weaker assumption of azimuthal symmetry, in other words terrain and atmospheric variations do not depend on the azimuth angle ϕ. Now it does not follow that the fields are independent of ϕ: in the case of a horizontal electric dipole source for example, the source distribution does depend on azimuth [49]. In what follows, we restrict ourselves to the case where the source is also independent of azimuth.

We can treat many radiowave propagation problems under these hypotheses. We first note that azimuthal variations of both terrain and atmospheric refractivity are generally slow. If the wavelength of interest is sufficiently small compared to the radius of the Earth, we can neglect lateral ground wave effects, in other words most of the energy between a transmitter and receiver is confined to the vicinity of the great circle plane containing the antennas [49]. We replace the problem with a symmetrical one by extending the atmospheric and terrain structures of the great circle plane containing the antennas by azimuthal symmetry, and applying the same treatment to the antenna patterns. Of course this approximation is not suitable if there is the possibility of energy being scattered back into the great circle plane by large terrain features. It is also clear that the symmetrization does not permit accurate modelling of terrain or sea scatter effects. It is however the only practical way of treating long-range propagation problems, and has been adopted by all the workers in this field.

Under our hypotheses, the solutions of Maxwell's equations can be written as a combination of horizontally and vertically polarized fields [49]. Indeed since the fields are independent of azimuth, the curl equations simplify to

$$\left\{\begin{array}{lcl}
\dfrac{1}{r^2\sin\theta}\dfrac{\partial(rE_\phi\sin\theta)}{\partial\theta} & = & ikZ_0H_r \\
\dfrac{1}{r\sin\theta}\dfrac{\partial(rE_\phi\sin\theta)}{\partial r} & = & -ikZ_0H_\theta \\
\dfrac{1}{r}\left(\dfrac{\partial(rE_\theta)}{\partial r}-\dfrac{\partial E_r}{\partial\theta}\right) & = & ikZ_0H_\phi \\
\dfrac{Z_0}{r^2\sin\theta}\dfrac{\partial(rH_\phi\sin\theta)}{\partial\theta} & = & ikn^2E_r \\
\dfrac{Z_0}{r\sin\theta}\dfrac{\partial(rH_\phi\sin\theta)}{\partial r} & = & ikn^2E_\theta \\
\dfrac{Z_0}{r}\left(\dfrac{\partial(rH_\theta)}{\partial r}-\dfrac{\partial H_r}{\partial\theta}\right) & = & -ikn^2E_\phi
\end{array}\right. \tag{4.13}$$

These equations split into two independent groups, respectively in (H_r, H_θ, E_ϕ) and (E_r, E_θ, H_ϕ). There is no difficulty splitting the divergence equations. In order to go further, we also need to split the boundary conditions at the air/ground interface. The boundary conditions state that the tangential fields must be continuous at the interface [21]. Now because of azimuthal symmetry, the tangential plane at a terrain point (r,θ,ϕ) is given by unit vectors $\mathbf{e}_\phi$ and $\boldsymbol{\tau}$, where the tangent vector $\boldsymbol{\tau}$ is of the form

$$\boldsymbol{\tau} = \cos\alpha\,\mathbf{e}_r + \sin\alpha\,\mathbf{e}_\theta \tag{4.14}$$

Hence continuity of the tangential components of the fields is the same as continuity of $E_\phi, \cos\alpha H_r + \sin\alpha H_\theta$ and $\cos\alpha E_r + \sin\alpha E_\theta, H_\phi$, and the continuity conditions indeed split into independent conditions on (E_ϕ, H_r, H_θ) and (E_r, E_θ, H_ϕ).

Our aim is now to obtain simple scalar wave equations for each type of field. We start with the case where E_r, E_θ and H_ϕ are all zero, corresponding to a vertical magnetic dipole source. Then the only non-zero component of $\mathbf{E}$ is the azimuthal component E_ϕ: the field is horizontally polarized everywhere, relative to the local radius to the Earth's centre. From the curl equations, we see that E_ϕ satisfies the scalar wave equation

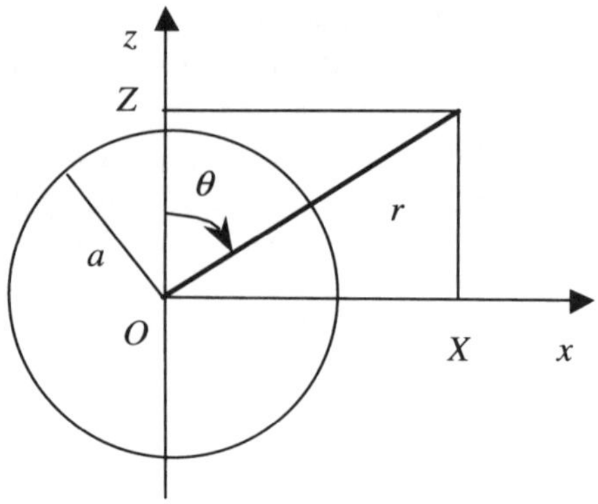

Figure 4.6 Transformation from (r, θ) to (X, Z) coordinates

$$\frac{1}{r}\frac{\partial^2(rE_\phi)}{\partial r^2} + \frac{1}{r^2}\frac{\partial\left(\frac{1}{\sin\theta}\frac{\partial(\sin\theta E_\phi)}{\partial\theta}\right)}{\partial\theta} + k^2n^2E_\phi = 0 \tag{4.15}$$

We factor out the cylindrical spread of the field by putting

$$E_\phi = \frac{1}{\sqrt{kr\sin\theta}}\psi_h \tag{4.16}$$

In terms of ψ_h, eqn. 4.15 becomes

$$\frac{\partial^2\psi_h}{\partial r^2} + \frac{1}{r}\frac{\partial\psi_h}{\partial r} + \frac{1}{r^2}\frac{\partial^2\psi_h}{\partial\theta^2} + \left(k^2n^2 - \frac{3}{4r^2\sin^2\theta}\right)\psi_h = 0 \tag{4.17}$$

We now make the change of variables

$$\left\{\begin{array}{lcl} X & = & r\sin\theta \\ Z & = & r\cos\theta \end{array}\right. \tag{4.18}$$

as illustrated in Figure 4.6.

We can rewrite eqn. 4.17 as

$$\Delta\psi_h(X,Z) + \left(k^2n^2 - \frac{3}{4X^2}\right)\psi_h(X,Z) = 0 \tag{4.19}$$

where the Laplacian refers to coordinates (X, Z). This is similar to the equation we derived in Chapter 2 in cylindrical coordinates. As then, we make the far field approximation, neglecting the term in $1/X^2$, and we obtain the usual wave equation

$$\Delta\psi_h(X, Z) + k^2n^2\psi_h(X, Z) = 0 \tag{4.20}$$

The situation is slightly more complicated for the case where H_r, H_θ and E_ϕ are zero, corresponding to a vertical electric dipole source. Then the electric field remains vertically polarized everywhere, in the sense that it is in the great circle plane. The only non-zero component of the magnetic field $\mathbf{H}$ is the azimuthal component H_ϕ, which satisfies the scalar wave equation

$$\frac{1}{r}\frac{\partial\left(\frac{1}{n^2}\frac{\partial(rH_\phi)}{\partial r}\right)}{\partial r} + \frac{1}{r^2}\frac{\partial\left(\frac{1}{n^2\sin\theta}\frac{\partial(\sin\theta H_\phi)}{\partial\theta}\right)}{\partial\theta} + k^2H_\phi = 0 \tag{4.21}$$

The appropriate substitution for this case is given by

$$H_\phi = \frac{n}{\sqrt{kr\sin\theta}}\psi_v \tag{4.22}$$

The function ψ_v satisfies the wave equation

$$\frac{\partial^2\psi_v}{\partial r^2} + \frac{1}{r}\frac{\partial\psi_v}{\partial r} + \frac{1}{r^2}\frac{\partial^2\psi_v}{\partial\theta^2} + (k^2n^2 + \kappa)\psi_v = 0 \tag{4.23}$$

with

$$\kappa = -\frac{1}{4r^2} + n\sqrt{\sin\theta}\left(\frac{\partial^2}{\partial r^2} + \frac{1}{r^2}\frac{\partial^2}{\partial\theta^2}\right)\left\{\frac{1}{n\sqrt{\sin\theta}}\right\} \tag{4.24}$$

Consider the case where n depends only on r, corresponding to a stratified atmosphere. Then

$$\kappa = \sin\theta\left\{\frac{\partial^2\log n}{\partial r^2} - \left(\frac{\partial\log n}{\partial r}\right)^2\right\} + \frac{3}{4r^2\sin^2\theta} \tag{4.25}$$

We see that the far-field approximation is not enough to ensure that κ can be neglected. We also need to constrain the variations of n. What is required is that the first and second height derivatives of $\log n$ remain small compared to the wavelength. In practice the variations of the atmospheric refractive index given by eqn. 4.1 are slow, and this is not a problem. By neglecting κ, we have replaced the vertical polarization problem too with the usual wave equation for ψ_v

$$\Delta\psi_v(X,Z) + k^2n^2\psi_v(X,Z) = 0 \tag{4.26}$$

4.3.1 Boundary conditions

Assume first that there are no terrain variations and the Earth's surface is perfectly spherical. For the case of horizontal polarization, the tangential field components are E_ϕ and H_θ. It follows from the curl equations that E_ϕ and its normal derivative $\partial E_\phi/\partial r$ must be continuous at the air/ground interface. For the case of vertical polarization, the tangential field components are H_ϕ and E_θ, and it follows from the curl equations that H_ϕ and $\partial_r(rH_\phi)/n^2$ must be continuous at the air/ground interface.

In the presence of terrain variations, the radial derivatives are replaced with local normal derivatives. Let $\boldsymbol{\nu}$ be the outer normal to the terrain surface (which is in the (r,θ) plane because of azimuthal symmetry). Using the notations of eqn. 4.14, wan can write the normal derivative of a function F as

$$\frac{\partial F}{\partial \boldsymbol{\nu}} = \cos\alpha\frac{\partial F}{\partial r} - \sin\alpha\frac{1}{r}\frac{\partial F}{\partial\theta} \tag{4.27}$$

It then follows from the curl equations that the continuity of E_ϕ and $\cos\alpha H_r + \sin\alpha H_\theta$ is equivalent to the continuity of E_ϕ and $\partial E_\phi/\partial\boldsymbol{\nu}$ for horizontal polarization. Hence ψ_h and its normal derivative must be continuous at the interface. Similarly, the continuity of H_ϕ and $\cos\alpha E_r + \sin\alpha E_\theta$ is equivalent to the continuity of H_ϕ and $\partial_{\boldsymbol{\nu}}(rE_\phi)/n^2$ for vertical polarization. If we neglect higher order terms in $1/a$, where a is the Earth's radius, we find that $n\psi_v$ and χ_v/n must be continuous at the interface, where

$$\chi_v = \frac{\partial\psi}{\partial\boldsymbol{\nu}} + \left(\frac{1}{a} + \frac{\partial n}{\partial\boldsymbol{\nu}}\right)\psi_v \tag{4.28}$$

If the refractive index of the ground is sufficiently large, we can use the Leontovich boundary conditions [49]. The continuity conditions are replaced by an approximate expression which does not require the knowledge of the fields below ground. The derivation of the Leontovich boundary conditions for the paraxial framework is given in Chapter 9.

We now wish to replace the wave equations we have obtained for ψ_h and ψ_v with parabolic equations. We could choose the horizontal direction X as the paraxial direction and factor out the vacuum phase variation in X as in the derivation of Chapter 2. However this is not recommended for radiowave propagation problems: while we needed the Cartesian coordinates (X, Z) to reduce the problem to the familiar wave equation, we have completely lost the underlying symmetries we had in spherical coordinates. We recall that boundary conditions have to be enforced at the Earth's surface, and that refractive index structures in the troposphere tend to be stratified along the surface: in fact we made use of these properties in the derivation of our wave equations. Hence the next stage is to rewrite our problem in terms of coordinates related to the Earth's surface.

4.4 Earth flattening transformation

It follows from the previous discussion that it would be very convenient to choose a coordinate system in which ranges are measured along an idealized spherical Earth and heights are measured along radials passing through the Earth's centre, as shown in Figure 4.7. Pekeris introduced this idea in [127]. It turns out that under this transformation, the wave equation is still approximately valid at low heights, provided the refractive index n is replaced by the modified refractive index

$$m(x, h) = n(x, h) + \frac{h}{a} \tag{4.29}$$

where a is the Earth's radius and h is the height above the Earth's surface. The modified refractivity M is related to m by

$$M = 10^6 \times (m - 1) \tag{4.30}$$

Modified refractivity is expressed in M-units.

In fact it is not possible to achieve this ideal Earth flattening transformation without distorting the wave equation, because it does not correspond to a conformal mapping in the complex plane. In order to obtain a conformal tranformation, we use a height coordinate z which is a logarithmic function of altitude h

$$z = a \log\left(1 + \frac{h}{a}\right) \tag{4.31}$$

We have shown above that our problem can be reduced to the usual wave equation

$$\Delta\psi(X, Z) + k^2 n^2(X, Z)\psi(X, Z) = 0 \tag{4.32}$$

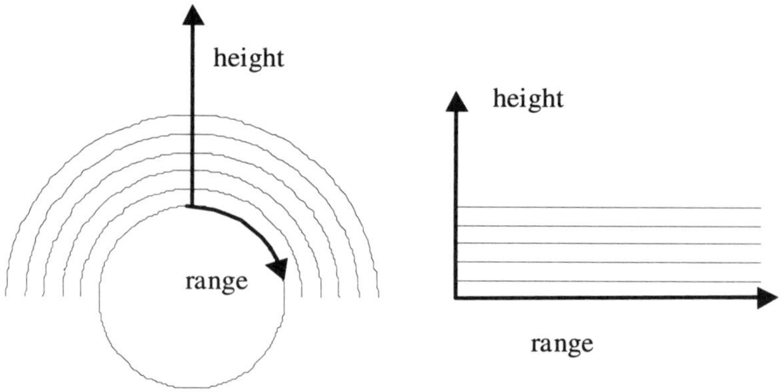

Figure 4.7 Earth flattening transformation

where ψ is related to the appropriate field component depending on the polarization. Putting $r = a + h$, we rewrite X and Z as

$$\begin{aligned} X &= (a+h)\sin\theta \\ Z &= (a+h)\cos\theta \end{aligned} \tag{4.33}$$

Our aim is to find a simple partial differential equation in terms of the logarithmic height z and the ground range $a\theta$ satisfied approximately by v. In order to do this, we look at the transformation defined in the complex plane outside the non-positive imaginary axis by

$$\xi = ia \log\left(\frac{\zeta}{ia}\right) \tag{4.34}$$

where we use the principal determination of the logarithm. If we put

$$\begin{cases} \zeta &= X + iZ \\ \xi &= x + iz \end{cases} \tag{4.35}$$

we can also look at this mapping as the change of coordinates $(X, Z) \mapsto (x, z)$. Using the notations of eqn. 4.33, we have

$$\zeta = i(a+h)\mathrm{e}^{-i\theta} \tag{4.36}$$

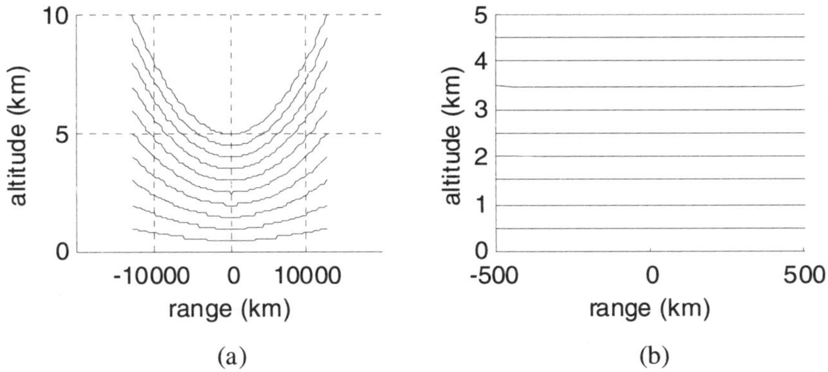

Figure 4.8 *Images of concentric circles under Earth flattening transformation for (a) whole hemisphere and (b) ranges from −500 km to 500 km*

and

$$\xi = a\theta + ia\log\left(1 + \frac{h}{a}\right) \tag{4.37}$$

or equivalently

$$\begin{cases} x = a\theta \\ z = a\log\left(1 + \frac{h}{a}\right) \end{cases} \tag{4.38}$$

Our transformation maps the circle of centre 0 and radius $a + h$ deprived of its South pole to the horizontal segment $-\pi a \leq x \leq \pi a, z = a\log(1 + h/a)$. Figure 4.8(a) shows the images of circles of radius $a + h$ for h varying from 0.5 to 5 km in 0.5 km steps. At low heights, the transformation clearly achieves an excellent approximation of the required flattening, since we have

$$0 \leq h - z \leq \frac{h^2}{2a} \tag{4.39}$$

This is illustrated by Figure 4.8(b), which shows the images of circular arcs with radius varying as in Figure 4.8(a), and ground ranges limited to −500 km to 500 km. The difference between h and z at 500 km range and 5 km height is less than 2 m.

Since the change of coordinates corresponds to a conformal mapping, the Laplacian in the new coordinate system is proportional to the Laplacian in the old coordinate system

$$\frac{\partial^2}{\partial x^2} + \frac{\partial^2}{\partial z^2} = \left|\frac{d\zeta}{d\xi}\right|^2 \left(\frac{\partial^2}{\partial X^2} + \frac{\partial^2}{\partial Z^2}\right) \tag{4.40}$$

The derivation of this property can be found in [57]. In the new coordinate system, the two-dimensional wave equation becomes

$$\frac{\partial^2 \psi}{\partial x^2} + \frac{\partial^2 \psi}{\partial z^2} + k^2 \tilde{n}^2 \left|\frac{d\zeta}{d\xi}\right|^2 \psi = 0 \tag{4.41}$$

where

$$\tilde{n}(x, z) = n(x, h) \tag{4.42}$$

Now we have

$$\left|\frac{d\zeta}{d\xi}\right| = \left|\mathrm{e}^{\xi/ia}\right| = \mathrm{e}^{z/a} \tag{4.43}$$

which gives the wave equation in the new coordinate system

$$\frac{\partial^2 \psi}{\partial x^2} + \frac{\partial^2 \psi}{\partial z^2} + k^2 \tilde{n}^2 \mathrm{e}^{2z/a} \psi = 0 \tag{4.44}$$

The corresponding modified refractive index $\widetilde{m}$ is given by

$$\widetilde{m}(x, z) = \tilde{n}(x, z)\mathrm{e}^{z/a} \tag{4.45}$$

At this point we have made no further approximations: eqn. 4.44 is valid at any height and range, excepting the half-line going through the South pole. We note that, if h is small compared to the Earth's radius, we have

$$z \sim h \tag{4.46}$$

so the logarithmic height z is equivalent to the linear height h for low altitudes. Since we can write the modified refractive index as

$$\widetilde{m}(x, z) = n(x, h)\left(1 + \frac{h}{a}\right) = m(x, h) \tag{4.47}$$

we conclude that our transformation is equivalent to the usual flat-Earth approximation at low heights, with the modified wave equation

$$\frac{\partial^2\psi}{\partial x^2} + \frac{\partial^2\psi}{\partial h^2} + k^2 m^2(x, h) = 0 \tag{4.48}$$

Since the usual flat-Earth approximation loses accuracy at larger heights, it is preferable to use the logarithmic height transformation when higher altitudes have to be modelled. This does not increase computational complexity. We note that the boundary conditions at the Earth's surface remain valid in the new coordinate system since the conformal transformation maps the image of the normal to the surface at a given terrain point to the normal to the transformed surface.

4.5 Two-dimensional wave equation summary

Granted certain types of symmetry, we have seen that we can reduce radiowave propagation problems to the two-dimensional scalar wave equation for a function ψ related to a field component

$$\frac{\partial^2\psi}{\partial x^2}(x, z) + \frac{\partial^2\psi}{\partial z^2}(x, z) + k^2 n^2(z)\psi(x, z) = 0 \tag{4.49}$$

This equation may be interpreted in three different ways, depending on the type of problem we wish to treat.

1. If we work with Cartesian coordinates (x, y, z) and the problem is independent of y, the function ψ is the transverse electric field E_y for horizontal polarization and the transverse magnetic field H_y for vertical polarization.

2. If we work with cylindrical coordinates (x, θ, z) and the problem does not depend on azimuth θ, the function ψ corresponds to azimuthal field components scaled for cylindrical spreading, $\sqrt{kx}E_\theta$ for horizontal polarization and $\sqrt{kx}H_\theta$ for vertical polarization.

3. Finally, if we work with flattened Earth coordinates (x, z) derived from spherical coordinates, we replace the refractive index n in eqn. 4.49 with the modified refractive index $\widetilde{m} = n\exp(z/a)$. The functions ψ_h and ψ_v corresponding respectively to horizontal and vertical polarization are given by

$$\psi_h = \sqrt{ka\sin\left(\frac{x}{a}\right)}\,\exp\left(\frac{z}{2a}\right)\,E_\phi \tag{4.50}$$

and

$$\psi_v = \frac{1}{n}\sqrt{ka\sin\left(\frac{x}{a}\right)}\,\exp\left(\frac{z}{2a}\right)\,H_\phi \tag{4.51}$$

The first interpretation is most commonly used for two-dimensional scattering problems, while the third is the usual one for tropospheric radiowave propagation problems.

4.6 PE for tropospheric radiowave propagation

The PE derivation for flattened Earth coordinates is the same as in section 2.2.2: we choose the positive x-direction as the paraxial direction and factor out the fast phase variation in x by putting

$$u(x,z) = \mathrm{e}^{ikx}\psi_j(x,z) \tag{4.52}$$

where $j = h$ or $j = v$ depending on polarization. Then we approximate eqn. 4.44 (or 4.48) as the product of two parabolic terms. Neglecting backscatter, we get the outgoing parabolic wave equation

$$\frac{\partial u}{\partial x}(x,z) = -ik\left\{1 - \sqrt{\frac{1}{k^2}\frac{\partial^2}{\partial z^2} + \widetilde{m}^2(x,z)}\right\}u(x,z) \tag{4.53}$$

Since the modified refractive index $\widetilde{m}$ becomes large at high altitudes, the narrow-angle version of the PE given by eqn. 2.21 becomes inaccurate at heights more than a few kilometres above the Earth's surface, even for propagation angles initially close to the horizontal. The error in the square-root approximation is of the order of $h^2/4a^2$, roughly 10^{-6} at 12 km altitude.

4.7 Path loss

For radiowave applications, parabolic equation results are usually given in terms of path loss. The transmission loss L of a radio link in a given environment is defined by the International Telecommunications Union as the ratio between the power radiated by the transmitting antenna and the power which would be available at the receiving antenna output if there were no losses in the radio frequency circuits [63]. The transmission loss is usually expressed in decibels. This definition takes the antenna radiation patterns into account as well as the propagation effects.

For many radio or radar applications, link performance evaluation is much simpler if the propagation and antenna effects can be decoupled. This leads to the following definition: the basic transmission loss L_b of a radio link in a given environment is defined as the ratio of the equivalent isotropic radiated power of the transmitter system to the power available from an isotropic receiving antenna [63]. In this definition, we replace the actual antennas by isotropic antennas with the same polarizations. Again system losses are not included in the definition and must be taken into account separately.

With this framework, the modelling of antenna patterns relies on a geometric optics picture: we look at a ray path between transmitter and receiver, and define the ray path transmission loss as

$$L_t = L_b - G_t - G_r \tag{4.54}$$

where G_t and G_r are the respective transmitter and receiver antenna gains in the ray path directions. This is often quite a useful way of doing things when ray-optics are valid. We must note however that in strong ducting environments, focusing and defocusing effects usually depend on the ray-path of interest, whereas by definition basic transmission loss does not. The definition is then not totally consistent since the loss due to the propagation medium should be ray-path dependent.

When the propagation medium is free space and the distance d between the terminals is large compared to the wavelength λ, we can use the far-field approximation to evaluate basic transmission loss. The free space basic transmission loss L_{bf} is given by

$$L = 20 \log \left(\frac{4\pi d}{\lambda} \right) \tag{4.55}$$

Neither transmission loss nor basic transmission loss are quite suitable for use with the parabolic equation method. The PE models the transmitter antenna pattern, so some of the difficulties of the basic transmission loss concept can be avoided. However what is computed is the field at all points of the computational grid, not transmission loss. In general further processing is needed to couple the field to the receiving antenna.

We work with the intermediate concept of path loss L_p, defined as the ratio between boresight equivalent isotropic radiated power for the actual antenna and the power received by an isotropic antenna with the same polarization, assuming that there are no system losses. By definition, the boresight equivalent isotropic radiated power (e.i.r.p.) of an antenna is the power P_{iso} required at the input of an isotropic antenna to produce the same far-field power flux-density in the boresight direction as the actual antenna radiating in vacuum. Note that our definition of P_{iso} does not take ground effects into account.

At distance r from the source, the power flux density S_{iso} radiated by our equivalent isotropic antenna is

$$S_{\text{iso}} = \frac{P_{\text{iso}}}{4\pi r^2} \tag{4.56}$$

Now the power flux density S_b in the boresight direction is related to the boresight far-field beam pattern for the electric field by

$$S_b = \frac{|B_{\text{max}}|^2}{2Z_0 r^2} \tag{4.57}$$

Since we are working with ratios, we may choose the value of Bmax. We work with the normalization

$$B_{\text{max}} = \frac{1}{\sqrt{2\pi}} \tag{4.58}$$

which corresponds to unity gain in the boresight direction. The boresight equivalent isotropic radiated power is then given by

$$P_{\text{iso}} = \frac{1}{Z_0} \tag{4.59}$$

where Z_0 is the impedance of a vacuum. We now need to relate the PE output to the path loss L_p. We first look at the case of horizontal polarization. Then we solve for the reduced function u corresponding to the azimuthal component E_ϕ of the electric field. The power flux density at the point of coordinates (X, Z) is

$$S = \frac{1}{2Z_0}|E_\phi(X, Z)|^2 \tag{4.60}$$

If we treat the receiving antenna as an isotropic point source, the received power P_r is given by

$$P_r(X, Z) = \frac{\lambda^2}{4\pi} S \tag{4.61}$$

In terms of the function ψ_h defined by eqn. 4.16, the power ratio is expressed as

$$\frac{P_{\text{iso}}}{P_r(X, Z)} = \frac{(4\pi)^2 X}{\lambda^3} |\psi_h(X, Z)|^{-2} \tag{4.62}$$

Finally, converting to flat Earth coordinates (x, z), we obtain the path loss in terms of the reduced PE field u as

$$L_p(x,z) = -20\log|u(x,z)| + 20\log(4\pi) + 10\log\left(a\sin\frac{x}{a}\right) - 30\log(\lambda) \quad (4.63)$$

If the range x is small relative to the Earth's radius, we can write

$$L_p(x,z) = -20\log|u(x,z)| + 20\log(4\pi) + 10\log(x) - 30\log(\lambda) \quad (4.64)$$

This formula also holds for the vertically polarized case. The derivation is similar, working with H_ϕ and replacing Z_0 by the admittance $1/Z_0$.

Results are often given in terms of propagation factor for radar application. The propagation factor F is the field relative to free space, expressed in dB. If we approximate the distance d between the terminals by range x, we get from eqns. 4.55 and 4.64

$$F(x,z) = 20\log|u(x,z)| - 10\log(x) - 10\log(\lambda) \quad (4.65)$$

4.8 Aperture field

In this section, we address the generation of an aperture field with the appropriate beam pattern. Microwave antennas are usually located several wavelengths above the ground. We can then neglect the interaction of the ground with the antenna, assuming that the aperture function in the absence of the ground is a good approximation to the actual aperture function of the antenna. In these circumstances, we can use the near-field/far-field relationship derived in Appendix B to generate the initial PE reduced field with the appropriate normalization. In the wide-angle case, we have

$$u(0,z) = \sqrt{2\pi}\mathrm{e}^{i\pi/4}\int_{-\frac{1}{\lambda}}^{+\frac{1}{\lambda}} \frac{B(\theta(p))}{\sqrt{\cos(\theta(p))}}\mathrm{e}^{2i\pi pz}\,dp \quad (4.66)$$

where

$$\sin\theta(p) = \lambda p \quad (4.67)$$

In practice, the $\mathrm{e}^{i\pi/4}$ factor is usually dropped since it does not affect amplitude calculations. We recall that the path loss normalization gives a boresight value of $1/\sqrt{2\pi}$ for the beam pattern function B. It is simpler to work with the normalized beam pattern function

$$B_n = \sqrt{2\pi}B \tag{4.68}$$

with value 1 at boresight, writing the aperture field as

$$u(0,z) = \int_{-\frac{1}{\lambda}}^{+\frac{1}{\lambda}} \frac{B_n(\theta(p))}{\sqrt{\cos(\theta(p))}} \mathrm{e}^{2i\pi pz} dp \tag{4.69}$$

In the narrow-angle case, this becomes

$$u(0,z) = \int_{-\infty}^{+\infty} B_n(\lambda p)\mathrm{e}^{2i\pi pz} dp \tag{4.70}$$

If we want to take ground reflection effects into account in the aperture field, we can introduce an image source and use the formula

$$u(0,z) = \int_{-\infty}^{+\infty} \left[B_n(\lambda p) + R(\lambda p)B_n(-\lambda p)\right] \mathrm{e}^{2i\pi pz} dp \tag{4.71}$$

where $R(\theta)$ is the reflection coefficient corresponding to grazing angle θ. For horizontal polarization and a perfectly conducting ground, this can be expressed as a sine transform, as we have seen in Section 3.6. In the narrow-angle case, if the beam pattern is known at zero elevation angle and zero antenna height, it is straightforward to obtain the aperture function for arbitrary elevation angle θ_0 and antenna height z_0 using the Fourier shift theorems. In the wide-angle case, the Fourier shift properties can be used to model a change of antenna height, but the relationship between p and $\theta(p)$ is no longer linear and it is not possible to model a change of elevation angle using Fourier shift properties.

Chapter 5
Rays and modes

5.1 Introduction

For many years, assessment of refractive effects on tropospheric links was based mainly on geometrical optics and mode theory methods. Although ray optics do not in general give a reliable estimate of field strength, they provide an excellent qualitative picture of propagation conditions, bearing in mind of course that diffraction phenomena are not represented by the geometrical optics approximation. Even now that PE techniques have become the dominant tool for radiowave applications, geometrical optics is still extremely useful. It provides a fast and accurate solution for propagation angles at which refractive effects are not severe, and this makes it invaluable in the construction of efficient hybrid propagation models, as described in Chapter 11. In many cases, ray-tracing also provides crucial information on the behaviour of the wavefront. This is used to model angle-dependent reflection effects, as we shall see in Chapter 10. In Section 1, we give a brief derivation of the ray equations in the simple case of range-independent media with a piecewise linear refractive index profile. More general information on geometrical optics and ray-tracing can be found in [21, 67, 25, 71] for example.

Mode theory was introduced by Watson [163] to solve the problem of diffraction by the Earth, and was later applied to more general problems by a number of authors [27, 161]. It relies on an eigenfunction decomposition of solutions of the wave equation in terms of normal modes. In cases where a few dominant modes are present, a very efficient solution is obtained at sufficiently large ranges. However numerical difficulties start appearing when large numbers of terms are required in the mode series. This happens when many significant modes are present, for example with elevated ducts, and also at ranges which are close to the horizon or in the line-of-sight region. A further disadvantage of mode theory is the difficulty of modelling range-dependent environments [162, 122]. For tropospheric propagation applications, mode theory models have now largely been replaced by parabolic equation techniques which do not suffer the same limitations. However mode theory results are still very much in use for validation purposes. In particular the MLAYER waveguide model [14, 123] provides a reliable reference solution for many test cases. In Section 2, we outline the basic principles of mode theory for the range-independent case and derive the mode theory solution for a piecewise linear refractive index profile. A more complete discussion of the modal series is given in Appendix C.

5.2 Ray-tracing

5.2.1 Basic principles

In this section, we describe a simple analytical ray-trace suitable for range-independent environments. The medium is represented as a superposition of horizontal layers, with the refractive index a linear function of height inside each layer. We assume a flat interface located at $z = 0$. We start with the two-dimensional wave equation for a function ψ associated with a field component

$$\frac{\partial^2 \psi}{\partial x^2}(x,z) + \frac{\partial^2 \psi}{\partial z^2}(x,z) + k^2 n^2(z)\psi(x,z) = 0 \tag{5.1}$$

For simplicity, we use the notation n whatever the origin of eqn. 5.1, but in most applications n will actually be the modified refractive index corresponding to Earth flattening (see Section 4.5). The source is located at range $x = 0$.

Geometrical optics are based on the assumption that ψ behaves locally like a plane wave and can be expressed as

$$\psi \sim \psi_0 \exp(ik\Phi(x,z)) \tag{5.2}$$

where the amplitude ψ_0 is a constant. Then substituting into the wave equation, we find that the phase Φ satisfies

$$\left|\frac{\partial \Phi}{\partial x}\right|^2 + \left|\frac{\partial \Phi}{\partial z}\right|^2 - n^2 = \frac{i}{k}\left(\frac{\partial^2 \Phi}{\partial x^2} + \frac{\partial^2 \Phi}{\partial z^2}\right) \tag{5.3}$$

We now make the high frequency approximation, assuming that the RHS of eqn. 5.3 can be neglected for large wavenumber k. This results in the well-known eikonal equation

$$\left|\frac{\partial \Phi}{\partial x}\right|^2 + \left|\frac{\partial \Phi}{\partial z}\right|^2 = n^2 \tag{5.4}$$

The wavefronts are curves along which the phase Φ is constant. By definition, rays are trajectories orthogonal to the wavefronts. It follows that the tangent $\boldsymbol{\tau}(s)$ along a ray parametrized by the arc length s must be parallel to the gradient of Φ. Since $\boldsymbol{\tau}$ is a unit vector, we get from eqn. 5.4 that

$$\boldsymbol{\tau}(s) = \frac{1}{n}\nabla\Phi \tag{5.5}$$

Since we consider a stratified medium where n depends on z only, we get

$$\frac{d}{ds}\left(n\frac{dx}{ds}\right) = 0 \tag{5.6}$$

$$\frac{d}{ds}\left(n\frac{dz}{ds}\right) = \frac{dn}{dz} \tag{5.7}$$

Eqn. 5.6 is equivalent to Snell's law, giving

$$n\cos\theta = C \tag{5.8}$$

along the ray, where θ is the local angle of the ray with the horizontal. The geometric radius of curvature $R(s)$ along the ray is defined by

$$\frac{1}{R(s)} = \left|\frac{d\boldsymbol{\tau}(s)}{ds}\right| \tag{5.9}$$

From eqn. 5.5, we have

$$\frac{1}{R(s)} = \frac{\nabla n \cdot \boldsymbol{\tau}(s)}{n} \tag{5.10}$$

Since n depends only on z, we get

$$\frac{1}{R(s)} = \frac{1}{n}\frac{dn}{dz}\cos\theta \tag{5.11}$$

At low altitudes, the refractive index (or modified refractive index) is close to 1. For rays launched with small elevation angles, $\cos\theta$ also remains close to 1. Thus we see from eqn. 5.11 that in regions where the refractive index gradient is constant, the radius of curvature of the ray is approximately constant; in other words the ray trajectories are approximately circular arcs.

In simple cases, the magnitude of the field is given by the intensity law of geometrical optics. This is based on the fact that the energy flux along a ray tube remains constant, and hence the field intensity must be proportional to the inverse of the cross-section of the tube. The intensity law can be expressed in terms of the principal radii of curvature of the wavefront along the ray [72, 26]. For the stratified medium considered here, one may derive a much simpler expression [25]. Let θ_0 be the launching elevation angle of the ray. Figure 5.1 shows the ray tube delimited by rays at elevation angles θ_0 and $\theta_0 + d\theta$. Let $Q = (x, z)$ and $Q' = (x', z)$ be the points where the two rays reach height z. If we fix height z, the range x is a function of the launch angle only. The one-dimensional cross-section in the (x, z) plane is equal to the length of segment $Q'Q''$. Its horizontal projection is QQ', which can also be written as

$$QQ' = \left| \frac{\partial x}{\partial \theta_0} \right| \tag{5.12}$$

The one-dimensional cross-section is then

$$\left| \frac{\partial x}{\partial \theta_0} \right| \sin \theta \tag{5.13}$$

where θ is the elevation of the ray at Q. Let ϕ be the azimuth angle. The cross-section area dA is given by

$$dA = \left| \frac{\partial x}{\partial \theta_0} \right| \sin \theta d\theta_0 \, x d\phi \tag{5.14}$$

Assume an isotropic source, with total radiated power P. The power radiated into the ray tube is

$$dP = \frac{P}{4\pi} \cos \theta_0 d\theta_0 d\phi \tag{5.15}$$

Hence the intensity I at point Q is

$$I = \frac{dP}{dA} = \frac{P_0 \cos \theta_0}{4\pi \left| \frac{\partial x}{\partial \theta_0} \right| \sin \theta} \tag{5.16}$$

Now the intensity I_0 at point Q in free space is given by

$$I_0 = \frac{P}{4\pi r^2} \tag{5.17}$$

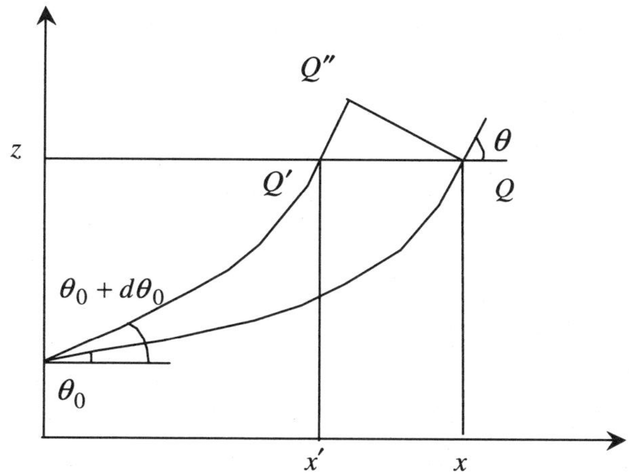

Figure 5.1 *Calculation of ray tube cross-section*

where r is the distance from the source to point Q. If follows that the field relative to free space (or propagation factor) F satisfies

$$|F|^2 = \frac{I}{I_0} = \frac{r^2 \cos\theta_0}{x\left|\dfrac{\partial x}{\partial \theta_0}\right|\sin\theta} \tag{5.18}$$

A large body of work has been published on extension of field-strength calculations to the singular case where the cross-section tends to zero, which is linked to the formation of a caustic (see for example [176]). We should also mention the geometrical theory of diffraction introduced by Keller [70, 69] which extends ray calculations to shadow regions to include diffraction and multiple diffraction. Detailed information on this topic may be found for example in [64, 22, 67].

5.2.2 Piecewise linear profiles

We now assume that in the horizontal slice between heights z_1 and z_2, with $z_1 < z_2$, the refractive index is of the form

$$n(z) = n_1 + \gamma(z - z_1) \tag{5.19}$$

where γ is non-zero. We derive simplified equations for the rays, using the fact that the refractive index n of the air is close to 1. We put $\alpha = \sin\theta$ and use the subscript i to denote values at height $z_i, i = 1, 2$. Neglecting terms in $(n-1)^2$, we have from eqn. 5.8

$$\alpha^2 \sim 2(n - n_1) + \alpha_1^2 \tag{5.20}$$

Hence we can express z as a function of α as

$$z = \frac{\alpha^2 - \alpha_1^2}{2\gamma} \tag{5.21}$$

We see that if the slope γ is positive, a ray launched at height z_1 with positive elevation α_1 will always reach height z_2. If γ is negative and α_1^2 is less than $|\gamma|(z_2 - z_1)$, the ray will turn around before if reaches height z_2. Now if we assume small elevation angles, we have $\tan\theta \sim \alpha$ and

$$\frac{dx}{d\alpha} \sim \frac{1}{\alpha}\frac{dz}{d\alpha} = \frac{1}{\gamma} \tag{5.22}$$

Hence if the ray does not turn around, we have

$$\alpha_2 = \sqrt{2(n - n_1) + \alpha_1^2} \tag{5.23}$$

and

$$x_2 - x_1 = \frac{\alpha_2 - \alpha_1}{\gamma} \tag{5.24}$$

By adding appropriate checks for possible surface reflections and turn-around in layers of negative slopes, we can trace rays in any piecewise linear refractive index environment. In order to add direct and reflected rays in phase, we shall require the optical path length d, which is the integral of nds along the ray. We have

$$nds = nd\alpha\sqrt{\left(\frac{dx}{d\alpha}\right)^2 + \left(\frac{dz}{d\alpha}\right)^2} \tag{5.25}$$

which becomes, neglecting terms of order greater than α^2

$$nds = \frac{1}{\gamma}\left(n_1 - \frac{\alpha_1^2}{2} + \alpha^2\right) d\alpha \tag{5.26}$$

By integration, we obtain the optical path length increment between heights z_1 and z_2 as

$$d_2 - d_1 = \frac{1}{\gamma}\left(\left[n_1 - \frac{\alpha_1^2}{2}\right](\alpha_2 - \alpha_1) + \frac{\alpha_2^3 - \alpha_1^3}{3}\right) \tag{5.27}$$

Provided a ray does not turn around, we can use the intensity ratio given by eqn. 5.18. For a ray launched at elevation angle α_0, the spreading increment dS between heights z_1 and z_2 is given by

$$dS = \frac{\partial x_2 - x_1}{\partial \alpha_0} = \frac{1}{\gamma}\left(\frac{\alpha_0}{\alpha_2} - \frac{\alpha_0}{\alpha_1}\right) \tag{5.28}$$

To see this, we note that between heights z_1 and z_2, we have

$$\alpha = \sqrt{\alpha_0^2 + b} \tag{5.29}$$

where b does not depend on α_0. Hence

$$\frac{\partial \alpha}{\partial \alpha_0} = \frac{\alpha_0}{\alpha} \tag{5.30}$$

and eqn. 5.28 follows using eqn. 5.22. We now obtain the propagation factor at range x by adding together the successive increments. Using the small angle approximation in eqn. 5.18, we have $r \sim x$, $\theta_0 \sim \alpha_0$ and $\cos\theta_0 \sim 1$, getting

$$F = f(\alpha_0)\sqrt{\frac{x}{\alpha S}}\mathrm{e}^{ikd} \tag{5.31}$$

where f is the antenna pattern factor, α is the elevation at the endpoint of the ray and d and S have been obtained by summing optical path length and spreading increments over the height segments of constant refractive index slope encountered along the ray. For a reflected ray, this expression must be multiplied by the complex reflection coefficient R corresponding to the grazing angle at the surface.

5.3 Mode theory

5.3.1 Basic principles

We start again with the two-dimensional wave equation, eqn. 5.1. As in the ray-tracing section, we assume that the propagation medium is stratified in height z and bounded by a plane interface located at $z = 0$. The boundary conditions on ψ are that ψ should be outgoing at infinity and satisfy a surface impedance condition on the interface, of the form

$$\frac{\partial \psi}{\partial z}(x,0) = ik\delta\psi(x,0) \tag{5.32}$$

where the surface impedance δ depends on polarization and the electrical properties of the ground. More information on surface impedance boundary conditions can be found in Chapter 9. The radiation condition implies in particular that ψ must tend to zero at infinity in every direction.

The basic idea of mode theory is to expand the solutions of eqn. 5.1 as

$$\psi(x,z) = \sum_{j=1}^{\infty} A_j f_j(z) \exp(ikC_j x) \tag{5.33}$$

where each term of the sum is itself an outgoing solution of the wave equation satisfying the surface impedance boundary condition at $z = 0$. In mode theory terminology, the C_js are the modes, the f_js the height-gain functions, and the A_js the excitation factors. Because we are dealing with a leaky waveguide, the usual mathematical tools for normal mode expansion are not valid, and in fact for general refractive index profiles, there is no guarantee that such an expansion should exist. Fortunately mode theory is valid for the cases of most practical interest, which are linked to radiowave propagation around the Earth, as we shall see in Appendix C.

Since each term in the sum must satisfy the wave equation, we see that a height-gain function f associated with mode C satisfies

$$f''(z) + k^2(n^2 - C^2)f(z) = 0 \tag{5.34}$$

The boundary condition at the interface becomes

$$f'(0) = ik\delta f(0) \tag{5.35}$$

Determination of the modes and modal functions is equivalent to an eigenvalue problem: we look at bounded functions which satisfy the boundary condition given by eqn. 5.35, and define the operator

$$Tf = f'' + k^2(n^2 - A)f \tag{5.36}$$

where A is a suitably chosen constant. The modes are related to the eigenvalues μ of T by

$$\mu = -k^2(A - C^2) \tag{5.37}$$

and the height-gain functions are associated eigenfunctions. We note here that the Sommerfeld radiation condition implies that if a mode C is different from 1, then the imaginary part of C must be positive. Hence there is a one-to-one correspondence between modes and eigenvalues.

It is interesting at this stage to consider the case of a constant refractive index. Assume n is constant equal to 1. We take $A = 1$. Then if μ is a non-zero eigenvalue, the eigenfunctions are of the form

$$f(z) = ae^{ikSz} + be^{-ikSz} \tag{5.38}$$

where

$$\mu = -k^2S^2 \tag{5.39}$$

and the imaginary part of S is non-negative. Now the radiation condition forces $b = 0$. We now see that it is impossible to impose the impedance boundary condition at $z = 0$ without forcing $a = 0$ as well, which contradicts the fact that an eigenfunction cannot be identically zero. A similar reasoning shows that zero cannot be an eigenvalue either. We conclude that in this case there are no modes at all.

Clearly we need to restrict mode theory to cases where the modal expansion makes sense, which implies in particular that T must have an infinite discrete sequence of eigenvalues. Even then, the convergence of the mode series to the desired solution of the wave equation has been an open problem for many years. Fortunately the modal expansion is meaningful for a large class of refractive index profiles. In what follows, we shall assume the squared refractive index profile is linearly increasing at large heights, with

$$n^2(z) = n^2(H) + \alpha(z - H), \quad z \geq H \tag{5.40}$$

where α is a positive constant. In fact this class of profiles is all we need to treat low altitude tropospheric radiowave propagation: if we are interested only in propagation below height H, we may assume that all refractive index perturbations are below H. Working with flat Earth coordinates, we replace the well-mixed atmosphere above H with a linearly increasing squared modified refractive index profile, as shown in Figure 5.2.

In Appendix C, functional analysis tools are used to show that for that type of profile, the solution of the wave equation can be approximated to arbitrarily

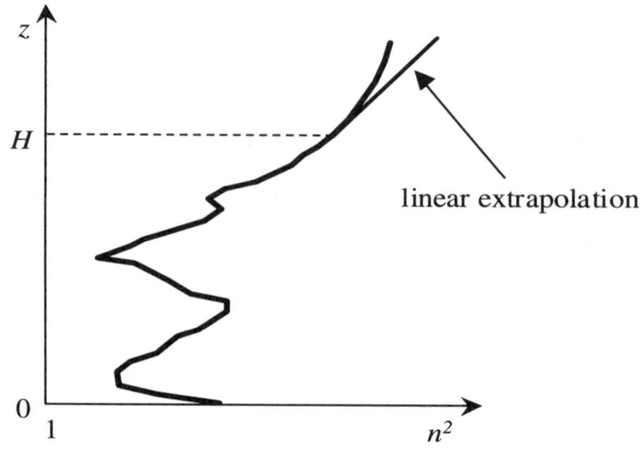

Figure 5.2 *Refractive index profile for mode theory, showing replacement with linear extrapolation above height H*

small error by a modal expansion of the form given by eqn. 5.33, where for fixed z, the sum converges for x large enough. The derivation requires the use of complex heights. The excitation factors are given by

$$A_j = \frac{\int_\Gamma \psi(0,z) f_j(z) dz}{\int_\Gamma f_j^2(z) dz} \tag{5.41}$$

where Γ is the arc in the complex plane shown in Figure 5.3, which tends to infinity along the direction $\pi/3$. We note that the height-gain functions are square-integrable on the arc Γ, but not on the positive real half-line. The normalization factors in the denominator depend on the behaviour of the eigenfunctions on the arc Γ.

In principle the integral in the numerator also requires the knowledge of the initial field $\psi(0,z)$ for complex heights z. In practice this is not necessary if height H is chosen sufficiently large to make sure that $\psi(0,z)$ has negligible energy above H. For the case of a dipole located at height z_t, the initial field $\psi(0,z)$ is a Dirac delta function and we have, for a receiver at height z_r,

$$\psi(x, z_r) = \sum_{j=1}^{\infty} \frac{f_j(z_t) f_j(z)}{\int_\Gamma f_j^2(z) dz} \exp(ikC_j x) \tag{5.42}$$

This representation is the most commonly used modal expansion. We now

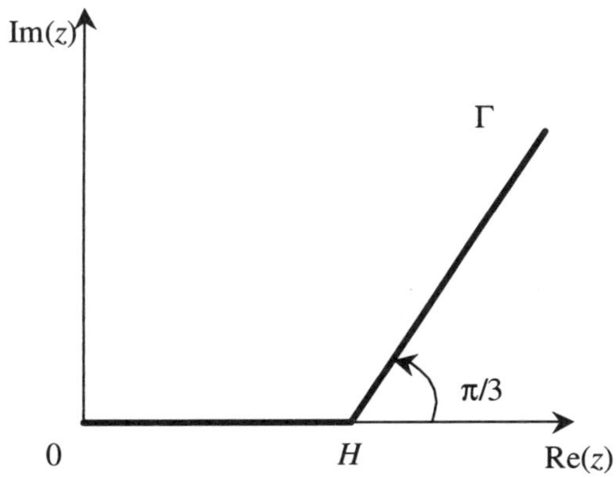

Figure 5.3 Integration contour for normalization of height-gain functions

turn to the calculation of the modes and eigenfunctions.

5.3.2 Piecewise linear profiles

For numerical implementation, it is easiest to approximate the squared refractive index profile with a piecewise linear function [105, 84], assuming a fixed positive slope above height H, as shown in Figure 5.4. We denote the breakpoints in height by $z_0 = 0, z_1, \ldots, z_L = H$. We have for $l < L$

$$n^2(z) = n_2(z_l) + \alpha_l(z - z_l), \quad z_l \leq z \leq z_{l+1} \tag{5.43}$$

and

$$n^2(z) = n^2(z_L) + \alpha_l(z - z_l), \quad z \geq z_L \tag{5.44}$$

The slopes α_l are real, with the constraint that α_L must be positive. To simplify the description, we assume the slopes are non-zero.

On each height segment $[z_l, z_{l+1}]$ of constant square refractive index slope, the eigenfunction equation is a variant of the Airy differential equation (see Appendix A)

$$w''(\zeta) = \zeta w(\zeta) \tag{5.45}$$

which has the two independent solutions w_1, w_2, related to the Airy function Ai by

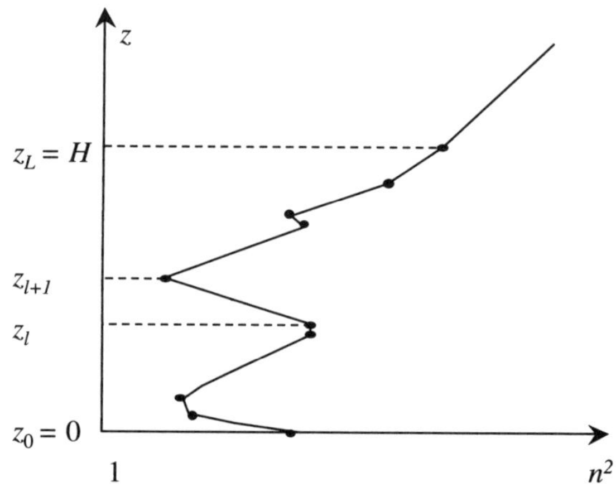

Figure 5.4 Piecewise-linear profile

$$w_1(\zeta) = \mathrm{Ai}(\zeta \mathrm{e}^{2i\pi/3}), \quad w_2(\zeta) = \mathrm{Ai}(\zeta \mathrm{e}^{-2i\pi/3}) \tag{5.46}$$

On each segment of constant slope, eigenfunctions associated with eigenvalue μ are of the form

$$f(z) = A_l\, w_1 \{\zeta_l(z,\mu)\} + B_l\, w_2 \{\zeta_l(z,\mu)\}, \quad z_l \leq z \leq z_{l+1} \tag{5.47}$$

with

$$\zeta_l(z,\mu) = \left(\frac{k^2}{\alpha_l^2}\right)^{1/3} \left(n^2(z_l) - \frac{\mu}{k^2} + \alpha_l(z - z_l)\right) \tag{5.48}$$

Using asymptotic expansions of Airy functions, one may show that w_1 is outgoing in ζ while w_2 is incoming (see Appendix A). Since the slope is positive above height z_L, the outgoing condition forces

$$f(z) = A_L w_1 \{\zeta_L(z,\mu)\}, \quad z \geq z_L \tag{5.49}$$

Starting from the top, we now match the functions at each break point z_l in order to obtain continuity of the functions and their derivatives. We put $A_L = 1$. The matching conditions at z_L yield the system

$$\begin{cases} A_{L-1}(\mu)w_1\{\chi_L(\mu)\} + B_{L-1}(\mu)w_2\{\chi_L(\mu)\} & = & w_1\{\eta_L(\mu)\} \\ A_{L-1}(\mu)w_1'\{\chi_L(\mu)\} + B_{L-1}(\mu)w_2'\{\chi_L(\mu)\} & = & \dfrac{\beta_L}{\beta_{L-1}} w_1'\{\eta_L(\mu)\} \end{cases} \tag{5.50}$$

where the quantities β_l, χ_l and η_l are defined by

$$\beta_l = \alpha_l \left(\frac{k^2}{\alpha_l^2}\right)^{1/3} \tag{5.51}$$

$$\chi_l(\mu) = \zeta_{l-1,\mu}(z_l) \tag{5.52}$$

and

$$\eta_l(\mu) = \zeta_{l,\mu}(z_l) \tag{5.53}$$

The determinant of this system is

$$\begin{aligned} D_L &= \beta_{L-1}\left[w_1\{\chi_L(\mu)\}\, w_2'\{\chi_L(\mu)\} - w_1'\{\chi_L(\mu)\}\, w_2\{\chi_L(\mu)\}\right] \\ &= \beta_{L-1} W(w_1, w_2) \end{aligned} \tag{5.54}$$

where $W(w_1, w_2)$ is the Wronskian of w_1 and w_2, which does not depend on χ_L. Since w_1 and w_2 are independent solutions of the Airy equation, their Wronskian is non-zero and this system has a unique solution $(A_{L-1}(\mu), B_{L-1}(\mu))$ for any given μ. Since the Wronskian is a constant, it is easy to show that $A_{L-1}(\mu)$ and $B_{L-1}(\mu)$ are entire functions of μ. We repeat the matching all the way down to the first segment, finding a unique solution for the coefficients $A_l(\mu)$ and $B_l(\mu)$ at each stage. Again $A_l(\mu)$ and $B_l(\mu)$ are entire functions of μ. The last step, which determines whether μ is an eigenvalue, is to ensure the boundary condition at $z = 0$ is satisfied. The modal condition is

$$\begin{aligned} &\beta_0\left[A_0(\mu)w_1'\{\eta_0(\mu)\} + B_0(\mu)w_2'\{\eta_0(\mu)\}\right] \\ &\qquad = ik\delta\left[A_0(\mu)w_1\{\eta_0(\mu)\} + B_0(\mu)w_2\{\eta_0(\mu)\}\right] \end{aligned} \tag{5.55}$$

Finally, we see that determination of the modes amounts to finding the zeros of a holomorphic function expressed in terms of Airy functions. In practice, only the modes with small imaginary part are of interest, since modes with large imaginary part correspond to terms in the mode sum which attenuate very quickly with range. However even the seach for these "locked" modes can be an extremely time-consuming undertaking, particularly at high frequencies when the number of low attenuation modes can become rather large. It is also a numerically tricky procedure, since the Airy functions range from exponentially small to exponentially large. More information on numerical conditioning and on root finding algorithms can be found in [105, 124, 119].

Once the required modes have been found, the normalization factors can be computed in closed form thanks to the properties of the Airy functions [123]. We note that if w satisfies the Airy differential equation, eqn. 5.45, we have

$$\left(\zeta w^2 - w'^2\right)'(\zeta) = w^2 \tag{5.56}$$

It follows that for eigenfunction f_j corresponding to the eigenvalue μ_j, the normalization factor is given by

$$\int_\Gamma f_j^2 = \sum_{l=0}^{L} \gamma_l \tag{5.57}$$

where

$$\gamma_L = -\frac{1}{\beta_L}\left\{\eta_L(\mu_j) f_j^2(z_L) + f_j'^2(z_L)\right\} \tag{5.58}$$

is the contribution from the interval $[z_L, \infty[$ and

$$\gamma_l = \frac{1}{\beta_l}\left\{\chi_{l+1}(\mu_j) f_j^2(z_{l+1}) + f_j'^2(z_{l+1}) - \eta_l(\mu_j) f_j^2(z_l) - f_j'^2(z_l)\right\} \tag{5.59}$$

is the contribution from interval $[z_l, z_{l+1}]$ for $l = 0, \ldots, L-1$.

Both the search for the roots of the modal equation and the computation of height-gain functions require accurate calculation of complex Airy functions. This is done with an appropriate combination of Taylor series for small values of the argument and asymptotic series for large values of the argument [14].

Chapter 6

Oversea propagation

6.1 Introduction

The simple narrow-angle split-step sine transform PE algorithm described in Chapter 3 is already powerful enough to treat many problems of oversea propagation. The assumption that the sea has infinite impedance is adequate at frequencies above 300 MHz or so regardless of polarization, provided we only consider small grazing angles on the surface and we neglect roughness effects. All the examples considered in this chapter are treated with a sine transform algorithm, neglecting polarization-dependent surface reflection effects. These will be examined in Chapters 9 and 10.

We look at commonly encountered ocean environments: evaporation ducts are considered in Section 6.2, and surface and elevated ducts given by bilinear and trilinear refractive index profiles are treated in Section 6.3. PE results are compared with ray-tracing and mode theory. Range-dependent environments are treated in Section 6.4 and examples involving measurements in situ are given in Section 6.5. The chapter ends with a discussion of absorption modelling in Section 6.6.

6.2 Evaporation ducts

In the boundary layer just above large bodies of water, turbulent transport processes result in a strong humidity gradient near the surface, giving rise to the evaporation duct. As the evaporation duct is present most of the time over the ocean, it has a major influence on shipborne radar and communications systems.

It is extremely difficult to measure vertical profiles of temperature of humidity at the very low heights which characterize the evaporation duct. One has to resort instead to models based on similarity theory [151], which parametrize the profiles in terms of bulk parameters. Stability length, the scaling

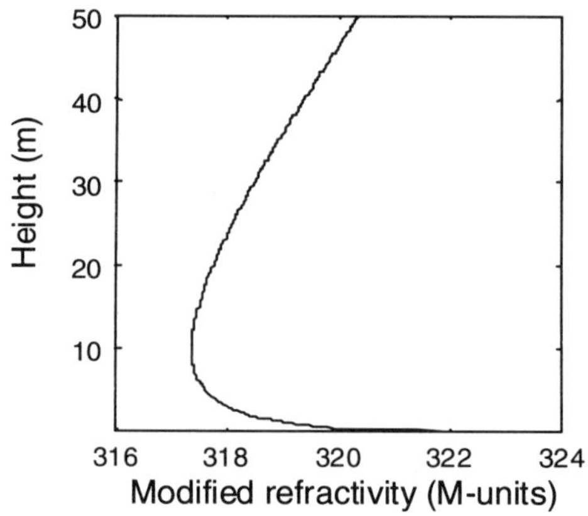

Figure 6.1 10 m evaporation duct profile

parameter for stability theory, is obtained as a function of four bulk parameters: sea temperature, and measurements of air temperature, humidity and wind speed at some fixed height. When the air and sea temperatures are almost equal, the boundary layer is neutrally buoyant and the stability length is $-\infty$. Then the model simplifies, allowing a parametrization in terms of duct height. The resulting modified refractivity profile is a log-linear function of height of the form

$$M(z) = M(0) + 0.125 \times \left(z - d\log\left(\frac{z}{z_0}\right)\right) \tag{6.1}$$

where d is the evaporation duct height and z_0 is the roughness length, which is usually very small. A typical value is $z_0 = 1.5 \times 10^{-4}$ m. Figure 6.1 shows the modified refractivity profile for a 10 m evaporation duct, with $M(0) = 330$ M-units. The modified refractivity curve exhibits the steep decrease close to the surface and very slow variations near the duct height which are typical of evaporation duct profiles.

Figure 6.2 shows evaporation duct effects on the field of a 3 GHz antenna located 25 m above the sea surface, for duct heights of 0, 10, 20 and 30 m. The contour plots of path loss are shown, using flat Earth coordinates. The zero duct height case is that of a standard atmosphere: since the modified refractivity gradient is positive, energy propagates away from the Earth, so the beam curves upwards in flat Earth coordinates. Because of diffraction around the Earth, there is no abrupt transition to zero beyond the radio-horizon, but a gradual decrease of signal. At that frequency, the 10 m duct does not have spectacular effects, but merely extends coverage slightly. Extension of

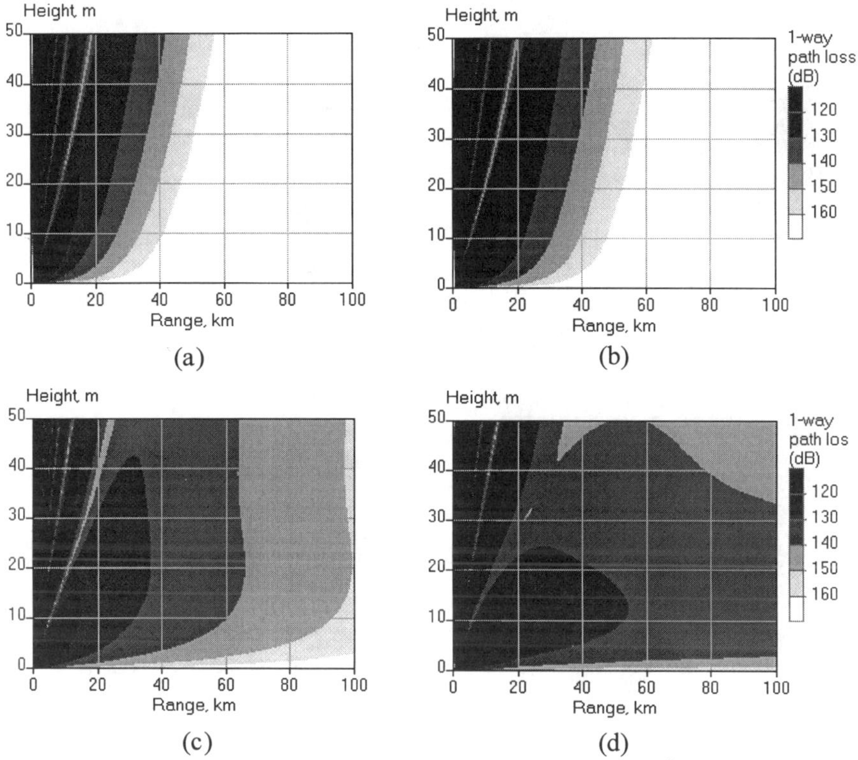

Figure 6.2 Evaporation duct effects at 3 GHz for (a) no duct and duct heights (b) 10, (c) 20 and (d) 30m

coverage is much more significant for the 20 m duct, and we can see energy begin to bend back towards the surface in the 30 m duct. Figure 6.3 shows the corresponding effects for a 10 GHz source. At 10 GHz, even the 10 m evaporation duct extends coverage substantially, and we can see how successive interference lobes get trapped in the duct as the height increases to 20 and 30 m.

Figures 6.4 and 6.5 show path loss against range on boresight, 25 m above sea level, for evaporation duct heights of 0, 10, 20 and 30 m, at 3 GHz and 10 GHz respectively. At long ranges path loss decreases very slowly in the stronger ducts, showing much increased radar performance at low altitudes.

It is interesting to compare the PE results with those of geometrical optics and mode theory. Figure 6.6 (a) shows ray-trace results for propagation in the standard atmosphere (no evaporation duct). Rays have been traced with

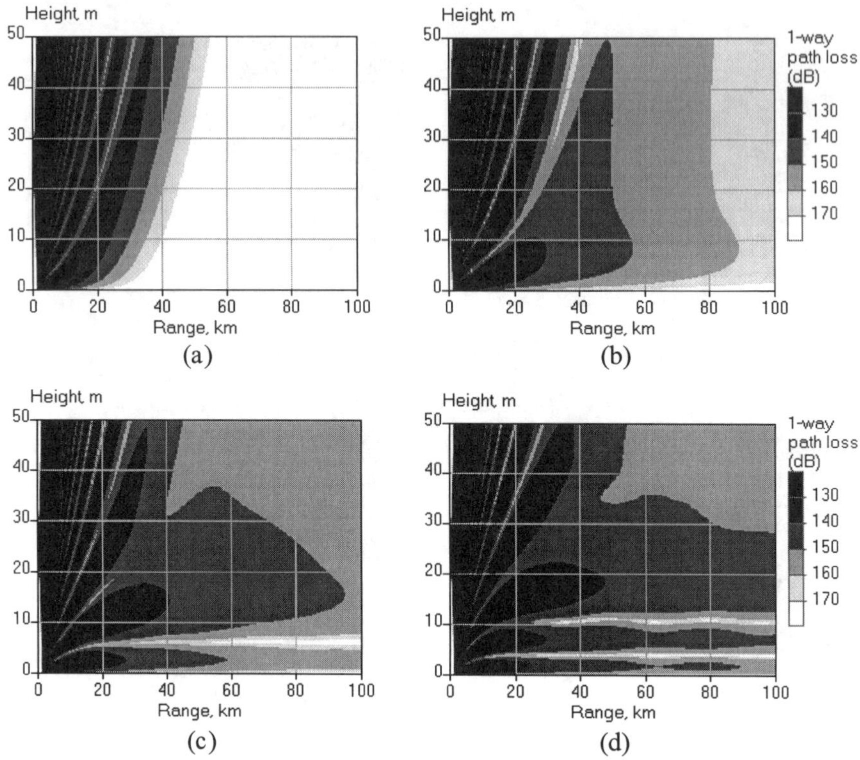

Figure 6.3 *Evaporation duct effects at 10 GHz for (a) no duct and duct heights (b) 10, (c) 20 and (d) 30m*

launch angles ranging from $-0.15°$ to $0.15°$, with separation of $0.005°$. Naturally, geometrical optics cannot represent energy propagating in the shadow of the Earth, hence there is an abrupt transition at the radio horizon. Figure 6.6 (b) shows ray-trace results for propagation in a 20 m evaporation duct. Ducting effects are largely unseen by the ray-trace.

By contrast, mode theory gives excellent results beyond the horizon for spherical diffraction and evaporation duct cases. It is particularly good for this type of problem because of the small number of modes which propagate without significant attenuation. Since the modal series does not converge close to the horizon and in the line of sight region, this method cannot be used there. The coherent mode sum results presented here have been obtained with the mode theory program MLAYER [123].

Figure 6.7 (a) shows mode theory results at 3 GHz for spherical diffraction,

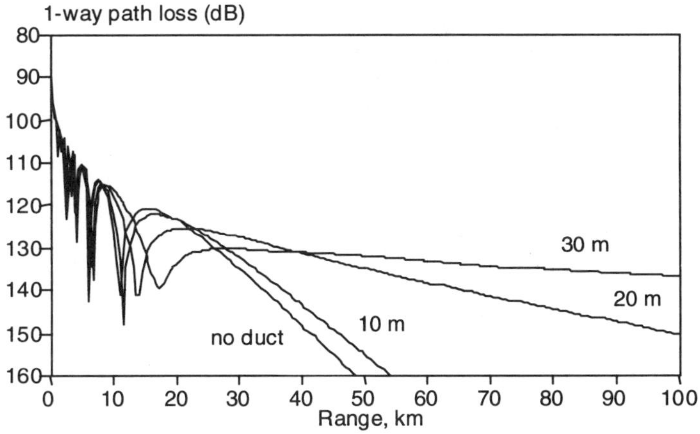

Figure 6.4 *Path loss against range on boresight (25 m above sea level) at 3 GHz for evaporation duct heights of 0, 10, 20 and 30 m*

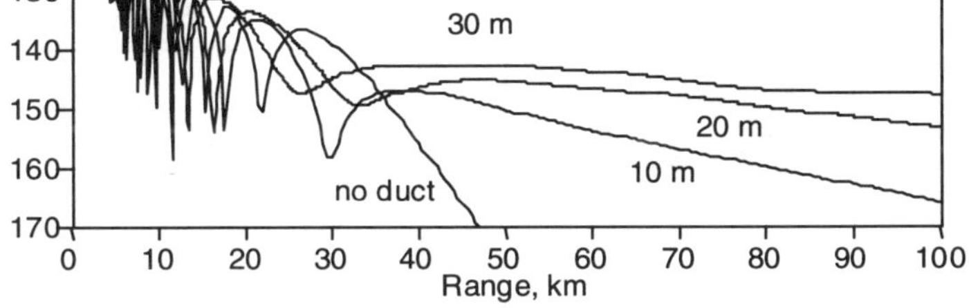

Figure 6.5 *Path loss against range on boresight (25 m above sea level) at 10 GHz for evaporation duct heights of 0, 10, 20 and 30 m*

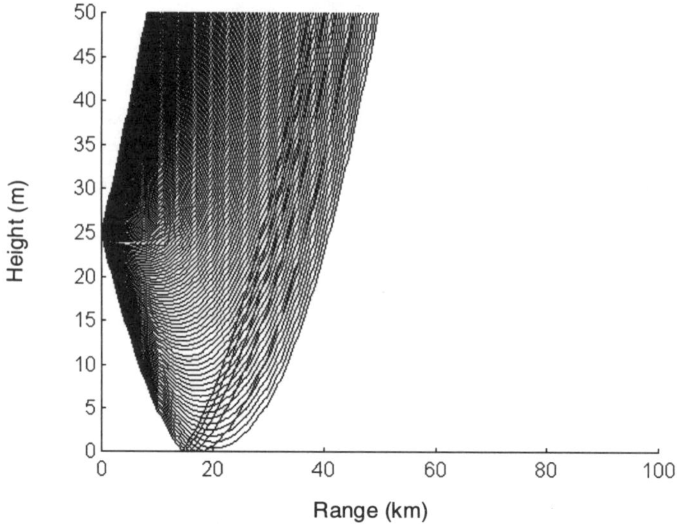

(a)

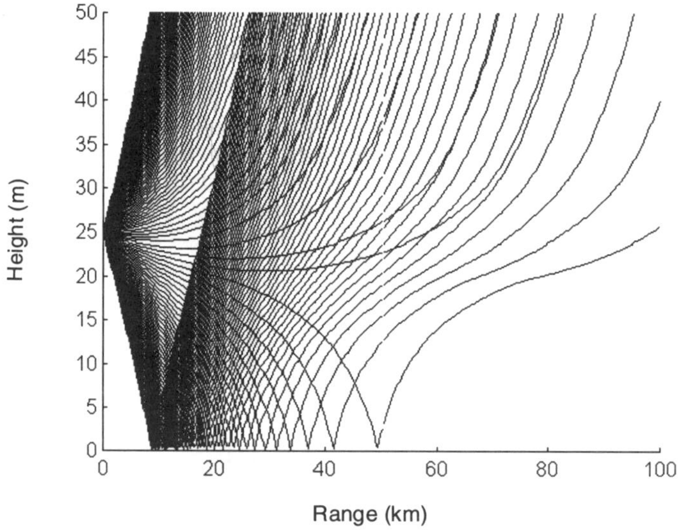

(b)

Figure 6.6 *Ray-tracing examples for (a) standard atmosphere and (b) 20 m evaporation duct*

Table 6.1 Refractivity for a surface duct modelled with bilinear profile

Height (m)	Refractivity (M-units)
0	330
100	320
1100	438

for a receiver at height 25 m. For comparison, the PE results are shown too, offset by 10 dB for clarity. The mode theory and PE results are identical beyond the horizon. Because of numerical reflections from the top of the integration domain, the PE results begin to show numerical problems when the field becomes very weak. This is not the case for the mode theory results: once the modes have been found, there is no numerical difficulty computing mode sums at arbitrary ranges. For this case, 6 modes were found with attenuation less than 5 dB/km. A similar comparison is shown in Figure 6.7(b) for the 20 m evaporation duct. Here 7 modes were found with attenuation less than 5 dB/km.

6.3 Bilinear and trilinear profiles

Ducting layers in the atmosphere can often be represented by simple bilinear or trilinear refractivity profiles. Surface layers are modelled as a bilinear function of height as shown in Figure 6.8(a). They always result in a surface-based duct which traps energy between the surface and the top of the ducting layer. Elevated layers require a trilinear profile, resulting in either a surface duct as in Figure 6.8(b), or an elevated duct, as shown in Figure 6.8(c). In the latter case, the lower boundary of the trapping layer is the height at which the refractive index has the same value as that at the top of the layer.

Because these ducts can involve very large numbers of significant modes at the frequencies of interest, the solution to this type of problem was elusive before the advent of PE models. Since the profile shape does not affect the performance of the PE model, there are no particular difficulties with bilinear or trilinear profiles.

This is illustrated by Figure 6.9, which shows path loss contours obtained using the bilinear profile given in Table 6.1, for 3 GHz and 10 GHz sources. The refractivity structure corresponds to a surface duct between 0 and 150 m. Ray-tracing results are also shown (c). Figure 6.10 shows analogous results for the trilinear profile given in Table 6.2, which corresponds to an elevated duct between 20 and 150 m.

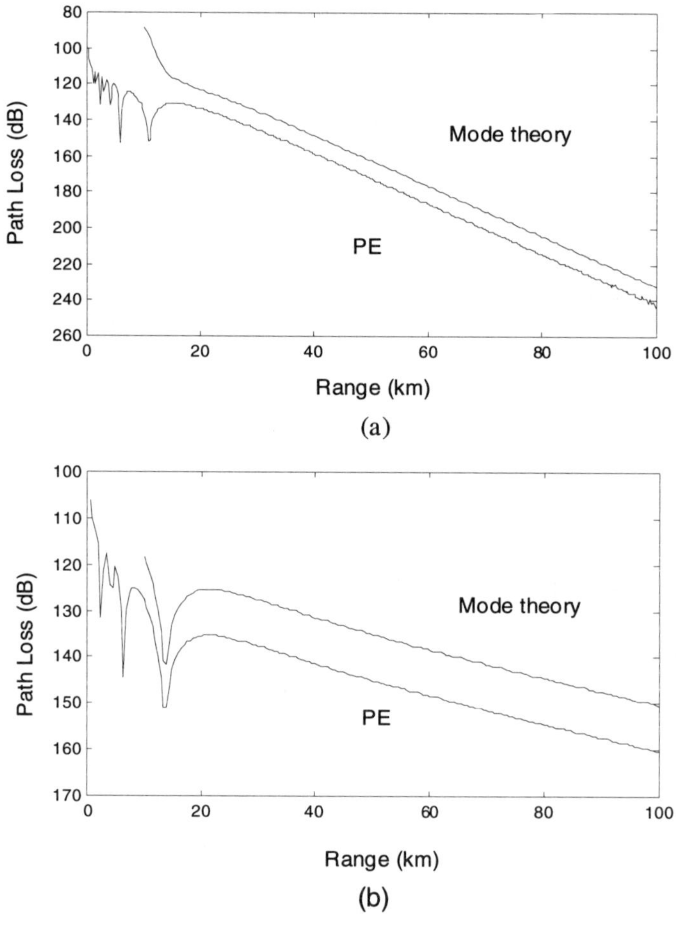

Figure 6.7 *Comparison of mode theory and PE results on boresight (25 m above sea level) for 3 GHz source in (a) standard atmosphere and (b) 20 m evaporation duct; the PE results are offset by 10 dB for clarity*

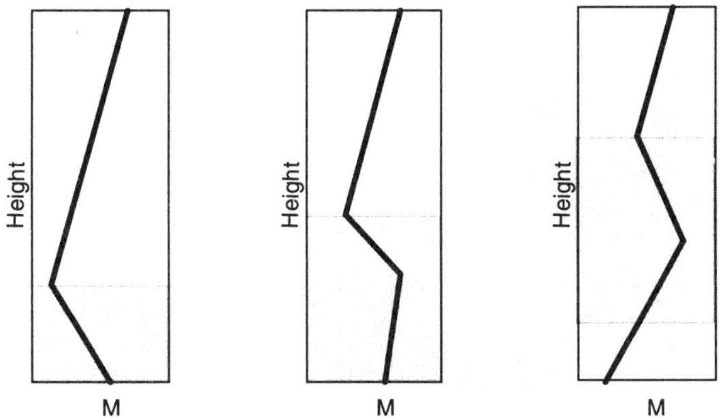

Figure 6.8 *Bilinear and trilinear modified refractivity profiles with (a) surface layer, (b) and (c) elevated layers; the ducting regions are indicated*

Table 6.2 *Refractivity for an elevated duct modelled with trilinear profile*

Height (m)	Refractivity (M-units)
0.0	330.0
100.0	341.8
150.0	331.8
1150.0	449.8

6.4 Range dependence

The split-step sine transform method can readily be adapted to deal with range-dependent environments, by treating the refractive index phase screen as a function of range. This amounts to solving a range-independent problem on each range interval $(x, x + \Delta x)$, switching to the relevant refractive index profile at each step. Naturally for a given range step the accuracy of the solution depends on the range variations of the refractive index. In typical tropospheric environments these are quite weak and there is no need to adapt the range step. There is essentially no increase in complexity in modelling range-dependent environments; in this respect the PE method is much more powerful than ray optics or mode theory.

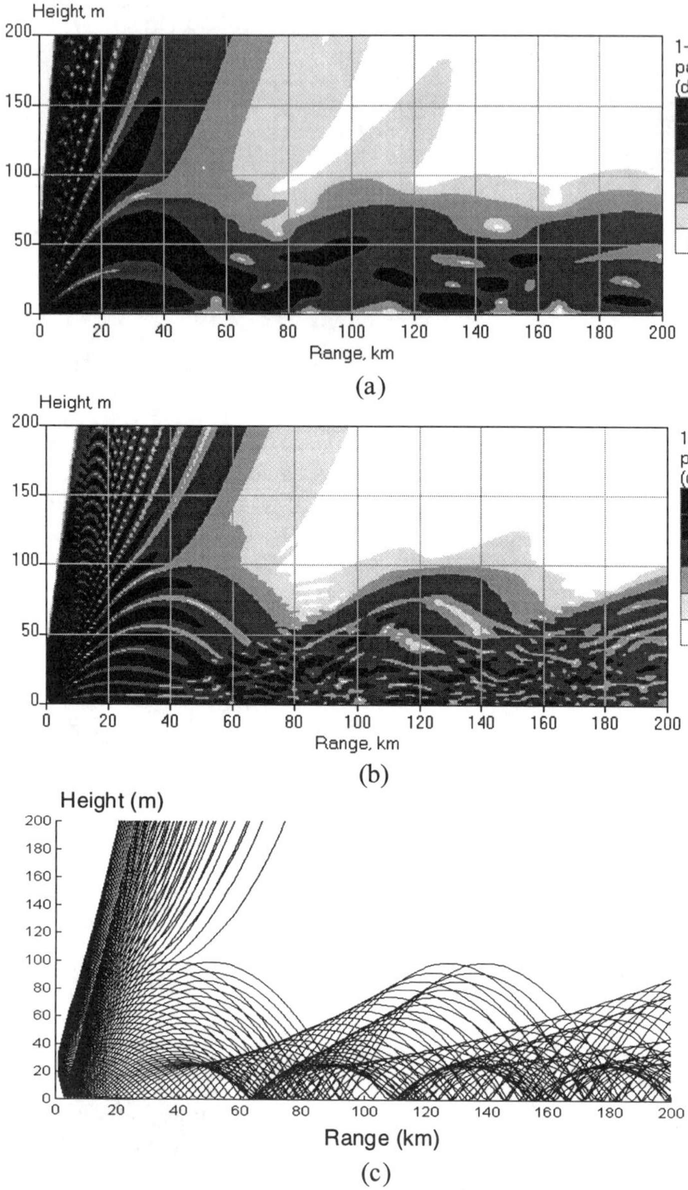

Figure 6.9 Bilinear surface duct modelling. PE results at (a) 3 GHz and (b) 10 GHz, and (c) ray-trace results

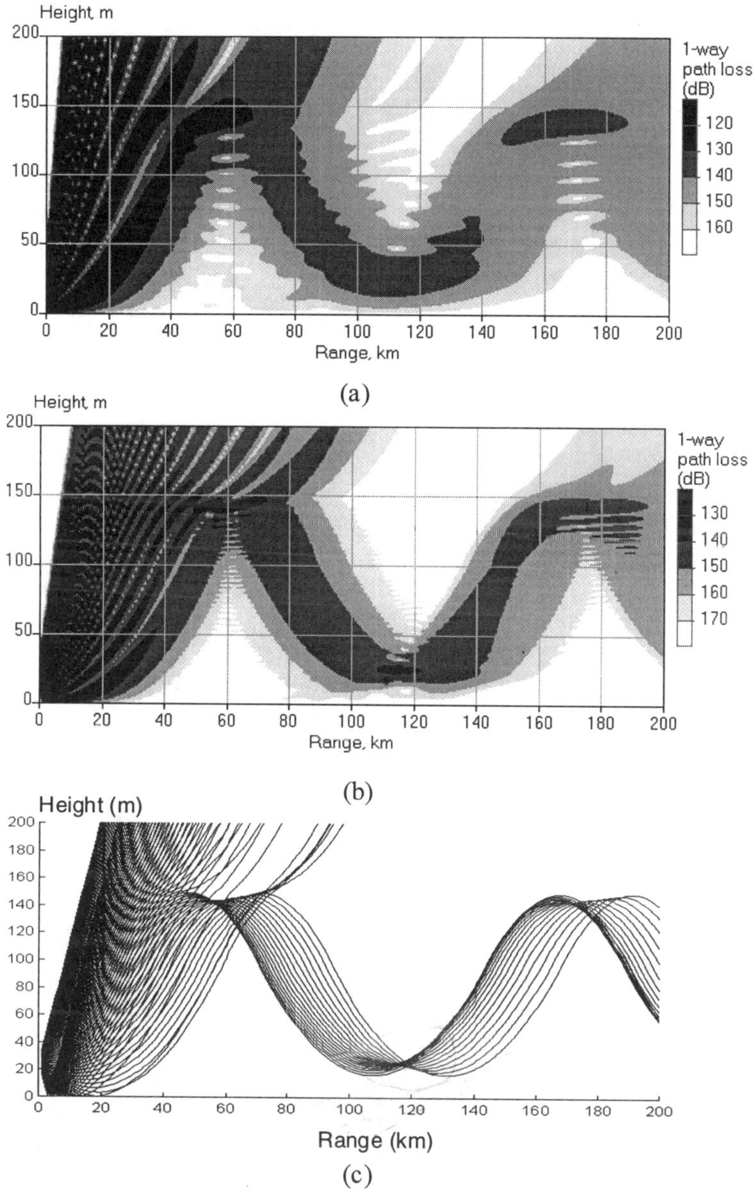

Figure 6.10 *Trilinear elevated duct modelling. PE results at (a) 3 GHz and (b) 10 GHz, and (c) ray-trace results*

Table 6.3 Modified refractivity values for sloping layer

Range: 0 km		Range: 100 km		Range: 200 km	
Height (m)	Refractivity (M-units)	Height (m)	Refractivity (M-units)	Height (m)	Refractivity (M-units)
0.0	300.0	0.0	300.0	0.0	300.0
50.0	310.0	150.0	315.0	100.0	310.0
100.0	300.0	250.0	310.0	150.0	300.0
200.0	311.8	350.0	321.8	250.0	311.8

Atmospheric layers are not necessarily parallel to the Earth's surface. Table 6.3 gives an example of sloping layer. Linear interpolation in both height and refractivity is used to link the profiles between ranges. The layer slopes up between 0 and 100 km, while increasing in thickness from 50 to 100 m and decreasing in strength from 10 M-units to 5 M-units. It then slopes down between 100 and 200 km, while becoming narrower and stronger. The consequences on propagation are illustrated in Figure 6.11, which shows a 3 GHz source in this environment, together with the profiles at 0, 100 and 200 km. Energy trapped in the duct at zero range travels up along the layer, with energy leakage as the duct weakens, then refocuses and travels downwards as the layer strengthens again.

6.5 Measured refractivity data

One of the great advantages of PE methods is that they can deal directly with measured radiosonde data, which often include multiple ducts or small scale features. Figure 6.12 shows a refractivity profile derived from measurements at Point Mugu on the Californian coast on 25 August 1993, during the VOCAR campaign [137]. A very strong surface-based duct was present, together with a weak elevated duct at about 1000 m. Table 6.4 gives the corresponding refractivity profile in tabular form. Figure 6.13 displays the path loss contours for a 3 GHz source in that environment. The antenna is located 25 m above sea level. This example was used as a test case for propagation model assessment [126]. PE models had no trouble dealing with this challenging case.

If two-dimensional measurements of refractivity are available, they too can be used directly as input for the PE model. A campaign of airborne measurements over the English Channel was carried out in 1989 [96]. The Meteorological Research Flight Hercules C130 aircraft carries a comprehensive set of meteorological instruments, permitting the detailed study of the two-dimensional structure of refractivity. Refractivity was measured directly with

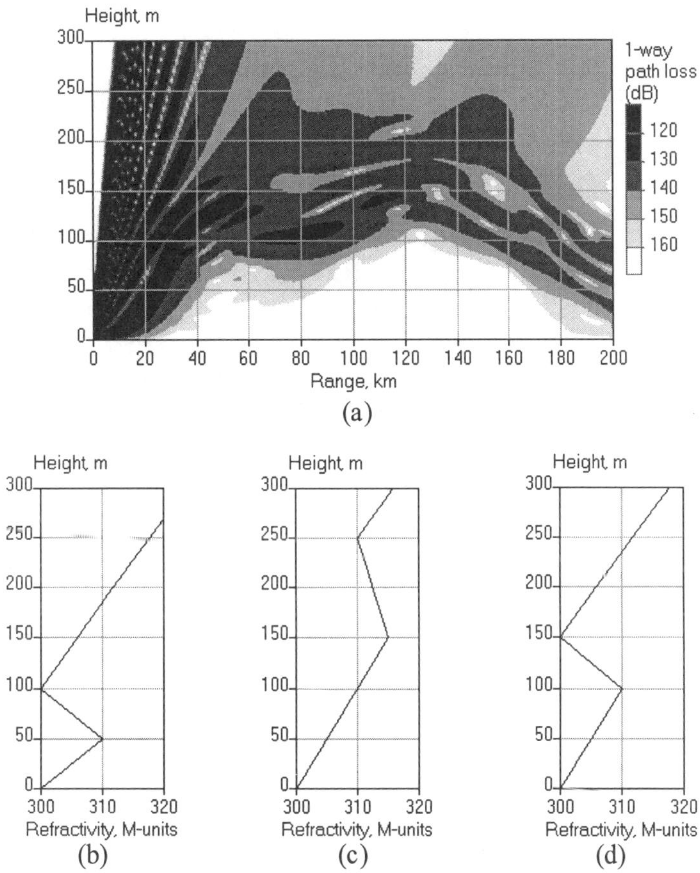

Figure 6.11 *Path loss contours for 3 GHz source in sloping duct (a), together with refractivity profiles at (b) 0, (c) 100 and (d) 200 km*

a refractometer and also calculated from pressure, temperature and relative humidity measurements using the Debye formula. In addition, the aircraft was fitted with a 400 MHz receiver, and a link was established with a transmitter at Bournemouth, on the British coast. The aircraft flew a sawtooth pattern in order to sample the range and height variations of the troposphere. The maximum height of interest was 1500 m, and the maximum distance was 250 km. Although it is clear that far more striking ducting effects would have been measured at a higher frequency, the 400 MHz beacon measurements did provide some opportunity of assessing PE-based forecasts.

Results from 18 July 1989 are given here. Conditions were anticyclonic,

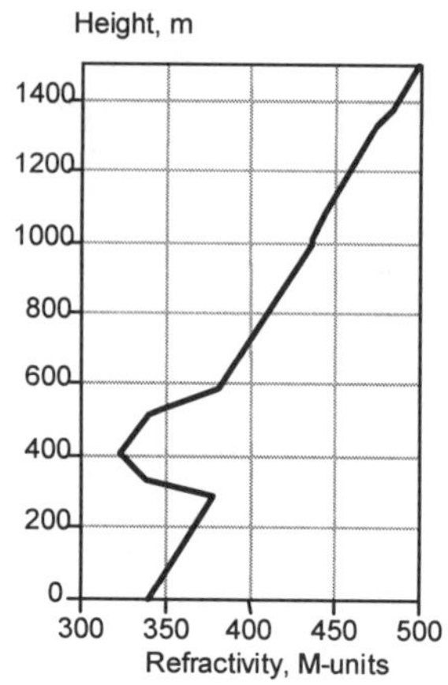

Figure 6.12 Radiosonde profile, 25 August 1993, Point Mugu

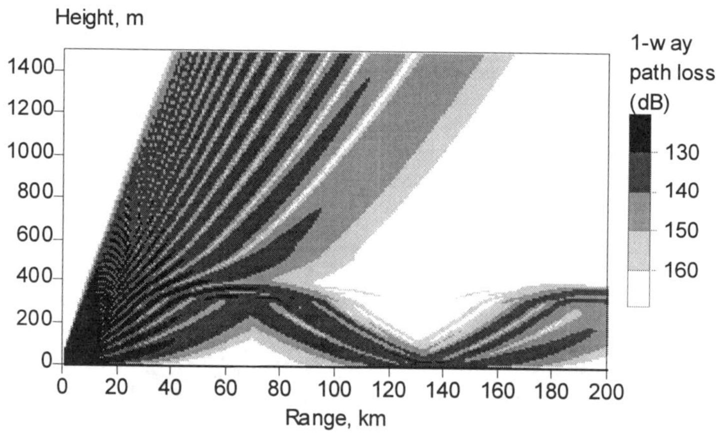

Figure 6.13 3 GHz source in environment of Figure 6.12

Table 6.4 Modified refractivity profile, Point Mugu, 25 August 1993

Height (m)	Refractivity (M-units)
0.0	339.8
292.0	378.1
332.0	337.9
403.0	322.7
514.0	338.6
590.0	381.1
998.0	435.4
1016.0	435.2
1083.0	443.7
1328.0	474.5
1379.0	484.3
2000.0	557.5

giving rise to a low level subsidence inversion. We use potential refractivity, which eliminates temperature variations due to pressure, making it easier to spot variations due to atmospheric layering [38]. It is obtained by replacing the temperature T in the Debye formula of eqn. 4.1 with the potential temperature θ which the air parcel would have if brought isentropically down to the surface. The potential temperature is given by

$$\theta = T\left(\frac{P_0}{P}\right)^{0.286} \tag{6.2}$$

where P_0 is the surface pressure [148]. Figure 6.14 shows contours of potential refractivity measured during the afternoon part of the mission, between 12:28 and 15:04 Greenwich Meridian Time. Over the two-hour period of interest, four saw-tooth traverses of the English Channel were made. Spatial sampling was sufficiently good to permit simple interpolation on a rectangular grid. In general, such detailed information is unavailable and one has to make do with sparse vertical refractivity profiles obtained from radiosondes. In that case, feature interpolation is often necessary in order to attempt adequate representation of range-dependent structures like sloping ducts [10].

Figure 6.15 shows the 400 MHz signal on one of the traverses. The measured signal is shown, together with the PE results based on the measured two-dimensional refractivity structure. For comparison, the PE results for a standard well-mixed atmosphere are also shown. The signal is measured well beyond the standard horizon, showing that anomalous propagation conditions

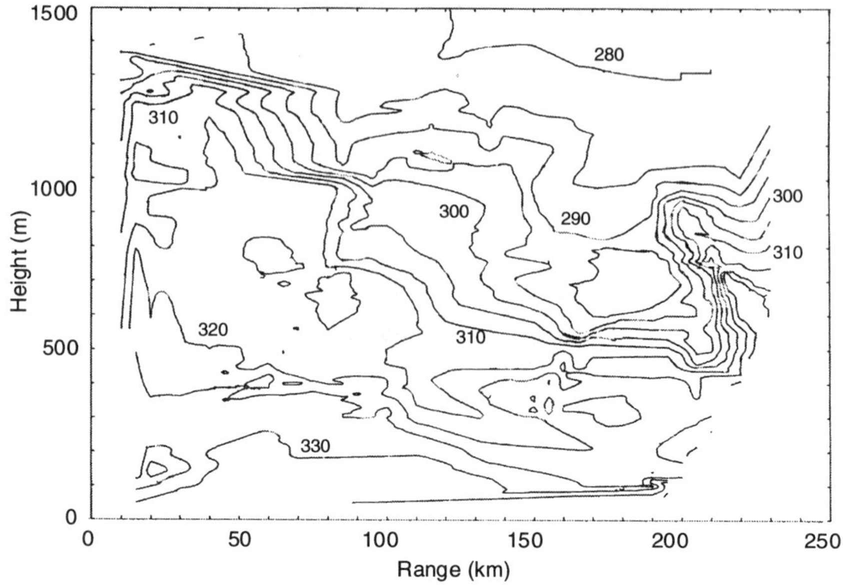

Figure 6.14 Potential refractivity contours for 18 July 1988, 12:28 GMT to 15:04 GMT

hold. Rather encouraging agreement is obtained between the PE results and the measurements, despite the various sources of experimental error [96].

6.6 Atmospheric absorption

Absorption effects can be incorporated in PE models by adding an imaginary part to the refractive index. If we denote this absorptive component by $a(x, z)$, the narrow-angle split-step PE becomes

$$u(x + \Delta x, z) = \mathrm{e}^{-\frac{ik\Delta x}{2}(n^2(x,z)-1)}\mathrm{e}^{-ka(x,z)\Delta x}e^{\frac{i\Delta x}{2k}\frac{\partial^2}{\partial z^2}} \cdot u(x, z) \qquad (6.3)$$

If atmospheric absorption does not vary with height, its effects are modelled very simply by multiplying the field at each range by the appropriate exponential, or equivalently by adding an absorption term to the path loss. For oversea propagation, however, evaporation ducts are accompanied by relatively strong variations of atmospheric absorption with height because of the large humidity gradients close to the surface. In that case differential absorption

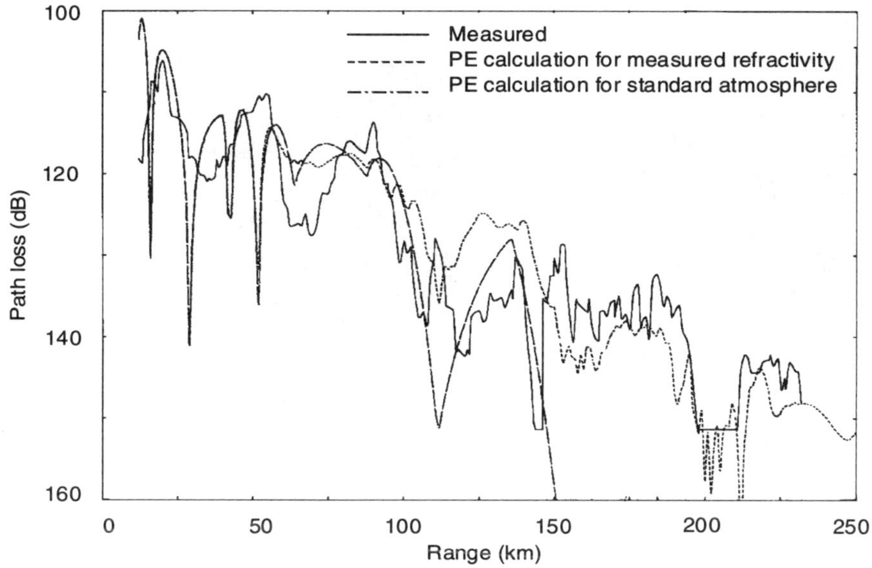

Figure 6.15 *400 MHz signal on one of the MRF flight traverses, 18 July 1989*

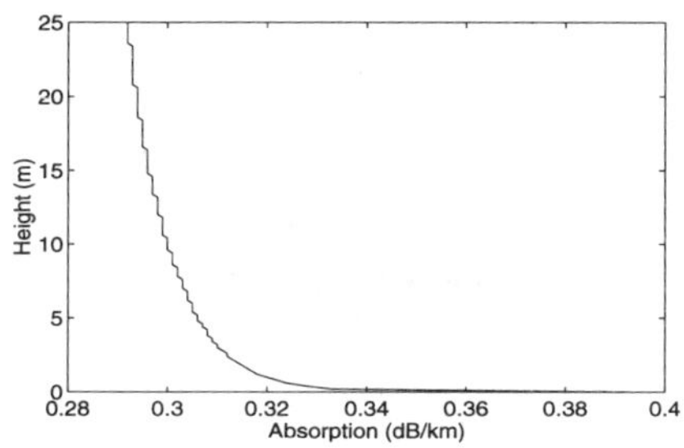

Figure 6.16 *Atmospheric absorption profile at 22 GHz in 20 m evaporation duct*

effects can be noticeable, and it is more accurate to use the modified PE given by eqn. 6.3.

Figure 6.16 shows the absorption profile in dB/km at 22 GHz for a 20 m evaporation duct, using bulk parameters of 18° Celsius for both sea and air temperatures, 71% for relative humidity and 10 m/s for wind speed. Absorption decreases quite sharply in the first few metres above the sea surface. Figure 6.17 shows path loss contours for a 22 GHz source located 15 m above sea level, using the variable absorption model of eqn. 6.3, and also using a con-

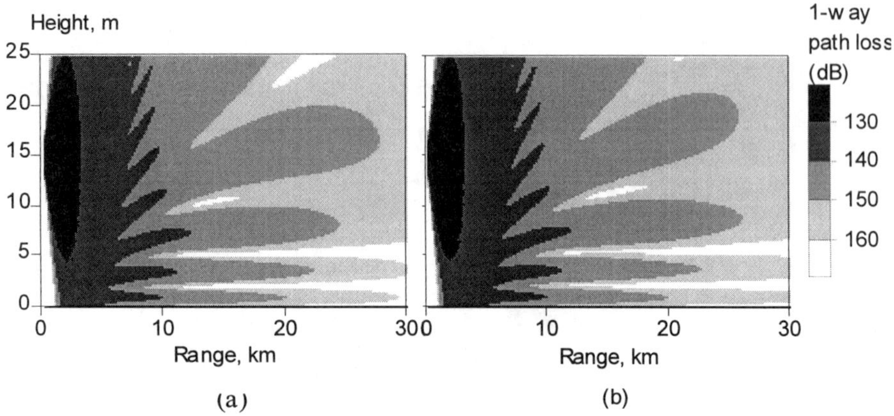

Figure 6.17 *Path loss contours for 22 GHz source in 20 m evaporation duct for (a) variable and (b) constant atmospheric absorption models*

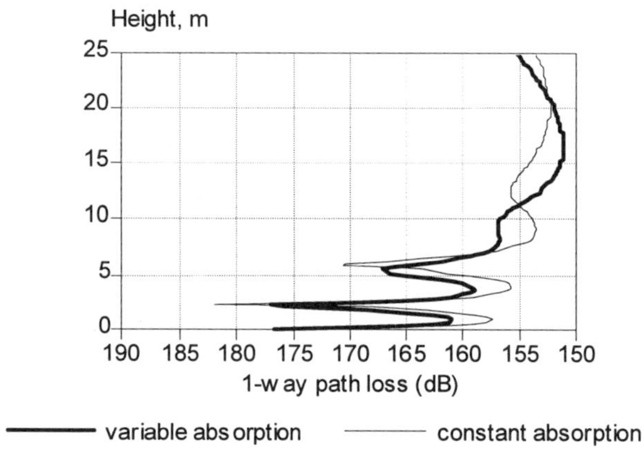

Figure 6.18 *Path loss against height for 22 GHz source in 20 m evaporation duct at range 30 km*

stant absorption of 0.25 dB/km, which is the boresight value for the profile of Figure 6.16. The lobing patterns are different for the two absorption models: in both cases several interference lobes are trapped in the evaporation duct, but with the variable absorption model the lobes close to the surface attenuate faster due to the higher absorption values lower down. This is confirmed by Figure 6.18 which shows path loss against height at a range of 30 km.

Chapter 7
Irregular terrain modelling

7.1 Introduction

Accurate modelling of radiowave propagation over irregular terrain is crucial for the planning of cellular communications networks and the prediction of radar performance in coastal environments. Many operational models are based on the very efficient simplified Deygout solution for multiple knife-edge diffraction [43, 44]. More sophisticated diffraction models representing the terrain profile as a succession of knife-edges, wedges, cylinders or flat strips are also available [159, 98, 143, 165]. Another important family of models uses an integral equation formulation [120, 118], which can be greatly simplified by making a paraxial approximation [60]. All of these techniques assume propagation in a homogeneous or linear atmosphere. By contrast parabolic equation methods model the combined effects of terrain diffraction and atmospheric refraction [85, 107, 11] while remaining straightforward to implement.

PE modelling of propagation over irregular terrain depends on the choice of coordinate system and terrain representation, as discussed in Section 7.2. The simplest technique models terrain as a sequence of horizontal steps, so that sloping terrain effects are modelled only approximately. The staircase method is extremely effective: the grid is essentially terrain-independent so that quite large range steps can be used. More accurate modelling of slopes can be obtained by using coordinate systems following terrain [65]. A simple and effective approach introduced in [102] represents terrain height as a continuous piecewise linear function of range, and uses a terrain-dependent reduced function. A more complicated conformal mapping method requiring information on terrain curvature is also available [11]. All these techniques can be implemented using straightforward extensions of the sine transform method described in the previous chapter. Alternatively, the grid can be adapted to terrain shape, still retaining a Cartesian coordinate system [85]. This requires a finite-difference implementation, which is presented in Section 7.3. Applica-

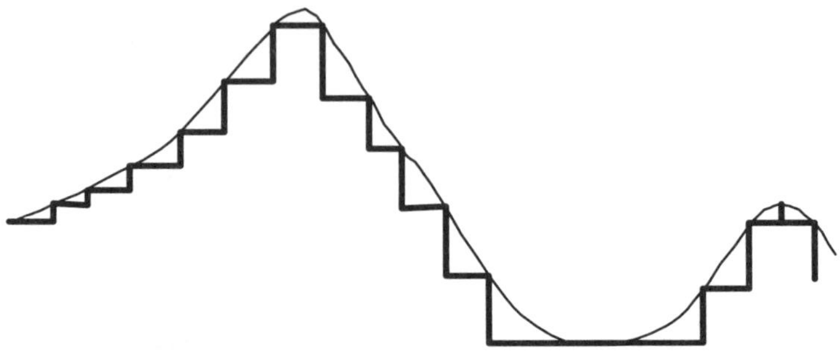

Figure 7.1 Staircase representation of terrain

tions to diffraction around the Earth, multiple edge diffraction and propagation over irregular terrain are given in Section 7.4.

In this chapter, we assume that the ground is perfectly conducting and that the incident wave is horizontally polarized. Generalization to surface impedance boundary conditions will be given in Chapter 9.

7.2 Terrain models

7.2.1 Staircase terrain modelling

This approach models terrain with successive horizontal segments, as shown in Figure 7.1. On each segment of constant height, the field is propagated in the usual way, applying the appropriate boundary condition at the ground. When the terrain height changes, corner diffraction is ignored and the field is simply set to zero on vertical terrain facets. The staircase treatment is shown in Figure 7.2. If the terrain goes up, as in Figure 7.2(a), the sequence of operations to compute the field at range $x + \Delta x$ is the following:

- propagate the field on horizontal segment S_1, ignoring the presence of the vertical boundary S_2; this is consistent with the paraxial approximation which neglects backscatter;
- truncate the field by setting it to zero on S_2; this is consistent with the assumption that the ground does not support propagation at the wavelength of interest.

If the terrain goes down, as in Figure 7.2(b), the sequence of operations is now:

- propagate the field on horizontal segment S_1, ignoring the presence of the vertical boundary S_2; hence we neglect backscattering due to corner diffraction;

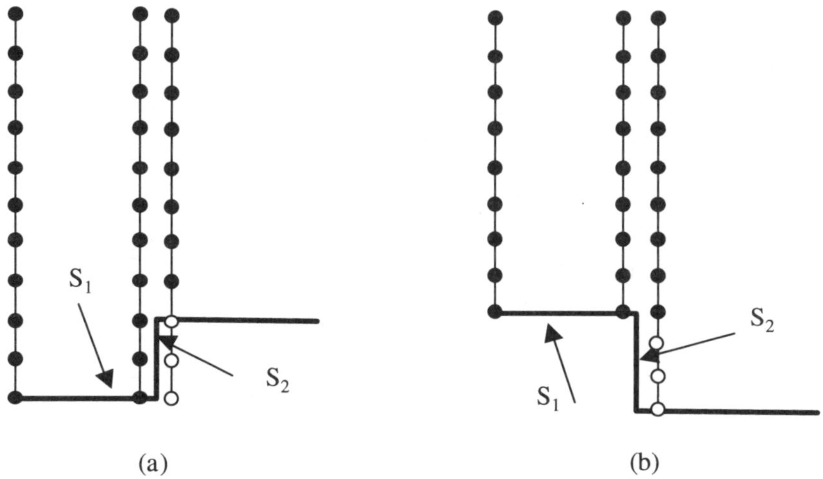

Figure 7.2 *Advancing the PE solution on staircase terrain for (a) ascending and (b) descending steps*

- pad the field by setting it to zero on S_2.

If the terrain really was a staircase, this model would be quite accurate, extending the PE solution for half-space screen diffraction described in Chapter 2. For smooth terrain, this representation is of course in error, since boundary conditions on sloping facets are not properly accounted for.

7.2.2 Piecewise linear terrain

Here the terrain is represented as a sequence of linear segments, as shown in Figure 7.3. The coordinate system measures height from the terrain. In other words we define new range and height variables by

$$\begin{cases} \xi = x \\ \zeta = z - h(x) \end{cases} \tag{7.1}$$

where $h(x)$ is the terrain height (see Figure 7.3).

Assume the terrain has slope α on segment $x_1 \le x \le x_2$. Then we have in the corresponding vertical slice

$$\begin{cases} \xi = x \\ \zeta = z - h(x_1) - \alpha(x - x_1) \end{cases} \tag{7.2}$$

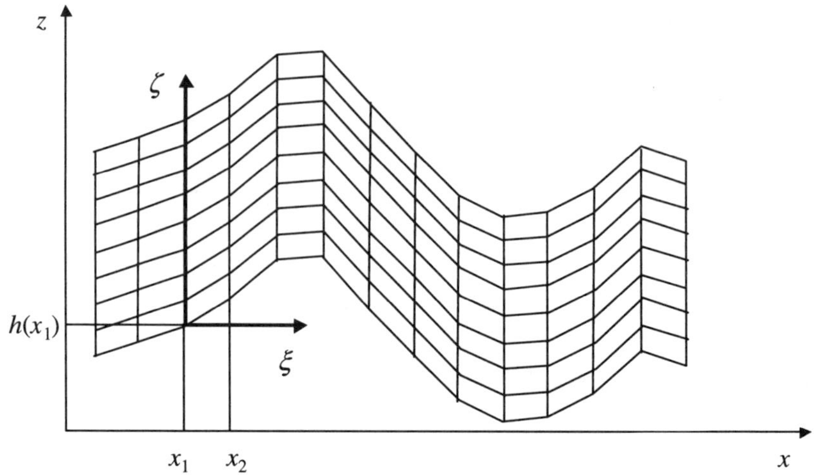

Figure 7.3 *Coordinate system for linear terrain*

We now look at the new function v defined by

$$v(\xi, \zeta) = \exp(ik\alpha\zeta)u(x, z) \tag{7.3}$$

This function compensates for the wavefront shift relative to the terrain. We have

$$\begin{aligned} \frac{\partial u}{\partial x} &= \exp(ik\alpha\zeta)\left\{\frac{\partial v}{\partial \xi} - \alpha\frac{\partial v}{\partial \zeta}\right\} \\ \frac{\partial u}{\partial z} &= \exp(ik\alpha\zeta)\left\{\frac{\partial v}{\partial \zeta} + ik\alpha v\right\} \\ \frac{\partial^2 u}{\partial z^2} &= \exp(ik\alpha\zeta)\left\{\frac{\partial^2 v}{\partial \zeta^2} + 2ik\alpha v\right\} \end{aligned} \tag{7.4}$$

Hence the SPE becomes in terms of v

$$\frac{\partial^2 v}{\partial \zeta^2} + 2ik\frac{\partial v}{\partial \xi} + k^2(n^2 - 1)v = 0 \tag{7.5}$$

We are back to the usual PE, but we now solve for a function which follows the terrain by applying appropriate angle shifts on successive segments. Implementation with a sine transform is straightforward. We denote by $(r_0, h_0), (r_1, h_1), \ldots$

the terrain break points and by $\alpha_1, \alpha_2, \ldots$ the successive terrain slopes. Suppose we have computed the field at range r_{m-1}. We then march the solution to range r_m with the sequence of operations outlined below.

1. First we shift the wave front, putting

$$v_m(r_{m-1}, \zeta) = u(r_{m-1}, h_{m-1} + \zeta) \exp(-ik\alpha_m \zeta) \tag{7.6}$$

2. We now propagate v_m to range r_m using the sine transform method. On interval $[r_{m-1}, r_m]$, the SPE algorithm is given by

$$v_m(\xi + \Delta\xi, \zeta) = \mathrm{e}^{-\frac{ik(n^2-1)}{2}\Delta\xi} \mathcal{S}^{-1} \left\{ \mathrm{e}^{-\frac{2i\pi^2 p^2}{k}\Delta x} \mathcal{S} \left\{ v_m(xi, \zeta') \right\} \right\} \tag{7.7}$$

   This propagates the field along the mth segment, automatically enforcing the boundary condition that the field should be zero at the ground.

3. At range r_m, we go back to u with the formula

$$u(r_m, h_m + \zeta) = v_m(r_m, \zeta) \exp(ik\alpha_m \zeta) \tag{7.8}$$

4. To continue the process, we shift the wave front again. In terms of v_m, the next function v_{m+1} is given by

$$v_{m+1}(r_m, h_m + \zeta) = v_m(r_m, \zeta) \exp(ik(\alpha_m - \alpha_{m+1})\zeta) \tag{7.9}$$

Propagation angles are now measured relative to the sloping terrain in each successive slice. Discretization in height proceeds along the lines of Section 3.3. From eqn. 7.3, we see that if we want the solution u to represent propagation angles up to $\theta_{\max}$ accurately, then v must represent propagation angles up to $\theta_{\max} + \alpha_{\max}$, where $\alpha_{\max}$ is the maximum terrain slope modulus. As a result the vertical grid resolution can increase substantially. Assuming small angles and filtering in half of the spectral domain as in Section 3.3, we have from Nyquist's criterion (eqn. 3.55)

$$\Delta\zeta \le \frac{\lambda}{4\left(\theta_{\max} + \alpha_{\max}\right)} \tag{7.10}$$

In hilly terrain, where terrain slopes can easily reach 5°, modelling propagation angles up to say 1° relative to terrain would require a height resolution of $\lambda/4$, which is impractical for large problems at microwave frequencies.

7.2.3 Conformal mapping

This approach assumes smooth terrain given by a function $h(x)$. We assume that h has a continuous second-order derivative in x. Again we define new range and height variables by

$$\begin{cases} \xi &= x \\ \zeta &= z - h(x) \end{cases} \tag{7.11}$$

Instead of considering a sequence of vertical slices following a piecewise linear terrain, we now look at a global change of function, given by

$$v(\xi, \zeta) = \exp\{i\theta(\xi, \zeta)\} u(x, z) \tag{7.12}$$

where the terrain-dependent phase function θ is defined by

$$\theta(\xi, \zeta) = k\zeta h'(\xi) + \frac{k}{2}\int_0^\xi h'^2(s)ds \tag{7.13}$$

The partial derivatives of u are expressed in terms of v by

$$\begin{aligned} \frac{\partial u}{\partial x} &= \mathrm{e}^{i\theta}\left\{\frac{\partial v}{\partial \xi} + iv\frac{\partial \theta}{\partial \xi} - \left[\frac{\partial v}{\partial \zeta} + iv\frac{\partial \theta}{\partial \zeta}\right] h'(x)\right\} \\ \frac{\partial u}{\partial z} &= \mathrm{e}^{i\theta}\left\{\frac{\partial v}{\partial \zeta} + iv\frac{\partial \theta}{\partial \zeta}\right\} \\ \frac{\partial^2 u}{\partial z^2} &= \mathrm{e}^{i\theta}\left\{\frac{\partial^2 v}{\partial \zeta^2} + 2i\frac{\partial v}{\partial \zeta}\frac{\partial \theta}{\partial \zeta} + v\left[i\frac{\partial^2 \theta}{\partial \zeta^2} - \left(\frac{\partial \theta}{\partial \zeta}\right)^2\right]\right\} \end{aligned} \tag{7.14}$$

In terms of function v, the SPE becomes

$$\begin{aligned} \frac{\partial^2 v}{\partial \zeta^2} & + 2ik\frac{\partial v}{\partial \xi} + 2i\left(\frac{\partial \theta}{\partial \zeta} - kh'(\xi)\right)\frac{\partial v}{\partial \zeta} \\ & + \left\{k^2(n^2-1) - 2k\frac{\partial \theta}{\partial \xi} + 2kh'(\xi)\frac{\partial \theta}{\partial \zeta} + i\frac{\partial^2 \theta}{\partial \zeta^2} - \left(\frac{\partial \theta}{\partial \zeta}\right)^2\right\} v = 0 \end{aligned} \tag{7.15}$$

With our choice of θ, we have

$$\frac{\partial \theta}{\partial \zeta} = kh'(\xi) \tag{7.16}$$

$$\frac{\partial \theta}{\partial \xi} = k\zeta h''(\xi) + \frac{k}{2}\left(h'(\xi)\right)^2 \tag{7.17}$$

So eqn. 7.15 simplifies to

$$\frac{\partial^2 v}{\partial \zeta^2} + 2ik\frac{\partial v}{\partial \xi} + k^2 m_h^2(\xi, \zeta)v = 0 \tag{7.18}$$

which is of the same form as the SPE, with a modified refractive index m_h defined by

$$m_h^2(\xi, \zeta) = n^2(\xi, h(\xi) + \zeta) - 2\zeta h''(\xi) \tag{7.19}$$

Hence we can solve for v with the split-step sine transform method. If only the amplitude of the field is required, we are finished since the ratio of u and v is of modulus 1. If the phase is required as well, the offset given by eqn. 7.13 must be calculated at each output range.

Eqn. 7.15 is more accurate than eqn. 7.5 if the second derivative of the terrain profile is known. In practice, this is not usually the case and we have no information on terrain slope and curvature. Then the best we can do is to approximate the second derivative at range r_{m-1} by

$$h''(r_{m-1}) \sim \frac{\alpha_m - \alpha_{m-1}}{\Delta x} \tag{7.20}$$

where we have used the notation of the previous section. Hence the extra multiplicative term due to the modification of the refractive index becomes

$$\exp\left\{ik(\alpha_{m-1} - \alpha_m)\zeta\right\} \tag{7.21}$$

which is exactly the same as the linear shift method of the previous section. The conformal framework is however more general, and can be used in particular to revisit the Earth flattening approximation. If we approximate the Earth's surface by the parabola

$$h(x) = -\frac{x^2}{2a} \tag{7.22}$$

where a is the Earth's radius, we find

$$m_h^2(\xi,\zeta) = n^2(\xi, h(\xi)+\zeta) + \frac{2\zeta}{a} \tag{7.23}$$

or, if we neglect terms of order more than 1 in ζ/a,

$$m_h(\xi,\zeta) \sim n(\xi, h(\xi)+\zeta) + \frac{\zeta}{a} \tag{7.24}$$

which is exactly the modified refractive index m defined in Chapter 4.

7.3 Finite-difference implementation

The Crank-Nicolson schemes described in Chapter 3 can be adapted for terrain modelling by following the ideas described earlier for sine transform implementation. Alternatively, sloping boundaries can be implemented by enforcing the boundary condition $u = 0$ directly. When the terrain is sloping upwards, we use eqn. 3.81 for points above the terrain. If the ground at the advanced range x_m is at height $j_m \Delta z$, we use the $N+1-j_m$ equations

$$\begin{cases} u_{j_m}^m = 0 \\ \alpha_j u_j^m + u_{j+1}^m + u_{j-1}^m = \beta_j u_j^{m-1} - u_{j+1}^{m-1} - u_{j-1}^{m-1} & , j_m < j < N \\ u_N = 0 \end{cases} \tag{7.25}$$

For terrain sloping downwards, we assume that the difference in terrain height between the current and advanced ranges is exactly one grid increment Δz. In order to use the implicit scheme, the second-order height derivative has to be approximated on the boundary at the current range x_{m-1}. We use the approximation

$$\frac{\partial^2 u}{\partial z^2}(x_{m-1}, j_{m-1}\Delta z) \sim \frac{\partial^2 u}{\partial z^2}(x_{m-1}, (j_{m-1}+1)\Delta z) \tag{7.26}$$

Putting $l = j_{m-1}$, this yields the one-sided approximation

$$\frac{\partial^2 u}{\partial z^2}(x_{m-1}, l\Delta z) \sim \frac{u_l^{m-1} + u_{l+2}^{m-1} - 2u_{l+1}^{m-1}}{\Delta z^2} \tag{7.27}$$

We can now write the system

$$\begin{cases} u_{l-1}^{m} = 0 \\ \alpha_l u_l^m + u_{l+1}^m + u_{l-1}^m = \gamma_1 u_l^{m-1} - \gamma_2 u_{l+1}^{m-1} - u_{l+2}^{m-1} \\ \alpha_j u_j^m + u_{j+1}^m + u_{j-1}^m = \beta_j u_j^{m-1} - u_{j+1}^{m-1} - u_{j-1}^{m-1} \quad , l < j < N \\ u_N = 0 \end{cases} \tag{7.28}$$

This is a straightforward modification of the right hand side of the Crank-Nicolson scheme.

The obvious drawback of this approach is that the range-step must be adapted to the slope. This model is more accurate than the staircase model, as knowledge of the next terrain point is incorporated into the forward stepping algorithm. It is however less precise than the shift approach, which incorporates knowledge of the slope itself in the algorithm.

7.4 Examples

7.4.1 Diffraction by the Earth

Diffraction by the spherical Earth can be treated in two ways: either by representing the Earth's surface explicitly as a sphere, or using the Earth flattening transformation of Chapter 4. Figure 7.4 shows path loss contours for a horizontally polarized 1 GHz transmitter located 25 m above the surface, with the Earth represented explicitly as variable terrain. Of course for this representation the refractive index is not modified. A linearly decreasing profile corresponding to a standard well-mixed atmosphere has been used, with a slope of -39 N-units/km. The Earth is treated as a perfect conductor, so the field is zero on the surface. As expected, energy propagates along trajectories curving downwards, with a radius of curvature larger than that of the Earth since the refractive index is decreasing, so that energy propagates away from the surface. Because of diffraction around the Earth's surface, there is no sharp transition at the radio-horizon, but a smooth decrease of signal strength beyond the horizon.

Figure 7.5 shows calculation results for the same parameters when the Earth flattening transformation is used. For this simulation the low altitude modified refractive index is used, giving a linearly increasing profile with a slope of 118 M-units/km. Again energy propagates away from the surface, but the conformal mapping has the effect of flattening the Earth and produces trajectories curving upwards.

For both simulations, a range step of 250 m and a height step of 1 m were used. The two methods give identical results, as shown in Figure 7.6, which

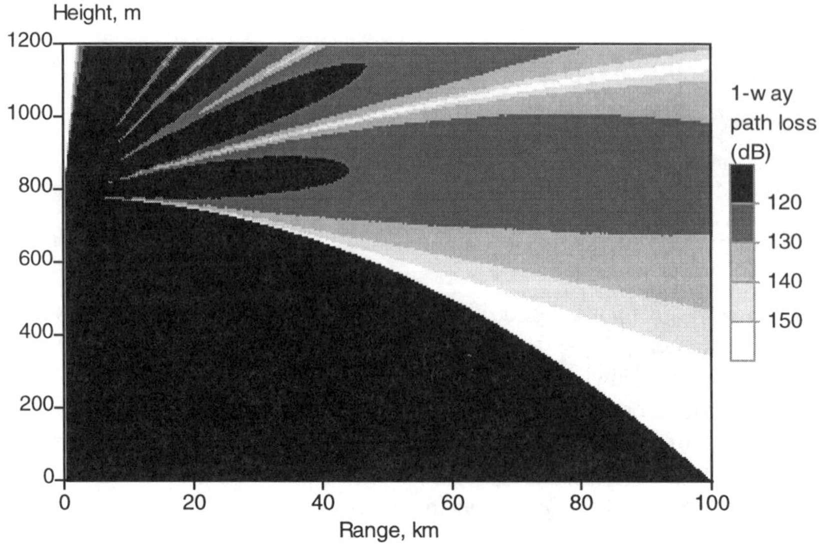

Figure 7.4 *Spherical diffraction for 1 GHz source 25 m above the ground, with explicit treatment of Earth's surface*

shows path loss as a function of range 25 m above the ground. A continuous line is used for the explicit Earth representation results, and a dotted line for the conformal mapping results. Agreement is very good, confirming the accuracy of the Earth flattening transformation.

7.4.2 Multiple knife-edge diffraction

We have seen in Chapter 2 that, for a single knife-edge obstacle, the PE formulation is formally equivalent to the Fresnel diffraction formula. This result can be generalized to multiple knife-edge diffraction. For the double knife-edge, small angle diffraction can be expressed by a Fresnel surface integral [114, 113]. This singular integral is not easy to compute directly and requires cumbersome matching of asymptotic expansions and Taylor series. By contrast the PE solution provides an easy automatic way to compute the same integral by solving numerically the partial differential equation of which it is the solution.

Consider the double knife-edge geometry shown in Figure 7.7. The transmitter is located at T and the receiver at R. The heights of the knife-edges K_1 and K_2 above the line of sight TR are denoted by h_1 and h_2.

A double application of the Huygens principle leads to the following expression of the field E relative to the free-space field E_0:

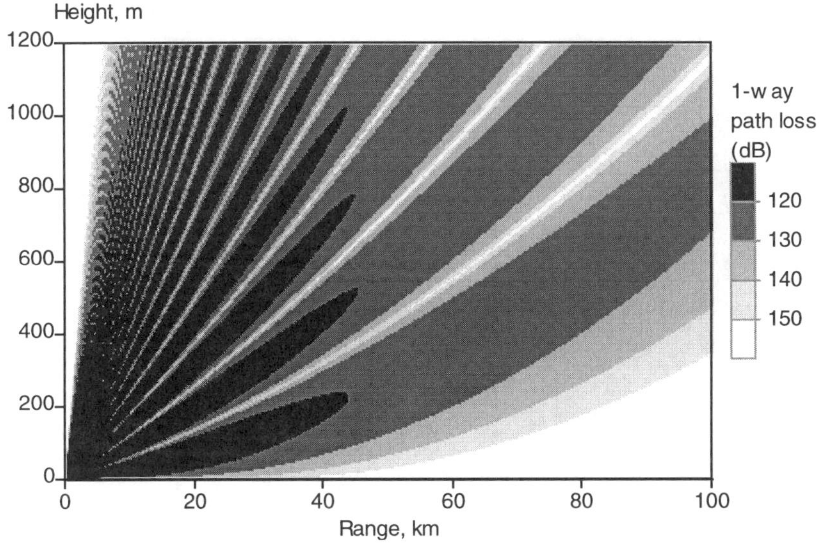

Figure 7.5 Spherical diffraction for 1 GHz source 25 m above the ground, with Earth flattening transformation

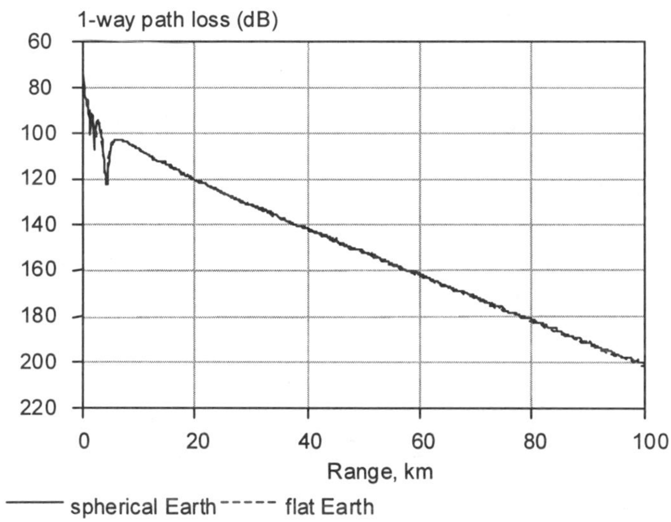

Figure 7.6 Comparison of path loss against range 25 m above the ground for flat and spherical Earth models

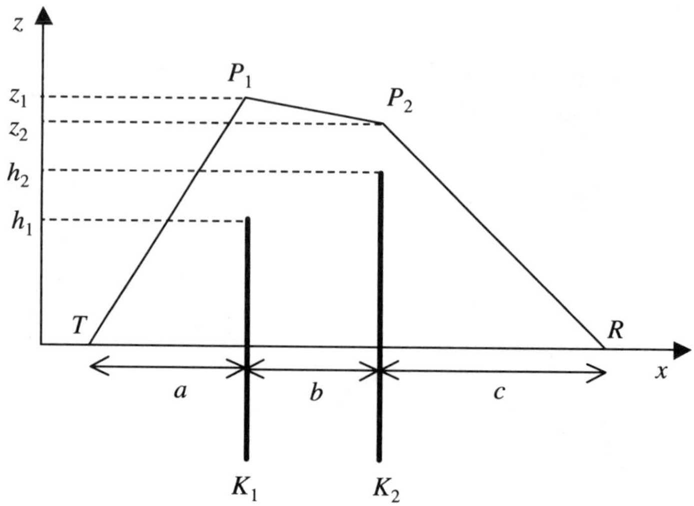

Figure 7.7 Double knife-edge geometry

$$\frac{E}{E_0} = \frac{\int_{h_1}^{+\infty}\int_{h_2}^{+\infty} \exp\left[ikd(z_1, z_2)\right]\, dz_1\, dz_2}{\int_{-\infty}^{+\infty}\int_{-\infty}^{+\infty} \exp\left[ikd(z_1, z_2)\right]\, dz_1\, dz_2} \tag{7.29}$$

where $d(z_1, z_2)$ is the path difference between TP_1P_2R and TR. This expression neglects edge currents and backscatter effects between the edges, like the parabolic wave equation. Now if we only consider paths with small diffraction angles, we have

$$d(z_1, z_2) \sim \frac{z_1^2}{2a} + \frac{(z_1 - z_2)^2}{2b} + \frac{z_2^2}{2c} \tag{7.30}$$

The resulting expression for the field is exactly equivalent to that obtained from the narrow-angle PE approximation. Indeed assuming a point source at T, the reduced narrow-angle PE field u at point P_1 is simply

$$u(a, z_1) = A\frac{z_1^2}{2a} \tag{7.31}$$

where A is a normalization constant. Using the results of Section 2.4.2, the reduced PE field at P_2 is

$$u(a+b,z_2) = A\int_{h_1}^{+\infty} \exp\left[ik\left(\frac{z_1^2}{2a} + \frac{(z_1-z-2)^2}{2b}\right)\right] dz_1 \tag{7.32}$$

The reduced field at R is now obtained by integrating over z_2, getting

$$\begin{aligned} u(a+b+c,0) \\ = A\int_{h_1}^{+\infty}\int_{h_2}^{+\infty} \exp\left[ik\left(\frac{z_1^2}{2a} + \frac{(z_1-z-2)^2}{2b} + \frac{z_2^2}{2c}\right)\right] dz_1 dz_2 \end{aligned} \tag{7.33}$$

As the reduced field in the absence of the two knife-edges is obtained by taking $h_1 = h_2 = -\infty$, the resulting expression for E/E_0 is exactly the same as eqn. 7.29, with the excess path $d(z_1, z_2)$ given by eqn. 7.30. Since the PE and Fresnel-Kirchhoff formulations are identical, the solution can be computed either by marching the PE with appropriate range steps and height discretization, or by computing the integral directly.

The method of [114, 113] consists in expressing the integral of eqn. 7.29 in terms of Fresnel surface integrals. In order to achieve this, the path difference d is expressed as a sum of two squares

$$d(z_1,z_2) = \frac{a+b}{2ab}\left(z_1 - \frac{az_2}{a+b}\right)^2 + \frac{(a+b+c)z_2^2}{2(a+b)c} \tag{7.34}$$

With the change of variables

$$\left\{\begin{aligned} s_1 &= \sqrt{\frac{2(a+b+c)}{\lambda ab}}\left(z_1 - \frac{a}{a+b}z_2\right) \\ s_2 &= \sqrt{\frac{2(a+b+c)}{\lambda(a+b)c}}z_2 \end{aligned}\right. \tag{7.35}$$

Eqn. 7.29 becomes

$$\begin{aligned} \frac{E}{E_0} = -\frac{i}{2}\int_{\sqrt{\frac{2(a+b)}{\lambda(a+b+c)}}h_2}^{+\infty} ds_2 \\ \times \int_{\sqrt{\frac{2(a+b)}{\lambda ab}}h_1 - \sqrt{\frac{2(a+b)}{b(a+b+c)}}s_2}^{+\infty} \exp\left[i\frac{\pi}{2}(s_1^2+s_2^2)\right] ds_1 \end{aligned} \tag{7.36}$$

It is customary to express the lower bounds as a function of the dimensionless Fresnel parameters p and q corresponding to the two knife-edges

$$p = h_1 \sqrt{\frac{2(a+b+c)}{\lambda a(b+c)}} \tag{7.37}$$

$$q = h_2 \sqrt{\frac{2(a+b+c)}{\lambda (a+b)c}} \tag{7.38}$$

and the angle α which represents knife-edge separation, given by

$$\tan \alpha = \sqrt{\frac{b(a+b+c)}{ac}} \tag{7.39}$$

The field relative to free-space is now expressed as

$$2i\frac{E}{E_0} = \int_q^{+\infty} ds_2 \int_{\frac{p-s_2\cos\alpha}{\sin\alpha}}^{+\infty} \exp\left[i\frac{\pi}{2}(s_1^2+s_2^2)\right] ds_1 \tag{7.40}$$

which by definition is the Fresnel surface integral

$$G(S) = \int_S \exp\left(i\frac{\pi}{2}\rho^2\right) dS \tag{7.41}$$

corresponding to the surface of integration S shown in Figure 7.8. References [114, 113] provide expansions of $G(S)$ for all values of p, q and α. The results can be extended to the case of an obstacle with a reflecting flat top, using similar techniques to compute the reflected field [164]. The diffracted field is expressed as the sum of two Fresnel surface integrals [164].

We now turn to the PE implementation of this type of problem. With the staircase terrain model, the treatment of vertical facets is straightforward. In particular single or multiple knife-edges can be modelled simply by introducing stairs of width zero. Alternatively, obstacles including vertical facets, for example buildings, can be modelled by a mixture of staircase and piecewise-linear treatment. This has been done in [102, 101] for flat-topped obstacles like that shown in Figure 7.9.

Depending on the sign of the slope of the obstacle top, different expressions are used in the Fresnel surface integral evaluation for the reflected field. Following [164, 101], we talk of illuminated, shadowed and tangential incidence for negative, positive and zero slopes respectively, as shown in Figure 7.10. Figure 7.11 from [101] shows a comparison of Fresnel surface integral results with those obtained with the PE. The angle α given by eqn. 7.39 is set to a fixed value of 28°, while the parameters p and q are allowed to vary by moving the receiver vertically. The illuminated and shadowed cases (curves I and S

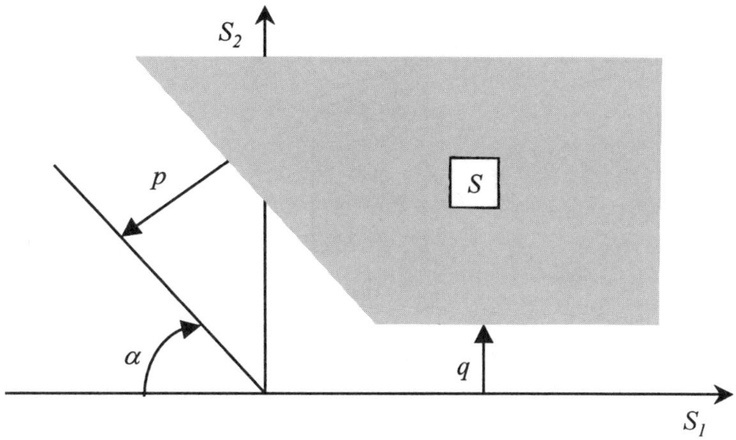

Figure 7.8 Area of integration for Fresnel surface integral

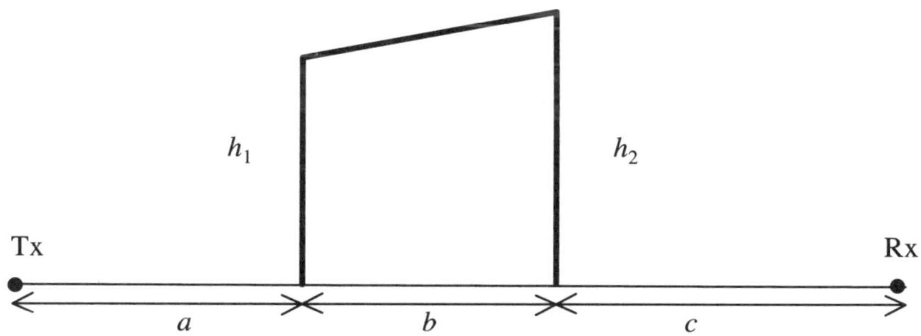

Figure 7.9 Flat-topped obstacle geometry

respectively) correspond to grazing angles of 2° and −2° of the incident field on the flat top. The curve labelled T corresponds to tangential incidence. Experimental results from [54] are also shown. Discrepancies between measured and modelled values could be due to the use of a perfectly flat and perfectly conducting obstacle for the theoretical models: this idealized representation does not account for the actual properties of the aluminium blocks used for the experiment [164].

Excellent agreement between the PE and Fresnel surface integral models is obtained, as expected since the two methods are simply different ways of calculating the same integral. The PE method is simpler to apply and easier to generalize. For example it is straightforward to treat the case of multiple knife-edges or multiple flat-top obtacles, whereas direct calculation of the integrals

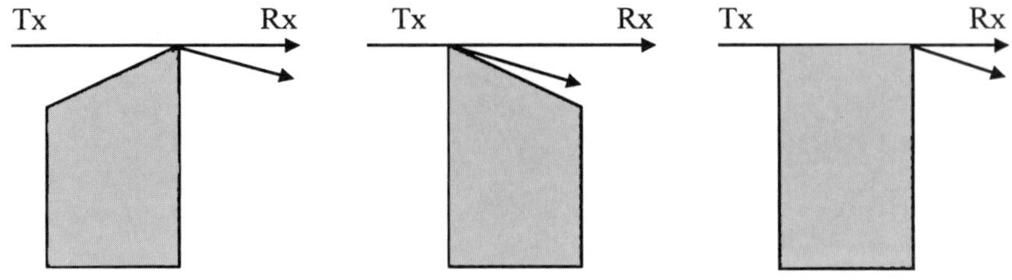

Figure 7.10 *Types of incidence on flat-topped obstacle: (a) illuminated, (b) shadowed and (c) tangential*

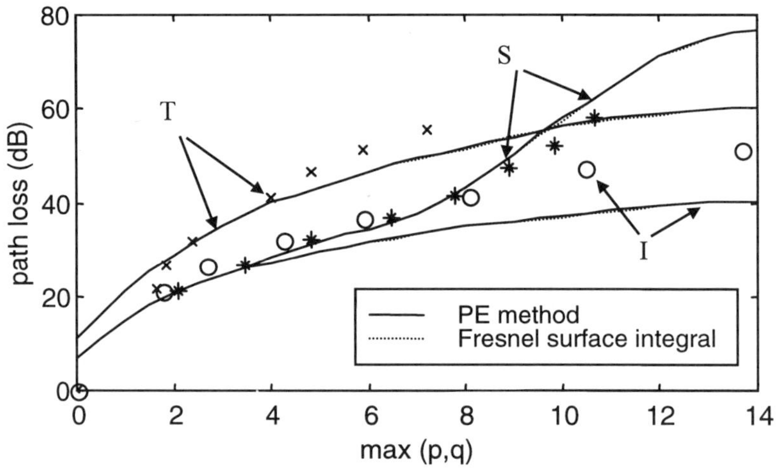

Figure 7.11 *Comparison of Fresnel surface integral, PE results and measurements [54] for illuminated (I), shadowed (S) and tangential (T) incidence on flat-topped obstacles [101]*

becomes far too cumbersome. An asymptotic solution based on path integral methods has been proposed in [81] to calculate the field diffracted by equally spaced parallel half-planes, at points located on the shadow boundary, as shown in Figure 7.12.

For n knife-edges between transmitter and receiver, the path loss L_n relative to free space (or diffraction loss) given by [81] is given by

$$L_n \sim 20\log(n+1) \tag{7.42}$$

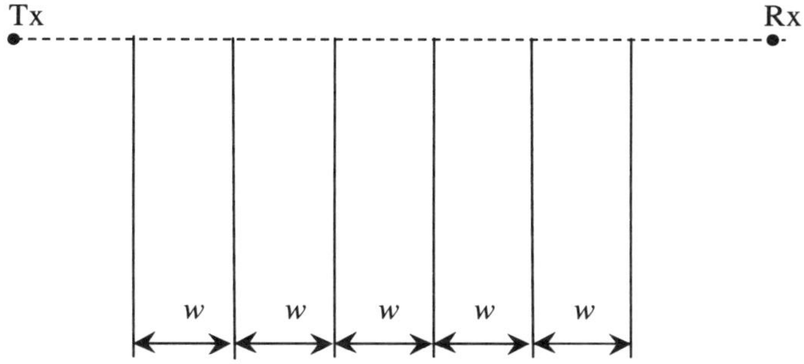

Figure 7.12 Geometry for equally spaced knife-edges

This corresponds to the leading term of an asymptotic expansion. For $n = 1$, this is the familiar result for single knife-edge diffraction. Figure 7.13 shows diffraction loss results for 1 to 10 edges, calculated from eqn. 7.42 and with the PE method [86]. The PE results were obtained with a wide-angle finite-difference code, using a frequency of 1 GHz and an edge separation of 45 m, corresponding to 150 λ. The PE results show a weak oscillation around the values predicted in [81], possibly because the PE representation includes higher order effects. In order to filter out these variations, the PE results have been averaged over 5 m. The averaged values are in excellent agreement with the asymptotic results, with a maximum error of 0.07 dB. The point PE values are also shown: the difference with the asymptotic values remains less than 0.5 dB, showing that the leading term of the asymptotic expansion provides a reasonable approximation of the diffracted field.

Multiple knife-edge models have often been used to represent idealized urban environments in microcellular network planning applications [139]. Of course the PE can be used to treat more realistic urban environments, such as that shown in Figure 7.14. A 900 MHz source located 35 m above the ground transmits over five buildings of various shapes. A wide-angle finite-difference PE code has been used in order to model deep shadow diffraction.

7.4.3 Combined effects of terrain and ducting

The planning of radio links requires accurate evaluation of coverage and interference. Until the development of PE methods, it was not possible to model simultaneously atmospheric refractive effects and diffraction by irregular terrain. This is of course straightforward with parabolic equation methods, which can deal easily with terrain and refractivity variations. The following examples have been obtained with the split-step sine transform method described in Section 3.3, using a staircase description of terrain.

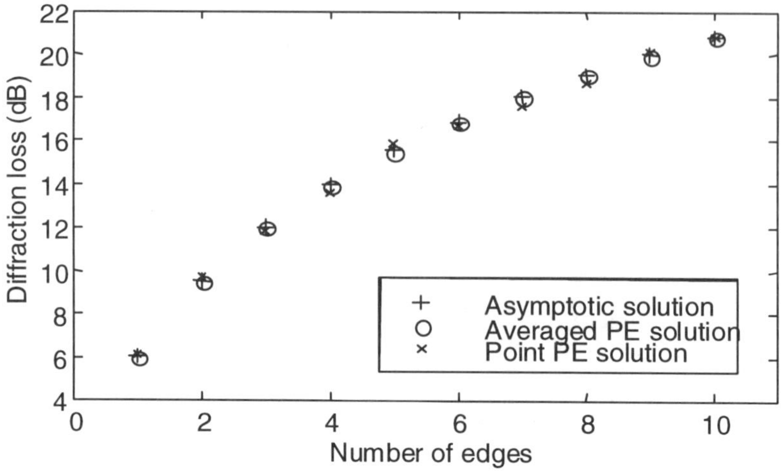

Figure 7.13 *Comparison of asymptotic and PE results for multiple knife-edge diffraction*

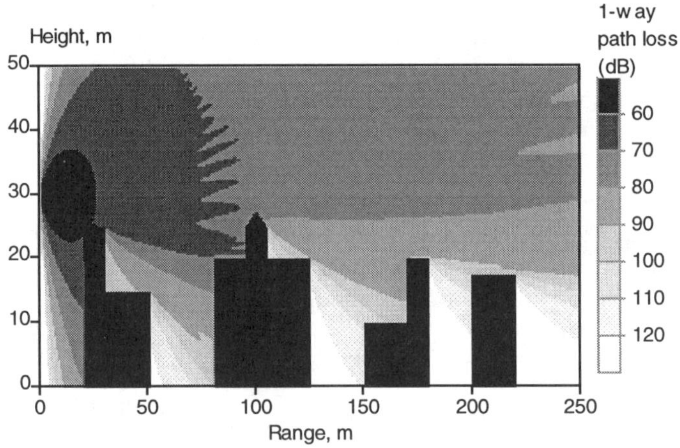

Figure 7.14 *Horizontally polarized 900 MHz source in urban environment*

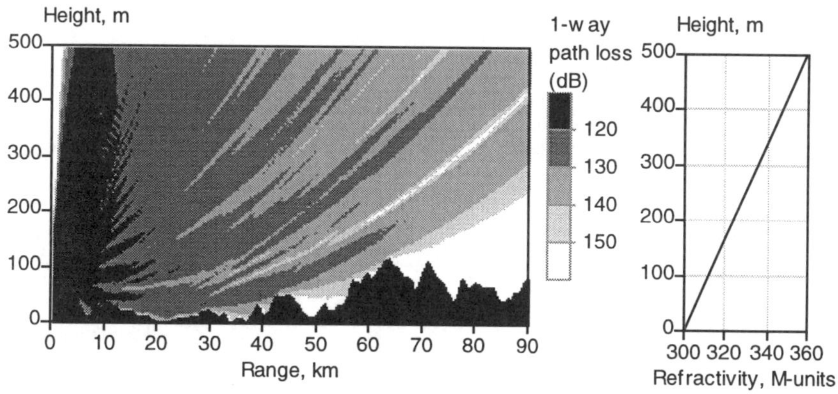

Figure 7.15 Diffraction by irregular terrain with standard atmosphere for horizontally polarized 1.8 GHz source

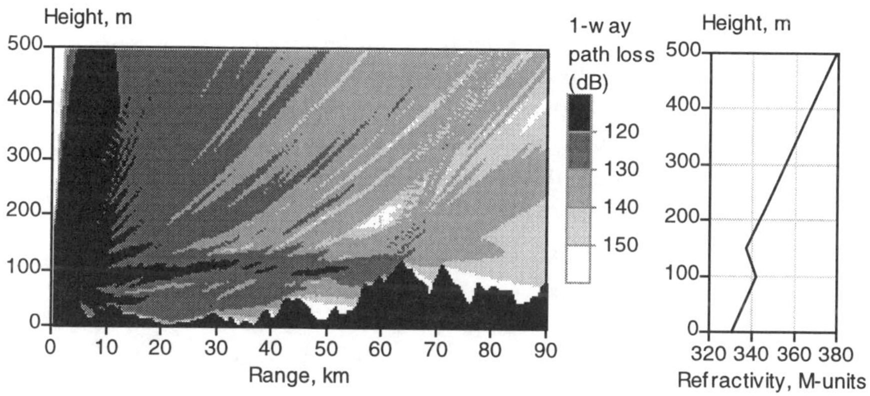

Figure 7.16 Diffraction by irregular terrain with elevated duct for horizontally polarized 1.8 GHz source

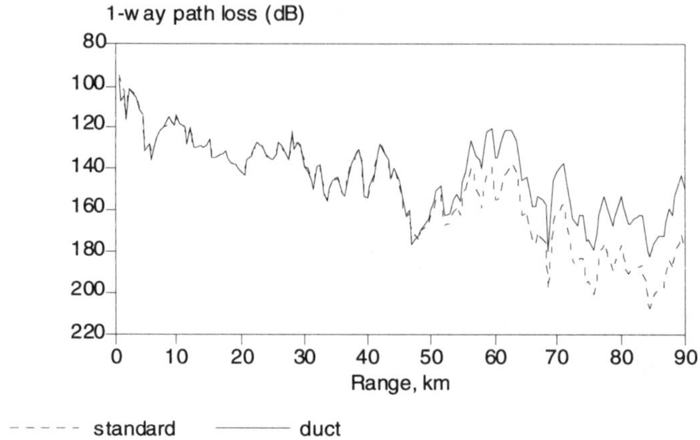

Figure 7.17 Comparison of 1.8 GHz results at height 2 m above terrain for standard atmosphere and elevated duct

Figure 7.15 shows path loss contours for a 1.8 GHz source located 15 m above ground level on hilly terrain with the corresponding modified refractivity profile. In this example a standard atmosphere has been used. Complex interference patterns are created by reflections from irregular terrain. Figure 7.16 shows path loss contours for the same antenna and terrain in the presence of an elevated duct, given by a trilinear modified refractivity profile. Energy trapped in the duct propagates to the dominant hill at 63 km, causing a much enhanced diffracted signal. This is apparent in Figure 7.17, which shows curves of path loss against height at a constant height of 2 m above ground level. Beyond the main hills, an enhancement of over 20 dB is observed.

Chapter 8
Domain truncation

8.1 Introduction

For many radiowave propagation problems the bottom boundary of the domain is the physical ground/air interface, often represented by a surface impedance boundary condition, while the top boundary is a computational artifact needed to limit the integration domain in height. The top boundary must represent at a finite distance the Sommerfeld radiation condition – it must be perfectly transparent, letting all the energy coming from below the boundary escape to infinity.

The earliest implementations of domain truncation in PE modelling were based on absorbing layers, which work well for small propagation angles but become quite expensive for larger angles from the horizontal. Absorbing layers are still the most popular truncating technique in view of their ease of implementation. We describe the properties of a simple absorbing layer in Section 8.2.

The idea of adding a lossy medium above the domain of interest is taken into a different direction with the perfectly matched layer of Bérenger [18], extending the propagation equations to a special medium which has the property of absorbing all plane waves irrespective of propagation direction. The perfectly matched layers (PML) were originally designed to provide efficient truncation for numerical solution of the full Maxwell equations. The method is at once beautiful and very powerful, and has subsequently found a variety of applications. In particular it has been adapted for the parabolic equation framework [158, 31]. In Section 8.3, we present a very simple approach which is adequate for scattering applications. Although this technique is not appropriate for long-range problems, it is extremely efficient for object scattering computations.

Next we consider non-local techniques which truncate the domain exactly. Non-local boundary conditions (NLBCs) have been derived by a number of authors for elliptic wave equation problems; an excellent review can be found in [51]. It turns out that NLBCs are also available for the parabolic wave equation. NLBCs in the elliptic case involve all the points on the boundary simultaneously, which makes them somwehat cumbersome to implement, whereas in the paraxial case NLBCs are causal and easy to apply. NLBCs for the paraxial framework were introduced by Papadakis [121] and independently by Baskakov and Popov [12] and Marcus [106, 107] for the case where the medium beyond the boundary is homogeneous and the initial field is screened. More general paraxial NLBCs were subsequently derived to deal with transmitting boundaries [42, 108] and linear media [88, 89, 91, 129, 152]. The NLBC framework is given in Section 8.4, and screeened NLBCs are described in Section 8.5. In this chapter we only consider NLBCs suitable for the narrow-angle two-dimensional PE. Wide-angle NLBCs and extension to three dimensions are treated in Chapters 12 and 13 respectively.

Finally we look at domain truncation techniques which allow the source to be **above** the parabolic equation domain, by adding a non-homogeneous term to the boundary condition to represent incoming energy. This is of great practical interest, since low altitude effects on the coverage of airborne sources can be computed in far less time than if the source had to be included in the domain. High sources can be treated with non-local boundary conditions [42, 89, 91], as described in Section 8.6. Perhaps more surprisingly, there is also a split-step/Fourier version to model incoming energy [93]. This is the topic of Section 8.7.

8.2 Absorbing layers

Absorbing layers are layers added above the region of interest, in which a filter is applied to the PE field in order to absorb the upgoing energy by the time it bounces back from the top of the domain. This is equivalent to adding a complex part to the refractive index, hence making the propagation medium lossy in the absorbing layer. In order for the filter not to affect propagation in the region of interest, it must be such that only negligible energy propagates back to the bottom of the layer. This method is applicable to both split-step/Fourier and finite-difference implementations of the PE.

A simple, but effective filter is given by the Hanning window, which is commonly used in signal processing. This is of the form

$$\phi(t) = \frac{1 + \cos(\pi t)}{2} \tag{8.1}$$

The Hanning window satisfies $\phi(0) = 1$, $\phi(1) = 0$, and the derivatives at the end points are zero, which ensures a smooth match with the rest of the domain. We can estimate the absorptive properties of the filter, assuming

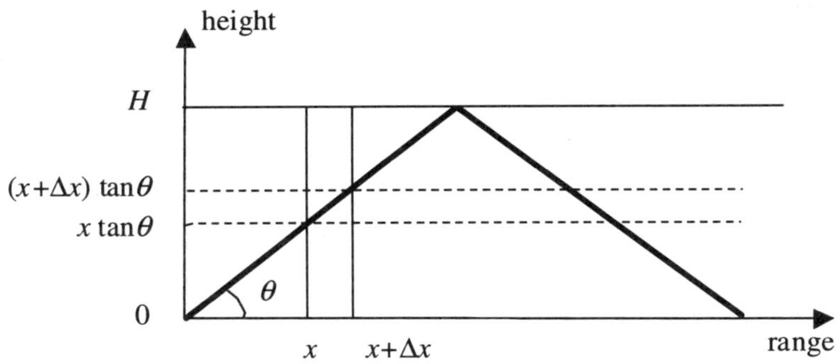

Figure 8.1 *Effect of absorbing layer on ray with elevation θ*

for simplicity that the propagation medium is a vacuum. We consider the geometry of Figure 8.1, with an absorbing layer of height H. The dB loss associated with a ray launched at angle θ from the horizontal after reflection from the top boundary is approximately

$$L = -\frac{40\,H}{\Delta x \tan\theta} \int_0^1 \log\phi(t)\,dt \tag{8.2}$$

where Δx is the integration range step. To see this, we note that the dB loss in the vertical slice between x and $x + \Delta x$ is

$$-20\log\phi(x\tan\theta/H) \sim -\frac{1}{\Delta x}\int_x^{x+\Delta x} 20\,\log\phi(\xi\tan\theta/H)\,d\xi \tag{8.3}$$

Eqn. 8.2 is obtained by summing over successive slices. At the top of the domain, truncation of the field amounts to the presence of a perfectly reflecting mirror. For the Hanning filter, we have [53]

$$-\int_0^1 \log\phi(t)dt = -2\int_0^1 \log\cos\frac{\pi}{2}t\,dt = 2\,\log 2 \tag{8.4}$$

For typical PE runs at microwave frequencies, the range step Δx is of the order of 500 m. Taking $H = 100$ m, we see that attenuation values reach at least 100 dB for propagation angles of less than 2.75°. If large propagation angles are involved, this method becomes rather expensive, as the ratio $H/\Delta x$ might have to be increased by several orders of magnitude in order to obtain sufficient attenuation.

Other filters defining "sponge" layers are described in the literature [62, 78].

8.3 Perfectly matched layer

Assume the propagation medium has refractive index 1 above height z_b. We construct a two-dimensional PML by replacing the height z with complex coordinate $\tilde{z}$ given by

$$\tilde{z} = z - i\int_0^z \gamma(\zeta)d\zeta \tag{8.5}$$

where

$$\begin{cases} \gamma(z) & = & 0, & z & \leq & z_b \\ \gamma(z) & > & 0, & z & > & z_b \end{cases} \tag{8.6}$$

Figure 8.2 shows the behaviour of the "stretched coordinate" $\tilde{z}$ [132]. We extend the SPE to this new height coordinate, getting

$$\frac{\partial^2 v}{\partial \tilde{z}^2} + 2ik\frac{\partial v}{\partial x} + k^2(n^2-1)v = 0 \tag{8.7}$$

for the extended function $v(x,\tilde{z})$. The solution of eqn. 8.7 coincides with the usual PE solution for heights $z \leq z_b$ and has the form of a damped wave above z_b. To see this, we look at an upward propagating plane wave with unit amplitude and propagation angle θ from the horizontal. In the PML medium, the solution has the form

$$v(x,\tilde{z}) = \exp\left(-ik\frac{x}{2}\sin^2\theta - ik\tilde{z}\sin\theta\right) \tag{8.8}$$

which is written in terms of z as

$$u(x,z) = \exp\left(-ik\frac{x}{2}\sin^2\theta - ikz\sin\theta\right)\exp\left(-k\sin\theta\int_0^z \gamma(\zeta)d\zeta\right) \tag{8.9}$$

We see that in terms of real height z, the solution is indeed a damped wave above z_b. We should stress here that the function $u(x,z)$ is not at all a solution of the parabolic wave equation above z_b. Rather, it expresses in terms of real heights the solution of the parabolic equation in stretched coordinates. We have replaced the original equation with a new equation with the property that the solution below z_b is unchanged, while the solution above z_b decays very quickly. We notice that, unlike the absorbing layer of Section 8.2, which is less

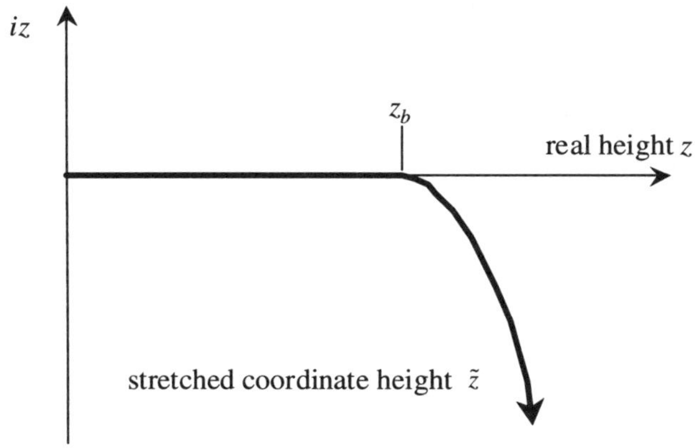

Figure 8.2 *Stretched height coordinate*

efficient at high propagation angles, the PML works less well at small propagation angles. For this reason it is not well suited to long-range propagation problems.

Finite-difference implementation of the PML is straightforward, writing eqn. 8.7 as

$$\frac{1}{1 - i\gamma(z)} \frac{\partial}{\partial z} \left(\frac{1}{1 - i\gamma(z)} \frac{\partial u}{\partial z} \right) + 2ik \frac{\partial u}{\partial x} = 0 \tag{8.10}$$

and using the standard finite-difference approximations of the derivatives. In an infinite PML medium, the choice of the damping function γ is arbitrary. In practice, the PML medium must be implemented with finite thickness, and some care must be taken to avoid spurious reflections at the top of the layer. In the initial Bérenger paper [18], the use of a geometrical progression is suggested, taking

$$\gamma(z) = \gamma_0 \left(g^{1/\Delta z} \right)^z \tag{8.11}$$

where Δz is the vertical resolution of the grid. Other possibilities are analysed in [31]. A very thin PML is usually adequate (typically of the order of 10 grid points), which makes this method particularly attractive for scattering problems.

8.4 Non-local boundary conditions

Non-local boundary conditions are based on the horizontal plane wave decomposition of the field at the boundary. Exact boundary conditions are written for each plane wave component, and a global boundary condition for the field is then constructed using an appropriate transform. The great advantage of this formulation is that no explicit knowledge of the spectral decomposition of the field is required. One drawback is that implementations so far have been limited to finite-difference schemes.

Some authors use the one-sided Fourier transform in range [121]. Here we choose the Laplace transform $\mathcal{L}$, with respect to range, defined by

$$\mathcal{L}\{u(x,z)\} = U(s,z) = \int_0^{+\infty} u(x,z)e^{-sx}dx, \ \mathrm{Re}(s) > 0 \tag{8.12}$$

We note that if u is outgoing at infinity then u is bounded, which implies that $U(s,z)$ is bounded in z for fixed s.

Assume that above the height z_b at which we want to truncate the PE domain, the refractive index is independent of range. Then for heights above z_b, the Laplace transform of the standard parabolic equation is

$$\frac{\partial^2 U}{\partial z^2}(s,z) + \left(2iks + k^2(n^2(z)-1)\right)U(s,z) = 2ik\varphi(z) \tag{8.13}$$

where $\varphi(z)$ denotes the initial field $u(0,z)$. For certain types of refractive index profiles, this equation can be solved in closed form. This gives a relation between U and its normal derivative which can be inverted to yield a perfectly transparent boundary condition at height z_b. In Chapter 9, we shall use similar techniques to derive generalized impedance boundary conditions at the air/ground interface. In Chapter 11, the Laplace transform method is used to extend the field upwards.

We shall assume in what follows that $n(z_b) = 1$. If we think in terms of the split-step PE solution, we see that this is not a loss of generality: shifting down the refractive index profile by $\beta = n(z_b) - 1$ amounts to multiplying the solution by the function $\exp(-ik\beta x/2)$. This does not affect the amplitude of the solution, and it is easy to correct the phase if required by applying the appropriate range-dependent offset. In fact this procedure can be generalized to the case where the refractive index values at z_b depend on range, provided the gradient above z_b is range-independent.

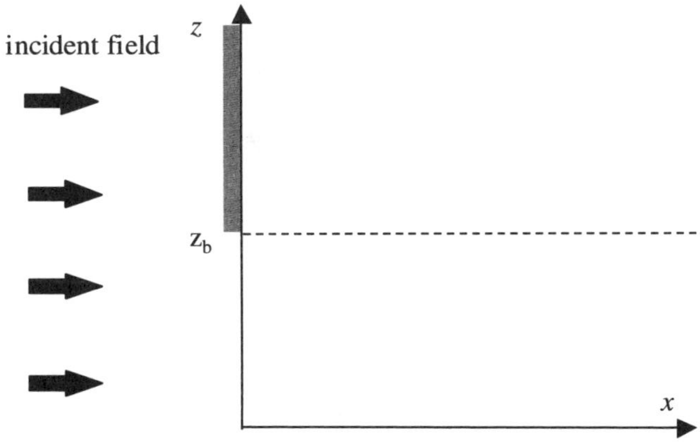

Figure 8.3 *Screened initial field*

8.5 Diffractive non-local boundary conditions

We first treat the case where the initial field $\varphi(z)$ is zero for z greater than z_b. This type of problem arises if we place an absorbing half-space screen at heights greater than z_b at range zero, as shown in Figure 8.3. The half-space screen can also be viewed as a "diffractive boundary" which is the terminology of [42]: the boundary condition represents energy diffracted around the edge of the screen into the region above z_b.

8.5.1 Homogeneous medium

We start with the case where $n(z) = 1$ for $z \geq z_b$. The bounded solution of (8.13) for the screened initial field is

$$U(s, z) = U(s, z_b)\, \mathrm{e}^{i\Omega(z-z_b)}, \; z \geq z_b \tag{8.14}$$

where $\Omega = (1+i)\sqrt{k}$. Differentiating with respect to z, we get for $z = z_b$

$$\frac{\partial U}{\partial z}(s, z_b) = i\Omega\sqrt{s}\, U(s, z_b) \tag{8.15}$$

Inverting this directly would lead to a non-integrable convolution kernel. Instead, following [107, 42], we write

$$\frac{\partial U}{\partial z}(s, z_b) = \frac{i\Omega}{\sqrt{s}}\, sU(s, z_b) \tag{8.16}$$

Inverting the Laplace transform and using the convolution theorem, we get the boundary condition

$$\frac{\partial u}{\partial z}(x, z_b) = \int_0^x \frac{\partial u}{\partial x}(\xi, z_b)\, g_0(x - \xi)\, d\xi \tag{8.17}$$

where the convolution kernel g_0 is given by

$$g_0(x) = \frac{i\Omega}{\sqrt{\pi x}} \tag{8.18}$$

This condition is non-local: at a given range, it depends on the field at all previous ranges. By contrast the usual local boundary conditions only involve the field and its normal derivative at the range of interest. We note that the NLBC is causal, in the sense that ranges beyond the current range are not involved.

8.5.2 *Linear medium*

For most radiowave propagation problems, the squared refractive index at the top of the region of interest increases with height. The NLBC derived for a homogeneous medium is not matched to the behaviour of the refractive index, and spurious reflections will occur if it is used. Instead we must derive a non-local boundary condition which matches the squared refractive index gradient at the top of the region of interest.

We assume that the squared refractive index above z_b is of the form

$$n^2(x, z) = 1 + \alpha(z - z_b), \ \ z \geq z_b \tag{8.19}$$

where α is non-zero. For a screened incident field, the Laplace transform of the SPE above z_b is then

$$\frac{\partial^2 U}{\partial z^2}(s, z) + \Big(2iks + k^2\alpha(z - z_b)\Big)\, U(s, z) = 0 \tag{8.20}$$

which is a variant of the Airy equation (see Appendix A).

Assume first that the slope α is positive, as will normally be the case. The solution of eqn. 8.20 which is bounded in z for fixed s is of the form

$$A(s) w_1 \left\{\zeta_s(z - z_b)\right\} \tag{8.21}$$

where

$$\zeta_s(z) = -\left(\alpha k^2\right)^{1/3} z - \frac{2iks}{(\alpha k^2)^{2/3}} \tag{8.22}$$

and w_1 is the "outgoing" Airy function related to the Airy function Ai by

$$w_1(z) = \mathrm{Ai}\left(ze^{2i\pi/3}\right) \tag{8.23}$$

Since the zeros of Ai are all real and negative [117], w_1 has no zeros in S and we can write

$$U(s, z) = U(s, z_b)\frac{w_1\left\{\zeta_s(z - z_b)\right\}}{w_1\left\{\zeta_s(0)\right\}} \tag{8.24}$$

By differentiating with respect to z, we obtain

$$\frac{\partial U}{\partial z}(s, z_b) = -(\alpha k^2)^{1/3} U(s, z_b)\frac{w_1'\left\{\zeta_s(0)\right\}}{w_1\left\{\zeta_s(0)\right\}} \tag{8.25}$$

The final stage is to find the inverse Laplace transform of $w_1'(\zeta_s(0))/sw_1(\zeta_s(0))$. This is done with factorization theorems for holomorphic functions [155], as described in Appendix A. Finally, we obtain the NLBC in the form

$$\frac{\partial u}{\partial z}(x, z_b) = \int_0^x \frac{\partial u}{\partial x}(\xi, z_b)\, g_\alpha(x - \xi)\, d\xi \tag{8.26}$$

where the convolution kernel is given by

$$g_\alpha(x) = -\left(\alpha k^2\right)^{1/3} \mathrm{e}^{2i\pi/3}\frac{\mathrm{Ai}'(0)}{\mathrm{Ai}(0)} - \frac{i\alpha k}{2}\sum_{j=1}^{\infty}\frac{\mathrm{e}^{s_j x}}{s_j} \tag{8.27}$$

and $(s_j)_{j=1}^{\infty}$ is the sequence of zeros of $w_1\left\{\zeta_s(0)\right\}$. These are related to the zeros $(a_j)_{j=1}^{\infty}$ of the Airy function Ai by

$$s_j = \frac{(\alpha k^2)^{2/3}}{2k}\mathrm{e}^{-i\pi/6} a_j \tag{8.28}$$

The convergence of the series is very fast for positive x since the zeros of the Airy function satisfy

$$a_j \sim -\left(\frac{3\pi}{8}(4j - 1)\right)^{2/3}, \quad j \to +\infty \tag{8.29}$$

and hence the terms in the series decrease exponentially. The kernel has an integrable singularity at zero.

For most applications, the medium above the boundary is upward transmitting, and hence α has a positive value. However it is interesting to consider the case of a ducting layer extending to infinity above z_b, as it demonstrates very strikingly the non-local nature of the boundary condition. A non-local boundary condition similar to eqn. 8.26 can be derived when α is negative, as shown by Thomson and Brooke in [152]. The appropriate solution of the Laplace transform of the SPE is then

$$U(s,z) = A(s)\mathrm{Ai}\left\{\zeta_s(z - z_b)\right\} \tag{8.30}$$

with

$$\zeta_s(z) = \left(|\alpha|k^2\right)^{1/3} z - \frac{2iks}{\left(|\alpha|k^2\right)^{2/3}} \tag{8.31}$$

The resulting non-local boundary condition is of the same form as eqn. 8.26, with the convolution kernel for negative α given by

$$g_\alpha(x) = -\frac{i|\alpha|k}{2}\sum_{j=1}^{\infty}\frac{\mathrm{e}^{s'_j x}}{s'_j} - \left(|\alpha|k^2\right)^{1/3}\frac{\mathrm{Ai}'(0)}{\mathrm{Ai}(0)} \tag{8.32}$$

where

$$s'_j = i\frac{(|\alpha|k^2)^{2/3}}{2k}a_j \tag{8.33}$$

For negative α, the series defining g_α does not converge absolutely, and convergence is much slower than for positive α. In both cases, the constant term in g_α acts like a Dirac function while the infinite series represents the continuous part of the kernel. The behaviour of the continous part of the kernel is radically different for positive and negative slopes. To illustrate this point, Figure 8.4 shows the variations of the modulus of the continuous part of the convolution kernel g_α for modified refractivity slopes of 100 M-units/km, -100 M-units/km and zero. The frequency is 3 GHz. The kernel corresponding to the positive slope decays extremely fast, much faster in fact that the kernel for a vacuum. This is easily understood in terms of geometrical optics: rays curve upwards in the positive slope environment and the energy travels away from the boundary faster than when rays are straight lines. The kernel for the negative slope decays rather slowly at short ranges, and then exhibits a number of peaks which represent energy returning at successive ranges after curving down in the infinite ducting layer.

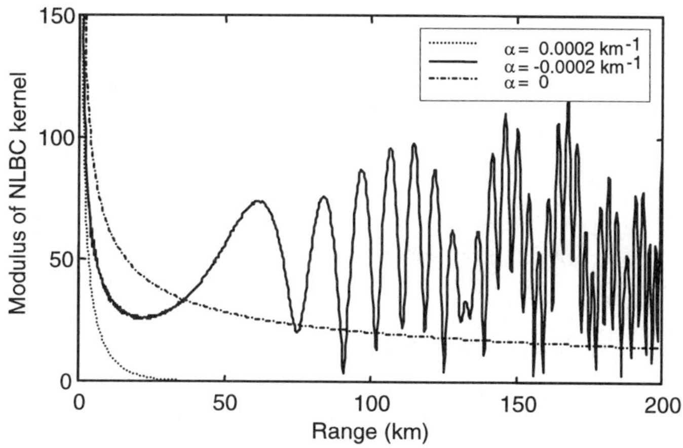

Figure 8.4 *Variation of modulus of NLBC kernel with range for different values of slope α*

8.5.3 Implementation

There is no difficulty implementing non-local boundary conditions for a homogeneous or linear medium within a finite-difference scheme [108]. We first discretize in range, putting

$$x_l = l\Delta x \tag{8.34}$$

where Δx is the range step, and

$$u_b^l = u\left(x_l, z_b\right) \tag{8.35}$$

Following [42], we make the approximation that $\dfrac{\partial u}{\partial x}(x, z_b)$ is constant on each range step interval, writing

$$\frac{\partial u}{\partial x}(x, z_b) = \frac{u_b^{l+1} - u_b^l}{\Delta x}, \; x_l < x < x_{l+1} \tag{8.36}$$

Then eqns. 8.17 and 8.26 are approximated by

$$\frac{\partial u}{\partial z}(x_m, z_b) = \sum_{l=0}^{m-1} \frac{u_b^{m-l} - u_b^{m-l-1}}{\Delta x} I_l \tag{8.37}$$

where

$$I_l = \int_{x_l}^{x_{l+1}} g_\alpha(\xi)\, d\xi \tag{8.38}$$

We now rearrange eqn. 8.37 to get

$$\Delta x \frac{\partial u}{\partial z}(x_m, z_b) = -u_b^0 I_m + u_b^m I_0 + \sum_{l=1}^{m-1} u_b^{m-l}\,(I_l - I_{l-1}) \tag{8.39}$$

The final step is to discretize in height z. We put

$$z_j = j\Delta z \tag{8.40}$$

and

$$u_j^m = u(x_m, z_j) \tag{8.41}$$

If z_N is the largest height of interest, the boundary z_b is located at

$$z_b = \frac{z_{N-1} + z_N}{2} \tag{8.42}$$

Using standard finite-difference approximations, we have

$$u(x_l, z_b) \sim \frac{u_N^l + u_{N-1}^l}{2} \tag{8.43}$$

and

$$\frac{\partial u}{\partial z}(x_m, z_b) \sim \frac{u_N^m - u_{N-1}^m}{\Delta z} \tag{8.44}$$

Finally, recalling that $u_N^0 = 0$ since the initial field is zero above z_b, we can write the discretized non-local boundary condition as

$$(1-a)u_N^m - (1+a)u_{N-1}^m = \sum_{l=1}^{m-1} b_l(u_N^{m-l} + u_{N-1}^{m-l}) \tag{8.45}$$

where

$$a = \frac{1}{2}\frac{\Delta z}{\Delta x} I_0 \tag{8.46}$$

and

$$b_l = \frac{1}{2}\frac{\Delta z}{\Delta x}(I_l - I_{l-1}) \tag{8.47}$$

For the vacuum NLBC with convolution kernel g_0, we have

$$a = \frac{i\Omega\Delta z}{\sqrt{\Delta x}} \tag{8.48}$$

and

$$b_l = \frac{i\Omega\Delta z}{\sqrt{\Delta x}}\left(\sqrt{l+1} + \sqrt{l-1} - 2\sqrt{l}\right) \tag{8.49}$$

For the linear case, we write the integral I_l as

$$I_l = G_\alpha(x_{l+1}) - G_\alpha(x_l) \tag{8.50}$$

where G_α is the indefinite integral of the convolution kernel g_α. For positive α, we have

$$G_\alpha(x) = -\Delta x\left(\alpha k^2\right)^{1/3} \mathrm{e}^{2i\pi/3}\frac{\mathrm{Ai}'(0)}{\mathrm{Ai}(0)} - \frac{i\alpha k}{2}\sum_{j=1}^{\infty}\frac{\mathrm{e}^{s_j x}}{s_j^2} \tag{8.51}$$

A similar expression holds for negative α. When x is zero, convergence of the series is extremely slow. Instead of computing the summation term by term, we use the identity

$$\sum_{j=1}^{\infty}\frac{1}{a_j^2} = \left(\frac{\mathrm{Ai}'(0)}{\mathrm{Ai}(0)}\right)^2 \tag{8.52}$$

which is derived in Appendix A.

This simple approach (discretization of the continuous non-local boundary condition) can lead to an unstable numerical scheme: the reason for this is that the discretized boundary condition is not matched to the finite-difference scheme used to solve the PE in the interior of the computational domain [12, 154]. This can be avoided by the use of a discrete non-local boundary condition specifically constructed to match the algorithm inside the domain. The reader is referred to the work of Arnold and Ehrhardt [3] for more information on this topic.

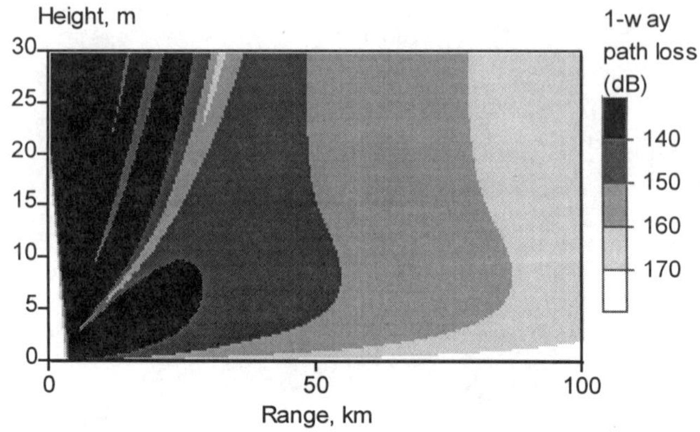

Figure 8.5 *Evaporation duct of height 10 m, 10 GHz antenna 25 m above sea level; Airy NLBC with slope 118 M-units/km*

8.5.4 Examples

We first look at an upwards transmitting case, considering 10 GHz propagation in a 10 m evaporation duct. The source is horizontally polarized and located 25 m above the ground. The beam pattern is Gaussian with a beamwidth of 0.4°. We truncate the domain with a non-local boundary condition at height 30 m. The initial field is negligible above that height, and we may use the screened version of the NLBC. Figure 8.5 shows path loss contours obtained with a finite-difference code when the slope α is matched to that of the refractivity profile at the top boundary, which is equal to 118 M-units/km. The Airy NLBC works very well, and there are no parasitic reflections from the top boundary. When the slope α is not matched to conditions at the top of the domain, distortions occur, due to the discontinuity in the refractive index gradient at the boundary. Figure 8.6 shows path loss contours obtained with $\alpha = 0$, corresponding to the transparent NLBC in vacuum: part of the energy is reflected back because of the gradient discontinuity.

Comparisons with a split-step Fourier code using an absorbing layer at the top of the domain confirm the accuracy of the Airy NLBC. The results at height 25 m shown in Figure 8.7 are identical for the Airy NLBC and the absorbing layer, while distortions are present for the vacuum NLBC.

We now consider a bilinear profile with a slope of 118 M-units/km betweeen 0 and 30 m and an infinite ducting layer of slope -100 M-units/km above 30 m. Figure 8.8 shows path loss contours for a horizontally polarized 3 GHz source

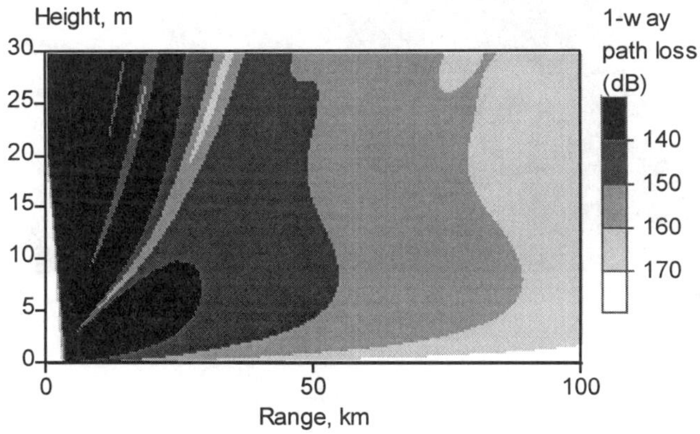

Figure 8.6 *Evaporation duct of height 10 m, 10 GHz antenna 25 m above sea level; Airy NLBC with* $\alpha = 0$

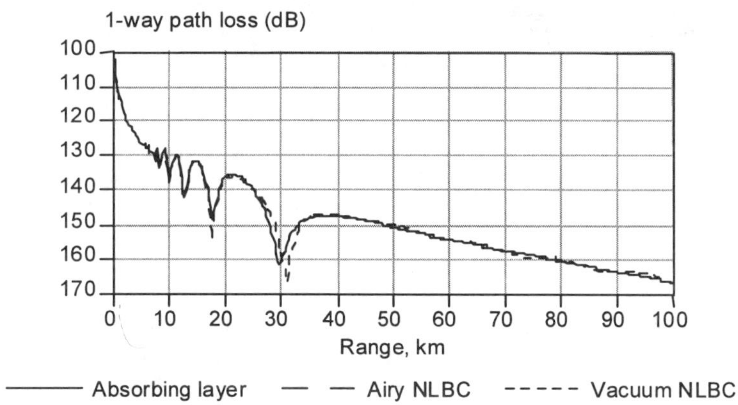

Figure 8.7 *Evaporation duct case. Path loss comparison at height 25 m for different truncation methods*

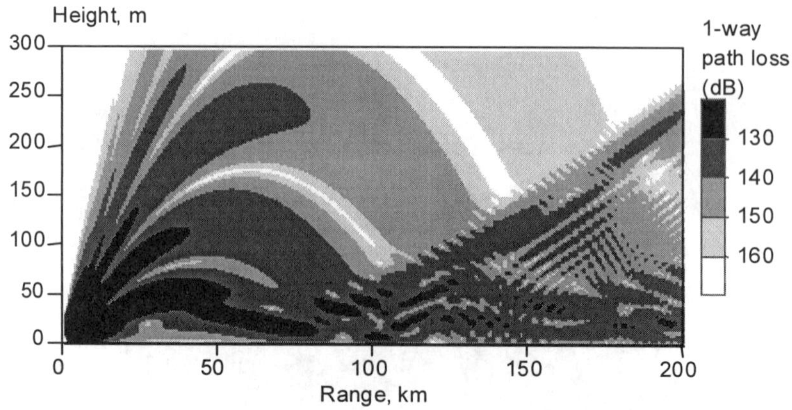

Figure 8.8 *Path loss contours obtained with full domain calculation*

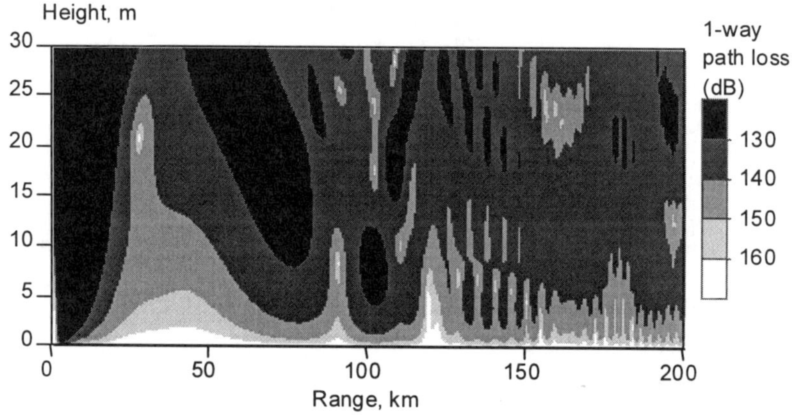

Figure 8.9 *Path loss contours obtained with ducting NLBC at 30 m*

at 20 m height. The results for this figure were obtained with a split-step Fourier code, adding an absorbing layer to the domain above 500 m. Figure 8.9 shows results obtained with a finite-difference code, truncating the domain at 30 m with the appropriate ducting NLBC. Energy coming back from above the domain is correctly represented.

In fact the results are identical to those obtained by considering the much larger domain of Figure 8.8, as shown in Figure 8.10, which gives path loss against range at a height of 20 m for both methods. The peaks in the convolution kernel g_α for $\alpha = -100$ M-units/km (see Figure 8.4) account for the

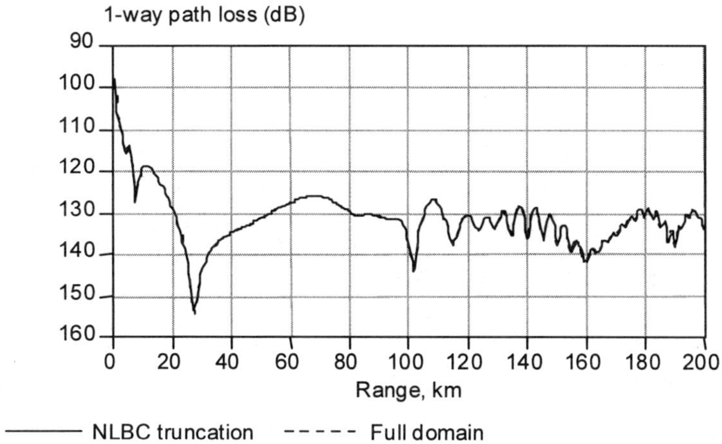

Figure 8.10 *Comparison of ducting NLBC results with full domain calculation*

energy returns due to the infinite ducting layer above the boundary.

8.6 Transmitting NLBCs

We now turn to the case where the initial field $\varphi(z)$ may have non-zero values above the boundary z_b. Dalrymple and Martin pointed out that a non-local boundary condition may be written for the field u in terms of a related diffractive boundary problem [42]. We denote by P the parabolic equation problem we wish to solve, which may involve terrain and refractive index variations below z_b. Above z_b, we assume the squared refractive index is constant or linear. Let P_0 be another parabolic equation problem, with the same environment as P above z_b. Suppose that we have found a solution $u_0(x, z)$ of problem P_0 in the domain $z \geq z_b$, such that $u_0(0, z) = \varphi(z)$ for $z \geq z_b$, in other words u and u_0 have the same initial field for $z \geq z_b$. Put $v = u - u_0$. Then since $v(0, z) = 0$ for $z \geq z_b$ and satisfies the parabolic equation for a constant or linear medium in the domain above z_b, it also satisfies the corresponding diffractive boundary condition (8.17) or (8.26) with the appropriate convolution kernel g. We then obtain the boundary condition for u in the form

$$\frac{\partial u}{\partial z}(x, z_b) = \int_0^x \frac{\partial u}{\partial x}(\xi, z_b) g(x - \xi) d\xi + I(x) \tag{8.53}$$

where the incoming energy term $I(x)$ is given by

$$I(x) = \frac{\partial u_0}{\partial z}(x, z_b) - \int_0^x \frac{\partial u_0}{\partial x}(\xi, z_b) g(x - \xi - t) d\xi \tag{8.54}$$

If the initial field is zero above z_b, we can simply take $P_0 = P$ everywhere, and $u_0 = 0$ in the whole domain. Then the incoming energy term is zero and eqn. 8.53 reduces to the diffractive NLBC. In the general case the auxiliary solution u_0 need not satisfy the same equation as u in the domain below z_b since P_0 may be different from P at heights below z_b. This has no bearing on the boundary condition, for which we only need information on $u - u_0$ below z_b.

From eqn. 8.53, we see that $I(x)$ is entirely determined by u. It follows that $I(x)$ does not depend on the choice of solution u_0, provided that u_0 satisfies the same equation as u above z_b and has the same initial field as u above z_b. This means that we may choose the refractive index structure and initial field below z_b as we see fit. Certain choices simplify the calculation of $I(x)$. Consider the case where $\varphi(z_b) = 0$. This property is quite frequently satisfied, in particular when we are concerned with high antenna modelling: then we may usually assume that the source is located sufficiently high above z_b to have a negligible aperture field below z_b. We now look at the problem P_0 obtained by making the refractive index symmetric with respect to z_b. For u_0, we take the solution obtained by making the initial field antisymmetric with respect to z_b, which is possible since $\varphi(z_b) = 0$. Then u_o must be antisymmetric with respect to z_b in the whole domain. It follows that both u_0 and its range derivative are identically zero on the z_b horizontal, and we have

$$I(x) = \frac{\partial u_0}{\partial z}(x, z_b) \tag{8.55}$$

Since u_0 satisfies some form of parabolic wave equation in the whole domain, we know that in principle $I(x)$ can be expressed as a function of the initial field. In the case where the refractive index is constant above z_b, this can be done explicitly: since the height derivative of u_0 satisfies the narrow-angle parabolic equation in vacuum, we can apply eqn. 2.43 from Chapter 2, getting

$$I(x) = -\frac{i\Omega}{\sqrt{\pi x}} \int_0^{+\infty} \varphi'(z + z_b)\, \mathrm{e}^{ikz^2/2x}\, dz \tag{8.56}$$

where we have used the fact that the height derivative of u_0 is symmetric with respect to z_b. This is a weighted Fresnel integral which represents the missing energy coming from the unscreened source into the domain of interest. In the far-field, the incoming energy term can be approximated by

$$I(x) \sim 2ik\frac{z_s - z_b}{x}\mathrm{e}^{ik(z_s-z_b)^2/2x}\,\psi\left(\frac{z_s - z_b}{x}\right) \tag{8.57}$$

where z_s is the source height and $\psi(\theta)$ is the far-field pattern, given in terms of the propagation angle θ from the horizontal.

We now treat the case of plane wave incidence for a constant refractive index medium above z_b. We cannot use the antisymmetric solution since the initial field is not negligible below z_b. Instead we simply take the initial field u_0 equal to that of u at all heights. Going back to eqn. 8.54, we can express $I(x)$ in terms of Fresnel integrals: for an incident plane wave of the form

$$\varphi_\theta(z) = \mathrm{e}^{ik\theta z} \tag{8.58}$$

we obtain

$$I(x) = \Omega\sqrt{\frac{2k}{\pi}}\theta\mathrm{e}^{-\frac{1}{2}ikx\theta^2}\int_{\theta\sqrt{\frac{kx}{2}}}^{+\infty}\mathrm{e}^{is^2}ds \tag{8.59}$$

This provides an alternative expression for the transmitting boundary condition derived in [42] for plane wave incidence.

8.6.1 *High antennas in linear media*

We consider a linear medium of the form given in eqn. 8.19, with positive slope α. It is then difficult to proceed further with the antisymmetric approach of eqn. 8.55. Instead, we define problem P_0 by extending the linear structure to the whole domain, and we look for the solution u_0 which has the same initial field as u for all values of z. This problem can be solved in closed form for an incident plane wave, and Fourier transform techniques provide the solution for an arbitrary initial field. For an incident wave of the form

$$\varphi_p(z) = \mathrm{e}^{2i\pi p(z-z_b)} \tag{8.60}$$

the solution $u_{p,0}$ of P_0 is given by

$$u_{p,0}(x,z) = \exp\left\{ik\left(\frac{\alpha x}{2}+\theta\right)(z-z_b) - \frac{ik}{3\alpha}\left(\left(\frac{\alpha x}{2}+\theta\right)^3 - \theta^3\right)\right\} \tag{8.61}$$

where

$$\theta = \lambda p \tag{8.62}$$

Letting α tend to zero, we obtain the familiar reduced plane wave solution of the standard parabolic equation. In the general case, let Φ be the Fourier transform of the incident field φ. We have

$$\varphi(z) = \int_{-\infty}^{+\infty} \Phi(p) \mathrm{e}^{2i\pi pz}\, dp \tag{8.63}$$

Using the spectral decomposition of φ, by superposition we obtain the solution u_0 as

$$u_0(x,z) = \int_{-\infty}^{+\infty} \Phi(p) u_{p,0}(x,z) \mathrm{e}^{2i\pi p z_b}\, dp \tag{8.64}$$

It is then straightforward to derive an expression for $I(x)$. In general, numerical implementation requires fast Fourier transforms. However u_0 can be expressed in closed form for the important case of a Gaussian source. For an aperture function of the form

$$\varphi(z) = \mathrm{e}^{-A(z-z_s)^2} \tag{8.65}$$

corresponding to a source at height z_s with a Gaussian beam pattern, the solution u_0 is given by

$$u_0(x,z) = \sqrt{\frac{k}{k+2iAx}}\, \mathrm{e}^{\frac{1}{2}ik\alpha x(z-z_b) - \frac{2}{3}ik\alpha^2 x^3 - kA\frac{(z_s - z + \alpha x^2/4)^2}{k+2iAx}} \tag{8.66}$$

It is straightforward to derive $I(x)$ from this expression.

8.6.2 Elevated duct example

To illustrate the high antenna procedure, we look at the case of the elevated duct given in Table 8.1, with a ducting layer of 5 M-units between heights of 150 m and 200 m. We consider a 3 GHz source located at a height of 900 m. The beam pattern is Gaussian, with half-power beamwidth 2° and elevation angle −0.83° from the horizontal. We apply the Airy NLBC with incoming energy term at a height of 200 m, which allows us to consider terrain variations below that height. The value of α is dictated by the profile above the 200 m boundary, as the incoming energy term must match propagation above the boundary. In this case α = 118 M-units/km. The discontinuity in refractive

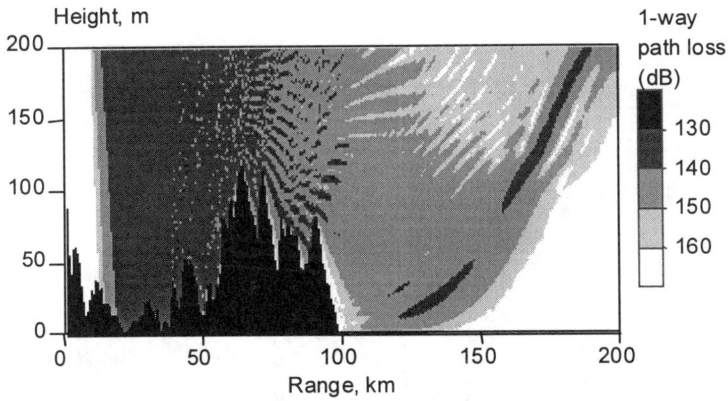

Figure 8.11 *Elevated duct example with 3 GHz source above the domain; the source altitude is 900 m; Airy NLBC at 200 m with* $\alpha = 118$ *M-units/km*

index slope at the boundary is an integral part of this particular case and does not introduce any spurious distortions.

Table 8.1 *M-profile for elevated duct NLBC example*

Height (m)	M (M-units)
0	-12.7
150	5
200	0
1200	118

In this case the initial field below 200 m is negligible, so all the incoming energy is arriving through the NLBC on the 200 m horizontal. Figure 8.11 shows the resulting path loss contours, demonstrating that the incoming energy is indeed well represented. Figure 8.12 shows PE results obtained by including the source in the integration domain, which now extends up to 1000 m. Seeing this full picture illustrates the difference in treatment between downgoing and upgoing energy in the NLBC method: downgoing energy, which affects results below the 200 m boundary, is represented by the incoming energy term in the NLBC, whereas upgoing energy, which does not affect the lower altitude results, is truncated exactly by the screened incident-field part of the NLBC.

Figure 8.13 gives path loss values against height at a distance of 150 km

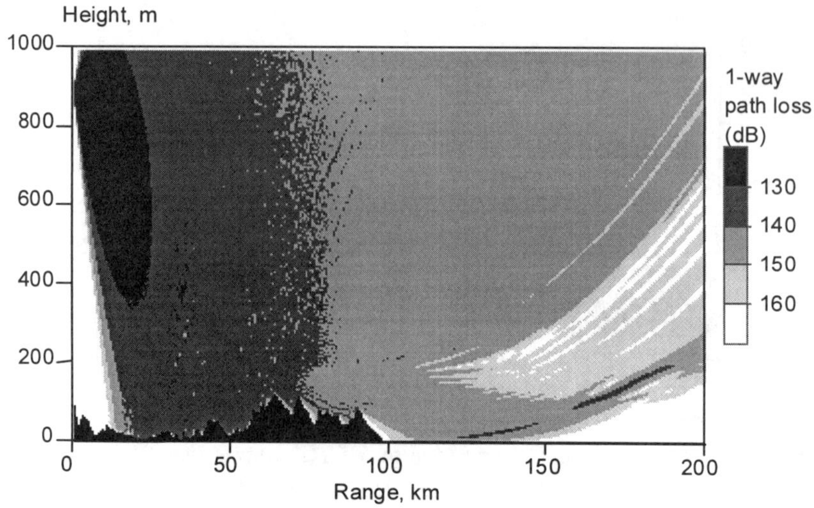

Figure 8.12 Elevated duct example with 3 GHz source inside PE domain, using split-step PE algorithm

from the source for the NLBC results with the source above the domain and the split-step Fourier algorithm with the source inside the domain. The two techniques give almost identical results.

8.7 Split-step incoming energy PE

The non-local methods described in Section 8.6 provide very efficient low altitude truncation for high antenna problems, but require finite-difference implementation. In fact the ideas of Section 8.6 can be adapted in order to treat antennas outside the PE integration domain with split-step techniques.

Assume the source is located above height z_b, with an initial field which is negligible below z_b. We call P the problem associated with the source, refractive index $n(x,z)$ and possible terrain variations below z_b. We assume that the medium is upwards transmitting above z_b, in the sense that no energy is reflected back below z_b by refractive index features above z_b. This is certainly the case when the squared refractive index is linearly increasing with height above z_b. More generally the medium is upwards transmitting if the refractive index is a smooth increasing function of height above z_b. Let P_0 be the problem associated with the same source and refractive index above z_b, but with an opaque medium below z_b, with boundary conditions to be decided later.

Let u and u_0 be the solutions of problems P and P_0, and let $v = u - u_0$. Since $u_0 = 0$ below z_b, v is equal to the required solution u in the domain

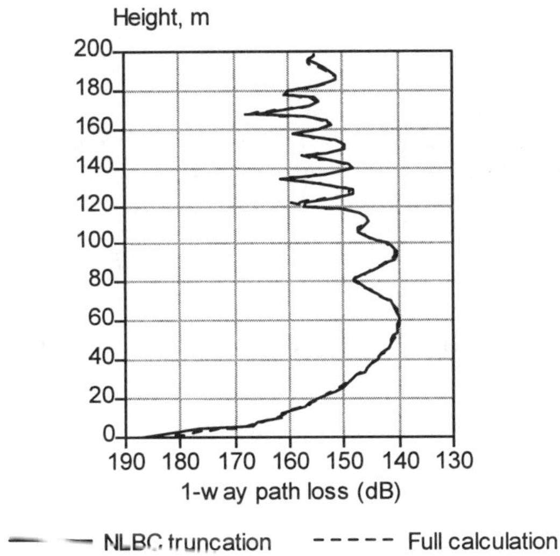

Figure 8.13 *Comparison between NLBC truncation and full domain calculation for high antenna example at range 150 km*

of interest. Because of the interface associated with problem P_0, v is not continuous, but we can still solve for v with a split-step formulation by taking the interface jumps into account.

For $z \neq z_b$, v satisfies the SPE

$$\frac{\partial v}{\partial x} = \frac{i}{2k}\frac{\partial^2 v}{\partial z^2} + \frac{ik}{2}(n^2 - 1)v \tag{8.67}$$

This is true for $z > z_b$ since P_0 and P have the same refractive index environment above z_b, and is also true for $z < z_b$ since $v = u$ in that region.

Assume for simplicity that the terrain associated with problem P is flat, located at $z = 0$, and perfectly conducting. If we denote by $V(x, p)$ the sine transform of v, we can write

$$\begin{aligned}\int_0^\infty \frac{\partial^2 v}{\partial z^2} \sin(2\pi pz)dz &= \left(\frac{\partial v}{\partial z}(x, z_b^-) - \frac{\partial v}{\partial z}(x, z_b^+)\right) \sin(2\pi p z_b) \\ &+ 2\pi p \left(v(x, z_b^-) - v(x, z_b^+)\right) \cos(2\pi p z_b) - 4\pi^2 p^2 V(x, p)\end{aligned} \tag{8.68}$$

where the + and − signs denote limiting values at the interface $z = z_b$ respectively above and below z_b. This expression simplifies if we take as the boundary condition for problem P_0 at the interface $z = z_b$

$$\frac{\partial v}{\partial z}(x, z_b^+) = 0 \tag{8.69}$$

In other words our opaque medium acts like a perfect conductor for vertically polarized incidence. Then with the usual split-step approximation, V satisfies

$$\frac{\partial V}{\partial x}(x,p) = -\frac{2i\pi^2 p^2}{k}V(x,p) + \frac{i\pi p}{k}\cos(2\pi p z_b)u_0(x, z_b^+) \tag{8.70}$$

since the jump at the interface is equal to u_0. The solution of this ordinary differential equation is

$$V(x+\Delta x, p) = \mathrm{e}^{-\frac{2i\pi^2 p^2 \Delta x}{k}} V(x,p) + J(x,p) \tag{8.71}$$

where the split-step PE incoming energy term $J(x,p)$ is given by

$$J(x,p) = \frac{i\pi p}{k}\cos(2\pi p z_b)\int_0^{\Delta x} u_0(x+t, z_b^+)\mathrm{e}^{\frac{2i\pi^2 p^2 (t-\Delta x)}{k}}\,dt \tag{8.72}$$

If we know the solution u_0 of problem P_0, we can solve for P below z_b. In general the integral cannot be given in closed form, and discretization is required. Assuming $u_0(x, z_b)$ constant over interval $(x, x+\Delta x)$, we approximate the incoming energy integral by

$$J(x,p) = -\cos(2\pi p z_b)\frac{\mathrm{e}^{-\frac{2i\pi^2 p^2 \Delta x}{k}} - 1}{2\pi p}u_0\left(x + \frac{\Delta x}{2}, z_b^+\right) \tag{8.73}$$

The incoming energy term can be computed either in z- or in p-space. The natural choice is to take z_b at the top of the region of interest, so that the usual filtering of upgoing energy does not affect propagation in that region.

In general the auxiliary solution u_0 is not available in closed form. Consider the simple case of a linearly increasing squared refractive index profile above z_b: then u_0 solves the problem of propagation around a perfectly conducting sphere for the case of vertical polarization, which has no easy solution if we are interested in values close to or beyond the horizon. Fortunately accuracy in computation of the incoming energy is most crucial in the line-of-sight region where rays actually penetrate the region below z_b. This allows the use of ray-tracing to compute u_0. Suitable filters must be applied to rays launched close to the horizon region to avoid spurious diffraction effects. The ray-trace method is schematically illustrated in Figure 8.14.

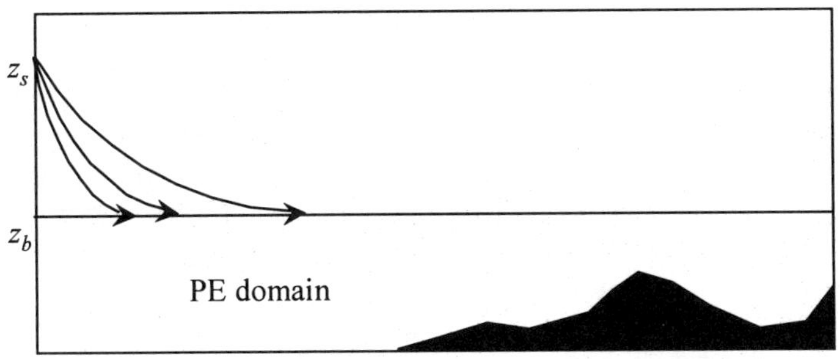

Figure 8.14 Calculation of auxiliary solution with ray-trace

The incoming energy split-step model works essentially like the ordinary split-step/Fourier algorithm:

1. Carry out sine transform of field at range x.
2. Multiply by free-space propagator.
3. Transform back to z-space.
4. Add incoming energy term.
5. Multiply by refractive index exponential.

When terrain variations are present, the above analysis must be modified slightly to account for offsets in the origin of the sine transform relative to z_b. If the ground between ranges x and ranges $x + \Delta x$ is at height h, we have to compute the sine transform f_h of

$$F_h(p) = \cos\left[2\pi p(z_b - h)\right] \frac{e^{-\frac{2i\pi^2 p^2 \Delta x}{k}} - 1}{2\pi p} \tag{8.74}$$

Using trigonometric identities, we obtain

$$f_h(z) = \frac{1}{2}\left\{f(z + z_b - h) + f(z - z_b + h)\right\} \tag{8.75}$$

where z is the height above ground and f is the sine transform of

$$F(p) = \frac{e^{-\frac{2i\pi^2 p^2 \Delta x}{k}} - 1}{2\pi p} \tag{8.76}$$

Hence the incoming energy term for varying terrain is obtained by suitable translations of f. Examples obtained with the split-step incoming energy PE algorithm will be given in Chapter 11, where it is used in conjunction with the horizontal PE method to extend coverage diagrams upwards for high antennas.

Chapter 9

Impedance boundary modelling

9.1 Introduction

Although the perfectly conducting ground model is adequate for many applications, it is not universally applicable. A more accurate model is required for surface wave propagation, for correct modelling of reflection effects at vertical polarization or for the modelling of propagation over rough surfaces.

In Chapter 8, we derived perfectly transparent non-local boundary conditions. These are actually a special case of non-local impedance boundary conditions describing the interface between two media. In Section 9.2, we give the derivation of these non-local boundary conditions in the paraxial framework and show how they can be approximated by the Leontovich boundary conditions [82] for most tropospheric propagation applications. Boundary conditions are also given in terms of the angle-dependent reflection coefficient. Much information on more general boundary conditions can be found in [141].

Finite-difference implementation of both local and non-local boundary conditions is straightforward. It is briefly described in Section 9.3. Because of the numerical advantages of the split-step/Fourier algorithm, it was soon realized that a version of it correctly propagating the impedance boundary condition would be extremely useful. Such an algorithm has to "mix" sine and cosine transforms, hence the "mixed Fourier transform" terminology. In Section 9.4, we turn to the mixed Fourier transform model developed by Kuttler and Dockery [77, 46], describing both the continuous and the discrete versions.

Applications to line-of-sight propagation of vertically polarized waves and to surface wave propagation are given in Section 9.5, showing that the discrete mixed Fourier transform indeed models these cases correctly. In this chapter, only smooth surfaces are considered. Rough surface modelling is treated in Chapter 10.

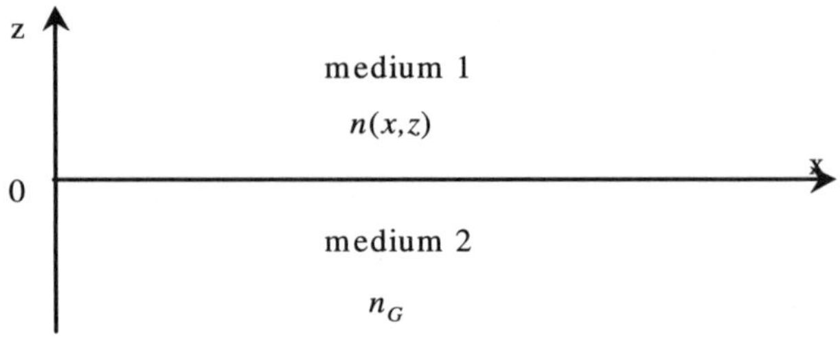

Figure 9.1 Horizontal interface between two media

9.2 Boundary conditions at a horizontal interface

We work with the two-dimensional wave equation for a field component ψ in Cartesian coordinates (x, z)

$$\frac{\partial^2\psi}{\partial x^2} + \frac{\partial^2\psi}{\partial z^2} + k^2n^2\psi = 0 \tag{9.1}$$

As we have seen in Chapter 4, this equation can arise either from a problem in Cartesian coordinates (x, y, z) which is independent of the transverse coordinate y, or from a tropospheric propagation problem where (x, z) are flattened Earth coordinates derived from spherical coordinates and n is the modified refractive index. In the first case, which we shall denote by (A), ψ is the tangential electric field E_y for horizontal polarization and the tangential magnetic field H_y for vertical polarization. In the second case, which we denote by (B), we have

$$\psi = \sqrt{ka\sin(x/a)\mathrm{e}^{z/a}}E_\phi \tag{9.2}$$

for horizontal polarization and

$$\psi = \frac{1}{n}\sqrt{ka\sin(x/a)\mathrm{e}^{z/a}}H_\phi \tag{9.3}$$

for vertical polarization, where E_ϕ and H_ϕ are the azimuthal transverse fields and a is the Earth's radius. In either case we consider a horizontal interface between two media located at height $z = 0$, as shown in Figure 9.1. In tropospheric problems with flattened Earth coordinates, the interface represents the Earth's surface.

Assuming that the conductivities of the two media are finite, it follows from Maxwell's equations that the tangential component of the fields must be continuous at the interface [25]. For case (A), the curl equation for the magnetic field becomes

$$H_x = -\frac{1}{ik}\frac{\partial E_y}{\partial z} \tag{9.4}$$

For horizontally polarized waves, the continuity conditions mean that ψ and the normal derivative $\partial_z \psi$ must be continuous across the interface. Similarly, for vertical polarization, ψ and $\partial_z \psi / n^2$ must be continuous across the interface.

From the discussions on boundary conditions in Chapter 4, it follows that for case (B), the same conditions hold for horizontal polarization, while for vertical polarization, $n\psi$ and χ/n must be continuous across the interface, where

$$\chi = \frac{\partial \psi}{\partial z} + \left(\frac{\partial n}{\partial z} - \frac{n-1}{a}\right)\psi \tag{9.5}$$

Medium 1 above the interface has a refractive index $n(x, z)$ which can vary with height and range. Assume that medium 2 is homogeneous, with refractive index n_G. The relative complex permittivity η of medium 2 is given by

$$\eta = n_G^2 = \epsilon_r + i\frac{\sigma}{kc\epsilon_0} \tag{9.6}$$

where ϵ_r is the relative permittivity of medium 2, σ its conductivity, ϵ_0 the permittivity of vacuum and c the velocity of light in vacuum. Keeping the same reference wavenumber k for both media, we define the reduced PE field

$$u(x, z) = \mathrm{e}^{-ikx}\psi(x, z) \tag{9.7}$$

In medium 1, u satisfies some version of the outgoing parabolic wave equation. It is immaterial for the derivation whether we choose the narrow-angle PE, some form of the wide-angle PE or the original square-root operator. In medium 2 we retain the square-root operator and write the outgoing parabolic wave equation as

$$\frac{\partial u}{\partial x}(x, z) = \left(i\sqrt{\frac{\partial^2}{\partial z^2} + k^2\eta} - ik\right) u(x, z) \tag{9.8}$$

We assume that the initial field is zero below the interface. Then the Laplace transform of eqn. 9.8 with respect to range x is

$$sU(s,z) = \left(i\sqrt{\frac{\partial^2}{\partial z^2} + k^2\eta} - ik \right) U(s,z) \tag{9.9}$$

Since medium 2 is homogeneous and contains no sources, the solution below the interface must remain bounded as z tends to $-\infty$, hence it is of the form

$$U(s,z) = \alpha(s)\mathrm{e}^{-ik\alpha(s)z} \tag{9.10}$$

where $\alpha(s)$ has a positive imaginary part. After some algebra with eqn. 9.9, we find

$$\alpha(s) = \sqrt{\eta + \left(\frac{s}{k} + i\right)^2} \tag{9.11}$$

Now the continuity conditions at the interface hold for the reduced field u. In the case of horizontal polarization, they also hold for the Laplace transform U by integrating over range. For vertical polarization, we need to make the further assumption that the refractive index at the surface is independent of range. In addition the surface refractive index gradient at the surface must be range-independent in case (B). From eqn. 9.10, we can express the normal derivative in medium 2 as

$$\frac{\partial U}{\partial z}(s,0) = -ik\alpha(s)U(s,0) \tag{9.12}$$

The continuity conditions then give an impedance type relationship between U and its normal derivative in medium 1 in the form

$$\frac{\partial U}{\partial z}(s,0) = (-ik\delta(s) + \beta)\, U(s,0) \tag{9.13}$$

where the generalized impedance function $\delta(s)$ is given by

$$\delta(s) = \sqrt{\eta + \left(\frac{s}{k} + i\right)^2} \tag{9.14}$$

for horizontal polarization and

$$\delta(s) = \frac{1}{\eta}\sqrt{\eta + \left(\frac{s}{k} + i\right)^2} \tag{9.15}$$

for vertical polarization, and β is zero except for the case of vertical polarization and flattened Earth coordinates, where

$$\beta = \frac{n(0)-1}{a} - \frac{n_G - 1}{\eta a} - \frac{\partial n}{\partial z}(0) \tag{9.16}$$

The first two terms on the RHS are always negligible, and tropospheric refractive index gradients are usually small, so we may neglect β. We note that the boundary condition given by eqn. 9.13 does not depend on the variations of medium 1, but only on the value of the refractive index and its height derivative at the boundary. Eqn. 9.13 is a generalized impedance boundary condition (GIBC). It depends on the Laplace variable s, which is related to the horizontal plane wave decomposition of the incident field as we have seen in Chapter 8.

9.2.1 Leontovich boundary conditions

We recall that the transform variable s is related to the spectral decomposition of the field with respect to the horizontal by

$$is = k(1 - \cos\theta) \tag{9.17}$$

where θ is the grazing angle. Values of s/k larger than unity correspond to evanescent waves and can be neglected. If the relative complex permittivity η of medium 2 has sufficiently large modulus, the impedance function $\delta(s)$ depends only weakly on propagation angle and we can treat the surface impedance as constant, writing

$$\alpha(s) \sim \alpha(0) \tag{9.18}$$

Inverting the Laplace transform, we obtain the Leontovich boundary condition

$$\frac{\partial u}{\partial z}(x,0) = (-ik\delta + \beta)u(x,0) \tag{9.19}$$

The surface impedance δ is given by

$$\delta = \sqrt{\eta - 1} \tag{9.20}$$

for horizontal polarization and by

$$\delta = \frac{\sqrt{\eta - 1}}{\eta} \tag{9.21}$$

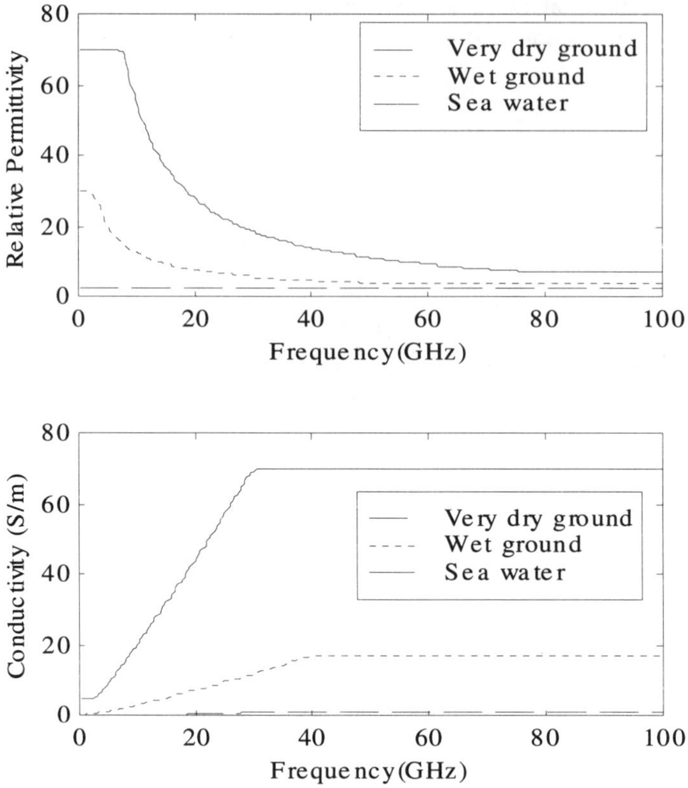

Figure 9.2 *Electrical constants of various types of ground as a function of frequency (a) relative permittivity and (b) conductivity*

for vertical polarization.

Figure 9.2 gives relative permittivity and conductivity as a function of frequency for very dry ground, wet ground and sea water [63]. Sea water values correspond to average salinity and a temperature of 20° C. Conductivity at the lower frequencies is of the order of 10^{-4} S/m for very dry ground and 10^{-2} S/m for wet ground.

Figure 9.3 gives the modulus of η for these types of ground. The Leontovich approximation is certainly justified for propagation over the sea, and also for propagation over wet ground for microwave frequencies, but it is less so for the very dry ground case if we are concerned with large propagation angles. We shall see below that the Leontovich approximation can still be used if propagation angles remain small.

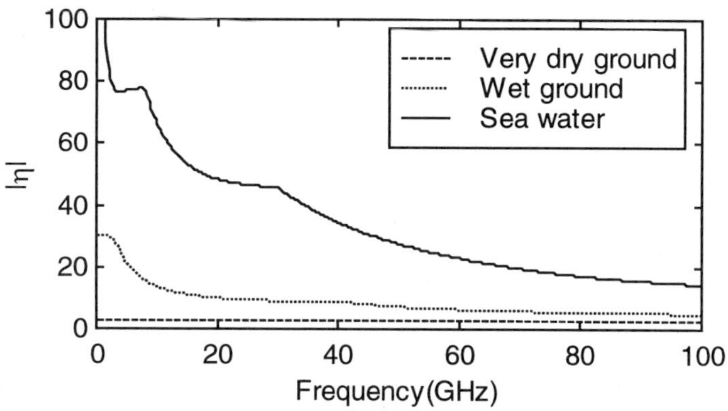

Figure 9.3 *Modulus of propagation constant η for various types of ground as a function of frequency*

9.2.2 Small angle GIBC

When η is not sufficiently large, the Leontovich approximation loses accuracy at large propagation angles. If propagation angles from the horizontal remain small, we may neglect the second-order term in s/k and use instead the approximation

$$\alpha(s) \sim \sqrt{\eta - 1 + \frac{2is}{k}} \tag{9.22}$$

which is valid for arbitrary values of η. The error is of the order of $\sin^4 \theta$ in the grazing angle, as for the narrow-angle PE approximation: if the SPE is used in medium 1, the small angle approximation of the impedance function is consistent with it. When medium 2 is vacuum, $\eta = 1$ and the GIBC reduces to the transparent boundary condition of Section 8.4, with a change of sign due to the fact we are now looking at the lower boundary of the domain.

For most applications, it is more convenient to work with the surface admittance, writing the boundary condition in the form

$$U(s,0) = -\frac{1}{ik\delta(s)} \frac{\partial U}{\partial z}(s,0) \tag{9.23}$$

After inversion of the Laplace transform, we get the boundary condition for u in the form

$$u(x,0) = \int_0^x \frac{\partial u}{\partial z}(\xi, 0)\, h(x - \xi)\, d\xi \tag{9.24}$$

where the convolution kernel h is the inverse Laplace transform of $-1/ik\delta(s)$. From [47], we have

$$h(x) = \frac{1+i}{2} \frac{\gamma}{\sqrt{k\pi x}} \mathrm{e}^{\frac{ik}{2}(\eta-1)x} \tag{9.25}$$

where $\gamma = 1$ for horizontal polarization and $\gamma = \eta$ for vertical polarization. Like the perfectly transparent boundary conditions of Chapter 8, this condition is non-local.

When is eqn. 9.24 well approximated by the Leontovich boundary condition? This is clear if the ground is conducting: the convolution kernel h then includes a damping factor d given by

$$d(x) = \exp\left(\frac{\sigma x}{\epsilon_0 c}\right) = \exp\left(-377\sigma x\right) \tag{9.26}$$

Even for quite poorly conducting ground, say with conductivity of the order of 10^{-4} S/m as in the very dry ground example of the previous section, the damping factor at 100 m is of the order of 0.02. Damping causes the convolution kernel to behave like a Dirac function, so that the Leontovich boundary condition is a good approximation of the non-local boundary condition. For pure dielectrics however there is no damping. In that case it is the high oscillatory behaviour of the kernel for large permittivities which makes it behave like a Dirac function.

9.2.3 Reflection coefficient

The generalized impedance boundary condition may also be considered in terms of reflection coefficients. This interpretation will be useful in Chapter 10, where we consider rough surfaces characterized by angle-dependent reflection coefficients. Assume that a thin homogeneous layer with refractive index 1 is located in medium 1 just above the interface. The solution of the Laplace transform of the SPE in the layer can be written as

$$U(s,z) = \alpha(s)\left(\mathrm{e}^{-iz\sqrt{2iks)}} + R(s)\mathrm{e}^{iz\sqrt{2iks)}}\right) \tag{9.27}$$

where $R(s)$ is the reflection coefficient corresponding to the spectral variable s. At the surface, we have

$$\frac{\partial U}{\partial z}(s,0) = -i\sqrt{2iks}\,\frac{1-R(s)}{1+R(s)}\,U(s,0) \tag{9.28}$$

which is equivalent to the surface impedance formulation of eqn. 9.13 when medium 2 is homogeneous if we take

$$R(s) = \frac{\sqrt{\frac{2is}{k}} - \delta(s)}{\sqrt{\frac{2is}{k}} + \delta(s)} \tag{9.29}$$

The reflection coefficient formulation is more general however: provided we know the reflection properties of the interface as a function of incidence angle, we can write a non-local boundary condition in the form of eqn. 9.24, where the convolution kernel h is given by the inverse Laplace transform of the admittance function H defined by

$$H(s) = \frac{i}{\sqrt{2is}} \frac{1 + R(s)}{1 - R(s)} \tag{9.30}$$

In the narrow-angle case, the spectral variable s is related to the grazing angle θ by

$$2is = k \sin^2 \theta \tag{9.31}$$

Then eqn. 9.29 is just the familiar relation between surface impedance and reflection coefficient.

9.3 Finite-difference implementation

We consider a local impedance condition of the form

$$\frac{\partial u}{\partial z}(x, 0) + \alpha u(x, 0) = 0 \tag{9.32}$$

Implementation of eqn. 9.32 with finite-difference schemes presents no difficulty: the first line of the matrix is modified to take the boundary condition into account. We write a finite-difference expression for the second order z-derivative at the boundary point $(x, 0)$ using a one-sided approximation. With the notation of Chapter 7, we have

$$\frac{\partial^2 u}{\partial z^2}(x, 0) \sim 2\, \frac{\frac{\partial u}{\partial z}(x, \Delta z/2) - \frac{\partial u}{\partial z}(x, 0)}{\Delta z} \tag{9.33}$$

We now approximate the z-derivative at $(x, \Delta z/2)$ by the usual central difference, and use the boundary condition to evaluate the z-derivative at the boundary point $(x, 0)$, getting

$$\frac{\partial^2 u}{\partial z^2}(x,0) \sim 2\,\frac{u(x,\Delta z) - u(x,0)(1-\alpha\Delta z)}{\Delta z^2} \tag{9.34}$$

Substituting in the SPE and averaging in range as in Chapter 3, we obtain the finite-difference equation

$$u_0^m(-1+\alpha\Delta z + b + a_0^m) + u_1^m = u_0^{m-1}(1-\alpha\Delta z + b - a_0^m) - u_1^{m-1} \tag{9.35}$$

where the coefficients b and and a_0^m are defined in eqns. 3.79 and 3.80 respectively. Since eqn. 9.35 only involves u_0^m and u_1^m, the resulting matrix is tridiagonal, and there is no increase in computational complexity.

9.4 Mixed Fourier transform

9.4.1 Continuous mixed Fourier transform

A continuous version of the mixed Fourier transform was derived by Kuttler and Dockery in [77]. The basic idea is to match the propagation algorithm to the boundary condition of eqn. 9.32. We look at the "diffractive" part of the split-step solution of the SPE, corresponding to a step in a homogeneous medium

$$\frac{\partial u}{\partial x} = \frac{i}{2k}\frac{\partial^2 u}{\partial z^2} \tag{9.36}$$

We have seen in Section 3.3 that the sine transform provides a solution matched to the Dirichlet condition $u(x, 0) = 0$. Similarly one could show that the cosine transform gives a solution satisfying the Neumann condition $\partial u/\partial z(x, 0) = 0$. Our aim is to find a transform which will automatically propagate the impedance boundary condition. We define the matched function v by

$$v = \frac{\partial u}{\partial z} + \alpha u \tag{9.37}$$

Now we notice that v also satisfies eqn. 9.36. The boundary condition states that v must be zero on the bottom interface $z = 0$, hence we can propagate v with the sine transform. In other words,

$$v(x + \Delta x, z) = \mathcal{S}^{-1}\left\{e^{-\frac{i\pi^2 p^2 \Delta x}{2k}} \mathcal{S}v(x, z)\right\} \tag{9.38}$$

We note that $U = \mathcal{S}v$ can be expressed in terms of u by

$$U(x, p) = \int_0^{+\infty} u(x, z) \left[\alpha \sin(\pi p z) - \pi p \cos(\pi p z)\right] dz \tag{9.39}$$

In order to find u at the advanced range, we could try to solve eqn. 9.37, which is an ordinary differential equation in terms of u. For the case of horizontal polarization, corresponding to $\mathrm{Re}(\alpha) < 0$, the only bounded solution is

$$u(x, z) = -\mathrm{e}^{-\alpha z} \int_z^{\infty} v(z')\mathrm{e}^{\alpha z'} dz' \tag{9.40}$$

However for the case of vertical polarization, corresponding to $\mathrm{Re}(\alpha) > 0$, the solutions are of the form

$$u(x, z) = \mathrm{e}^{-\alpha z} \left[\int_0^z v(z')\mathrm{e}^{\alpha z'} dz' + C\right] \tag{9.41}$$

and they are bounded for any choice of C. Now C satisfies

$$C + \int_0^{\infty} v(z)\mathrm{e}^{-\alpha z} dz = 2\alpha \int_0^{\infty} u(x, z)\mathrm{e}^{-\alpha z} dz \tag{9.42}$$

So we see that unless the integral on the RHS of eqn. 9.42 is known, we cannot determine u without ambiguity. This leads us to the following definition of the forward mixed transform: we put $\mathcal{M}u = \{U, K\}$ where

$$U(x, p) = \int_0^{+\infty} u(x, z) \left[\alpha \sin(\pi p z) - \pi p \cos(\pi p z)\right] dz \tag{9.43}$$

and $K(x)$ is defined by

$$K(x) = \begin{cases} 0 & \text{for } \mathrm{Re}(\alpha) \leq 0 \\ 2\alpha \int_0^{+\infty} u(x, z)\mathrm{e}^{-\alpha z} & \text{for } \mathrm{Re}(\alpha) > 0 \end{cases} \tag{9.44}$$

Instead of using eqns. 9.40 and 9.41, we can write the inverse transform as

$$u(x, z) = K(x)\mathrm{e}^{-\alpha z} + \int_0^{+\infty} U(p, z) \frac{\alpha \sin(\pi p z) - \pi p \cos(\pi p z)}{\alpha^2 + p^2} dz \tag{9.45}$$

The proof, which is based on contour integration, can be found in [155]. We now use the mixed transform pair to propagate u. The following differential equations are satisfied:

$$\frac{\partial U}{\partial x}(x,p) = -\frac{i\pi^2 p^2}{2k} U(x,p) \tag{9.46}$$

and

$$K'(x) = \frac{i\alpha^2}{2k} K(x) \tag{9.47}$$

The latter equation is obtained using the fact that u satisfies the SPE, together with repeated integration by parts. The boundary condition ensures that the integrated part vanishes. It is now straightforward to propagate U and K and to apply the inverse transform at the advanced range, getting the formula

$$\begin{aligned} u(x+\Delta x, z) &= \mathrm{e}^{\frac{i\alpha^2 \Delta x}{2k}} K(x) \mathrm{e}^{-\alpha z} \\ &+ \int_0^{+\infty} U(x,z)\, \mathrm{e}^{-\frac{i\pi^2 p^2 \Delta x}{2k}} \frac{\alpha \sin(\pi p z) - \pi p \cos(\pi p z)}{\alpha^2 + p^2}\, dz \end{aligned} \tag{9.48}$$

It is easy to check that u indeed satisfies the SPE. Since v is the sine transform of U, v vanishes on $z = 0$ and the boundary condition is automatically satisfied.

Eqn. 9.48 can of course be implemented numerically with discrete sine and cosine transforms. However each step then takes approximately twice the time required for the Neumann or Dirichlet case (corresponding to a perfectly conducting boundary), since two sine and two cosine transforms are needed. Another drawback, reported in [46], is that it becomes difficult to maintain accuracy when the poles $\pm i\alpha$ are close to the real line. These difficulties were solved by the introduction of the discrete mixed Fourier transform (DMFT) in [46]. In Chapter 3, we saw that it was possible to bypass the continuous integrals by discretizing the PE partial differential equation directly. This is the approach we use here.

9.4.2 Discrete mixed Fourier transform

As in Chapter 3, we start by discretizing in height z, putting

$$f_j(x) = u(x, j\Delta z),\ j = 0, \ldots, N \tag{9.49}$$

We want to define the DMFT in such a way that the impedance boundary condition will be automatically propagated. We put

$$r = \sqrt{1 + (\alpha \Delta z)^2} - \alpha \Delta z \tag{9.50}$$

for vertical polarization (corresponding to the case where α has a positive real part) and

$$r = -\sqrt{1 + (\alpha \Delta z)^2} - \alpha \Delta z \tag{9.51}$$

for horizontal polarization (corresponding to the case where α has a negative real part). We note that r and $-r^{-1}$ are the roots of the quadratic equation

$$r^2 + 2r\alpha\Delta z - 1 = 0 \tag{9.52}$$

Our choice of r according to the sign of the real part of α guarantees that $|r|$ is less than 1, which is required for the numerical stability of the algorithm.

The DMFT of $f = (f_0, \ldots, f_N)$ is given by

$$\begin{aligned}
F_0 &= A \sum_{m=0}^{N}{}' r^m f_m \\
F_j &= \sum_{m=1}^{N-1}{}' \left[\alpha \sin\left(\tfrac{\pi j m}{N}\right) - \frac{\sin\left(\frac{\pi j}{N}\right)}{\Delta z} \cos\left(\tfrac{\pi j m}{N}\right) \right] f_m \\
& \qquad\qquad j = 1, \ldots, N-1 \\
F_N &= A \sum_{m=0}^{N}{}' (-r)^{N-m} f_m
\end{aligned} \tag{9.53}$$

where the prime superscript on the sum indicates that the first and last terms are weighted with the coefficient 1/2 and

$$A = \frac{2(1 - r^2)}{(1 + r^2)(1 - r^{2N})} \tag{9.54}$$

The inverse discrete matched transform is given by

$$f_m = r^m F_0 + \frac{2}{N} \sum_{j=1}^{N-1} \frac{\alpha \sin\left(\frac{\pi j m}{N}\right) - \beta_j \cos\left(\frac{\pi j m}{N}\right)}{\alpha^2 + \beta_j^2} F_j + (-r)^{N-m} F_N, \; m = 0, \ldots, N \tag{9.55}$$

where

$$\beta_j = \frac{\sin\left(\frac{\pi j m}{N}\right)}{\Delta z}, \; j = 1, \ldots, N-1 \tag{9.56}$$

We can greatly simplify the inverse procedure by looking at the DMFT in terms of the discrete matched function g defined by

$$g_m = \frac{f_{m+1} - f_{m-1}}{2\Delta z} + \alpha f_m, \; m = 1, \ldots, N-1 \tag{9.57}$$

It is easy to check that $(F_1, \ldots, F_{N-1})$ is exactly the discrete sine transform of g. We have

$$F_j = \sum_{m=1}^{N-1} g_m \sin\left(\frac{\pi j m}{N}\right), \; j = 1, \ldots, N-1 \tag{9.58}$$

To obtain the inverse transform, we first apply the inverse discrete sine transform to $(F_1, \ldots, F_{N-1})$, getting f. We now have to retrieve f from g, F_0 and F_N. We first note that since r and $-1/r$ are the roots of the quadratic equation, eqn. 9.52, any vector of the form

$$\phi_m = B_1 r^m + B_2(-r)^{N-m}, \; m = 0, \ldots, N \tag{9.59}$$

satisfies the homogeneous difference equations

$$\phi_{m+1} - \phi_{m-1} + \alpha\phi_m = 0, \; m = 1, \ldots, N-1 \tag{9.60}$$

This implies that if we have found a particular vector ψ satisfying the inhomogeneous difference equations

$$\frac{\psi_{m+1} - \psi_{m-1}}{2\Delta z} + \alpha\psi_m = g_m, \; m = 1, \ldots, N-1 \tag{9.61}$$

then the vector

$$f_m = \psi_m + B_1 r^m + B_2(-r)^{N-m}, \; m = 0, \ldots, N \tag{9.62}$$

also satisfies the inhomogeneous difference equations. We note that since

$$\sum_0^N{}' r^m = \frac{1}{A} \tag{9.63}$$

and

$$\sum_0^N{}' (-r)^{N-m} r^m = 0 \tag{9.64}$$

we have

$$A \sum_0^N{}' f_m r^m = A \sum_0^N{}' \psi_m r^m + B_1 \tag{9.65}$$

and

$$A \sum_0^N{}' f_m (-r)^{N-m} = A \sum_0^N{}' \psi_m (-r)^{N-m} + B_2 \tag{9.66}$$

Once a particular solution ψ has been determined, the appropriate choice for B_1 and B_2 is simply

$$B_1 = F_0 - A \sum_0^N{}' \psi_m r^m \tag{9.67}$$

$$B_2 = F_N - A \sum_0^N{}' \psi_m (-r)^{N-m} \tag{9.68}$$

A simple way to obtain ψ is to set $\psi_0 = \psi_N = 0$ and to solve the resulting system of equations with a double pass method, decomposing the corresponding matrix into the product of upper and lower triangular factors. The first pass is given by the forward recursion

$$\eta_m - r\eta_{m-1} = 2g_m \Delta z, \ m = 1, \ldots, N-1 \tag{9.69}$$

The second pass is given by the backward recursion

$$\psi_{m+1} + \frac{1}{r}\psi_m = \eta_m, \ m = N-1, \ldots, 0 \tag{9.70}$$

In order to solve the PE with the DMFT, we have to define a discretized version of the PE and to examine the transformed equation. We have to define the discretized version of the second-order partial z-derivative, which we denote by D^2. For simplicity we drop the variable x in the derivation that follows. The difficulty is of course to define D^2 at the boundaries of the domain, since inside the domain the usual central difference approximation holds.

We work on vectors $f = (f_0, \ldots, f_N)$ which satisfy a discretized version of the impedance boundary condition, in the form

$$Df_0 = -\alpha f_0 \tag{9.71}$$

We also assume that

$$Df_N = 0 \tag{9.72}$$

which corresponds to a field which is constant for $z \geq N\Delta z$. Put $h = D^2 f$. Using one-sided approximations at the edges of the domain and central differences inside the domain, we obtain the following expression for h:

$$\begin{aligned} h_0 &= \frac{2(f_1 - f_0(1-\alpha))}{\Delta z^2} \\ h_m &= \frac{f_{m+1} + f_{m-1} - 2f_m}{\Delta z^2}, \quad m = 1, \ldots, N-1 \\ h_N &= \frac{2(f_{N-1} - f_N)}{\Delta z^2} \end{aligned} \tag{9.73}$$

The discretized version of the homogeneous PE is

$$D^2 f + 2ik\frac{\partial f}{\partial x} = 0 \tag{9.74}$$

We now look at the transformed equation. Let $H = Mh$. After a certain amount of algebra, we find

$$\begin{aligned} H_0 &= \frac{1}{\Delta z^2}\frac{(r-1)^2}{r}F_0 \\ H_j &= \frac{2}{\Delta z^2}\left(\cos\frac{\pi j}{N} - 1\right)F_j, \; j = 1, \ldots, N-1 \\ H_N &= -\frac{1}{\Delta z^2}\frac{(r+1)^2}{r}F_N \end{aligned} \tag{9.75}$$

Hence the DMFT F satisfies the system of ordinary differential equations

$$\begin{aligned}
F_0'(x) &= \frac{i}{2k\Delta z^2}\frac{(r-1)^2}{r}F_0(x) \\
F_j'(x) &= \frac{i}{k\Delta z^2}\left(\cos\frac{\pi j}{N}-1\right)F_j(x),\ j=1,\ldots,N-1 \\
F_N'(x) &= -\frac{i}{2k\Delta z^2}\frac{(r+1)^2}{r}F_N(x)
\end{aligned} \tag{9.76}$$

The solution at the advanced range is given by

$$\begin{aligned}
F_0(x+\Delta x) &= \exp\left\{\frac{i\Delta x}{2k\Delta z^2}\frac{(r-1)^2}{r}\right\}F_0(x) \\
F_j(x+\Delta x) &= \exp\left\{\frac{i\Delta x}{k\Delta z^2}\left(\cos\frac{\pi j}{N}-1\right)\right\}F_j(x) \\
& \qquad\qquad j=1,\ldots,N-1 \\
F_N(x+\Delta x) &= \exp\left\{-\frac{i\Delta x}{2k\Delta z^2}\frac{(r+1)^2}{r}\right\}F_N(x)
\end{aligned} \tag{9.77}$$

The DMFT is exactly matched to the impedance boundary condition, in the sense that if we start with a vector $f(x)$ satisfying the impedance boundary condition, the inverse transform of the advanced solution $F(x+\Delta x)$ will automatically satisfy it too.

We note that since the imaginary part $(r+1)^2/2$ is positive, $F_N(x)$ increases exponentially with range. In order to avoid numerical problems, it is best to enforce $F_N(x)=0$ at each range.

To obtain the complete split-step DMFT solution, the DMFT algorithm is combined with multiplication by the refractive index exponential, as in Chapter 3. Assume for example that we are using the simplest split, as given in eqn. 3.12. Then the DMFT solution is obtained with the following steps:

1. Start with solution $f=(f_0(x),\ldots,f_N(x))$ at range x.
2. Compute matched function $g=(g_0(x),\ldots,g_N(x))$.
3. Compute discrete sine transform of $F(x)$ of $g(x)$ (which is also the DMFT of $f(x)$).
4. Advance the DMFT function to range $x+\Delta x$ using the system of eqns. 9.77.
5. Compute inverse discrete sine transform to get $g(x+\Delta x)$

6. Retrieve intermediate solution $f_i(x + \Delta x)$ from $g(x + \Delta x)$, $F_0(x + \Delta x)$, $F_N(x + \Delta x)$.

7. Multiply by refractive index exponential to get solution $f(x + \Delta x)$ at advanced range.

The computing cost is increased slightly by comparison with the perfectly conducting case, since the matched function must be computed at each range, and the field at the advanced range must be retrieved from the matched function. However the FFT computational burden is exactly the same, i.e. two sine transforms per step. The DMFT is hence extremely efficient.

We note that the DMFT can be adapted for range-dependent impedance, simply by implementing the appropriate variations of the coefficient α with range. Possible backscatter due to interface discontinuities is of course neglected with this approach.

9.5 Examples

9.5.1 Line-of-sight propagation

Line-of-sight propagation provides a simple validation of the DMFT method, by comparison with a two-ray geometrical optics model. Figure 9.4 shows propagation factor results for a 1 GHz source located at height 25 m above mean sea level, propagating over the sea surface. The electrical constants of the sea surface are taken to be $\epsilon_r = 80$, $\sigma = 4$ S/m. The refractive index is assumed to be unity for this simulation. The DMFT parabolic equation results are in excellent agreement with the geometrical optics results for both vertical and horizontal polarizations.

9.5.2 Surface wave

Surface wave propagation is a more demanding problem, since we are now considering diffraction around the Earth while enforcing the impedance boundary condition at the surface. Figure 9.5 shows ground wave results obtained by Dockery and Kuttler [46] with their parabolic equation model TEMPER, together with those obtained by Barrick [8] using a surface wave program developed by Berry and Christman [19]. This example involves a vertically polarized 10 MHz source, located at a height of 30 m, radiating over the sea ($\epsilon_r = 80$, $\sigma = 4$ S/m). The atmosphere is assumed to be linear, with a modified refractivity gradient of 118 M-units/km. It is reassuring to note the excellent agreement, better than 0.1 dB at all ranges, between these very different models.

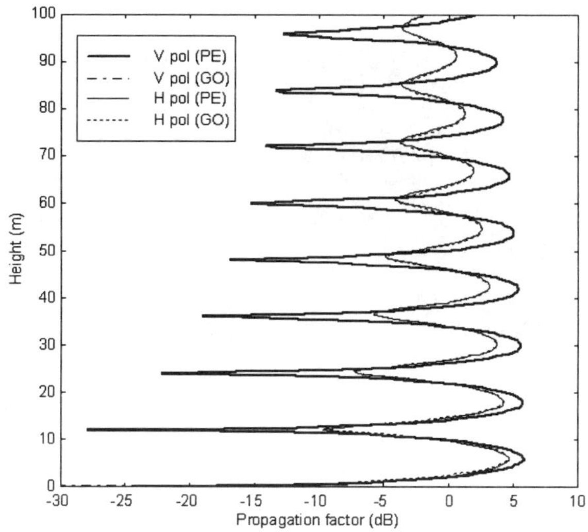

Figure 9.4 *Line-of-sight propagation over the sea at 1 GHz for vertical and horizontal polarizations, distance 2 km from source*

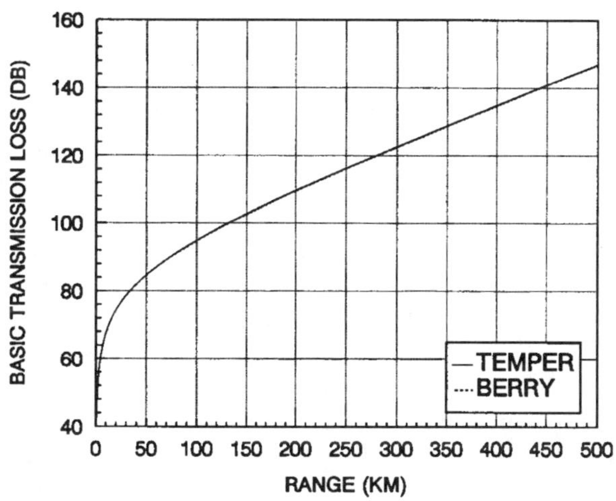

Figure 9.5 *Surface wave propagation over the sea at 10 MHz, receiver height = 4 m ©IEEE [46]*

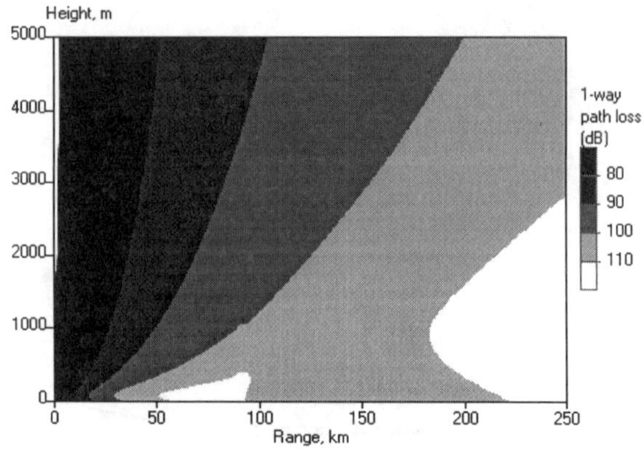

Figure 9.6 Recovery effect at 3.13 MHz

9.5.3 Recovery effect

Propagation over inhomogeneous flat ground provides a further test of surface wave modelling with the parabolic equation. We consider a path with two segments having different ground constants and apply the relevant impedance boundary condition on each segment. If the path starts over poorly conducting ground and continues over a good conductor, we expect to see an increase of the signal at the transition between the two segments, due to the fact that less energy is now lost inside the ground. This is the well-known recovery effect, which has been observed experimentally by Millington and Isted [112].

The example shown here allows comparisons with the measurements of [112] to be made. The source is vertically polarized and located at ground level; the transmitting frequency is 3.13 MHz. The path starts overland and continues oversea, crossing the coast line at a range of 86 km. Ground constants are $\epsilon_r = 5$, $\sigma = 0.01$ S/m for the overland part of the path, and $\epsilon_r = 81$, $\sigma = 4$ S/m for the oversea part. Figure 9.6 shows path loss contours obtained with the discrete mixed Fourier transform PE. The change of surface electrical properties at the land/sea boundary causes a redistribution of energy closer to the ground. Figure 9.7 shows the surface values of the field, with a marked recovery at the land/sea boundary. The measured values from [112] are indicated by crosses. This type of problem is well-suited to treatment by integral equation methods, using the compensation theorem [100]. Compensation theorem results from [167] are in excellent agreement with the PE model.

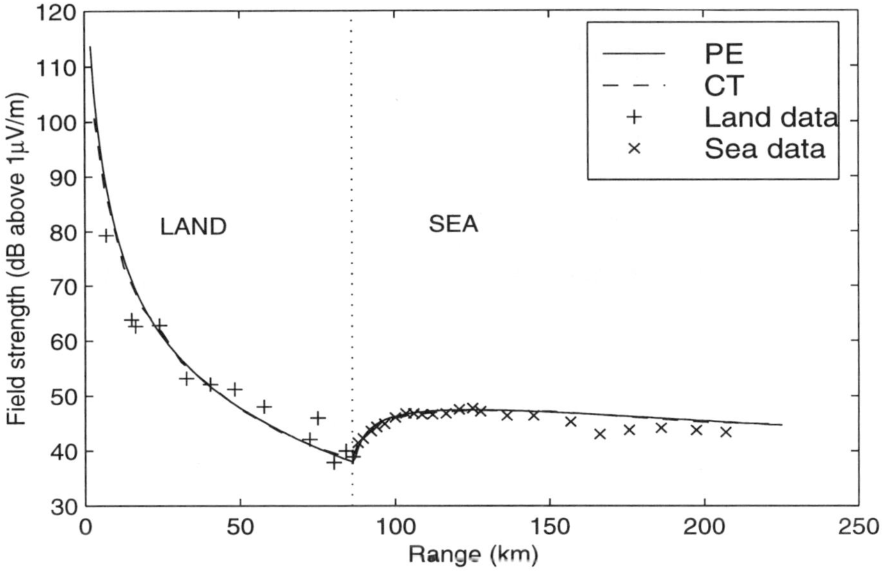

Figure 9.7 Recovery effect: comparison of PE results with measurements [112] and integral equation results [100]

Chapter 10

Propagation over the rough sea surface

10.1 Introduction

For parabolic equation modelling of forward scatter over a rough surface, we are concerned with the average field, corresponding to specular reflection. Most applications involve a non-homogeneous propagation medium, a situation which is extremely difficult to treat. Although some progress has been made on rough surface scattering in idealized ducting environments [147, 156, 136], there is no rigorous solution in general. Moreover in the case of propagation over the rough sea surface, which constitutes the most important application, insufficient information is available on the complex boundary layer phenomena which cause both the refractive effects and the rough surface formation. Further complication is added by the fact that it is usually very small grazing angles which are of interest for long-range radiowave applications [9].

Despite these difficulties, it is desirable to have at our disposal some engineering model which will at least provide an idea of rough surface effects on long-range radiowave propagation. This is achieved via the derivation of an effective surface impedance, or an effective reflection coefficient, taking surface roughness into account. One can then use the discrete mixed Fourier transform introduced in Chapter 9.

For HF applications, effective surface impedance is computed using the small perturbation approximation, while for microwave applications, the Kirchhoff approximation which treats the surface as locally plane is applied. Some background on the representation of rough surfaces is given in Section 10.2. The Phillips ocean wave spectrum, which is commonly used in radiowave applications, is described in Section 10.3, and Section 10.4 recalls the definition of the Rayleigh roughness parameter. We then turn to HF surface wave propagation in Section 10.5. Microwave applications require the estimation of surface grazing angle for use with the DMFT, as described in Section 10.6.

10.2 Random rough surfaces

In order to be able to use spectral representation theorems, we consider random rough surfaces $Z(x, y)$ which are stationary in the wide sense [68]: each random variable $Z(x, y)$ has a second moment, the average height $\overline{Z(x, y)}$ is the same at all points of the surface, and the correlation function $\overline{Z(x_1, y_1)Z(x_2, y_2)}$ depends only on the difference vector $(x_2 - x_1, y_2 - y_1)$. If the height distribution at all points of the surface is Gaussian, these assumptions imply strict stationarity of the random process. However stationarity in the wide sense is satisfied by much more general random surfaces: the height distribution need not be Gaussian, and does not even have to be the same at each point of the surface.

From now on, we assume that the mean height of the rough surface is zero at all points. We put

$$K(x, y) = \overline{Z(0, 0)Z(x, y)} \tag{10.1}$$

The correlation function $K(x, y)$ is then also the covariance function of the surface.

Surfaces which are stationary in the wide sense have a spectral representation in terms of power spectrum [61, 68]. The spectral representation is crucial for rough surface scattering calculations. The spectral density or power spectrum $W(p, q)$ is the Fourier transform of the correlation function of the surface height [61]. We note that the rms height h of the surface is given by

$$h = \sqrt{K(0, 0)} = \left\{ \int_{-\infty}^{\infty} \int_{-\infty}^{\infty} W(p, q)\, dp\, dq \right\}^{1/2} \tag{10.2}$$

10.3 Sea surface spectrum

The semi-isotropic Phillips spectrum was proposed in [128] to represent fully developed seas. In the half-space from which the wind is blowing, it is identically zero. In the half-space into which the wind is blowing, it is given by

$$W(p, q) = \begin{cases} \dfrac{B}{\pi(p^2 + q^2)^2} & \text{for} \quad (p^2 + q^2)^{1/2} > \dfrac{g}{U^2} \\ 0 & \text{for} \quad (p^2 + q^2)^{1/2} \leq \dfrac{g}{U^2} \end{cases} \tag{10.3}$$

where g (m/s^2) is the acceleration of gravity at the Earth's surface, U (m/s) is the wind speed and the constant B is usually taken to be 0.005. Taking $g = 9.81$ m/s^2, the rms wave height h (m) for the Phillips spectrum is given by

$$h = 0.0051\, U^2 \tag{10.4}$$

Table 10.1 gives the correspondence between sea state and wind speed for sea states 2 to 5, together with the rms wave height derived from the Phillips spectrum.

Table 10.1 Sea states, wind speed and rms wave height

Sea state	Wind speed	rms wave height
2	10 knots (5.14 m/s)	0.135 m
3	15 knots (7.72 m/s)	0.304 m
4	20 knots (10.29 m/s)	0.540 m
5	25 knots (12.86 m/s)	0.843 m

We note here that in order for the random rough surface to be physically meaningful, the realizations of the surface should be continuous, and that in order to apply the Kirchhoff method, second order derivatives should exist [13]. This is true if the correlation function has reasonable continuity properties at zero [68], which is not the case for the Phillips spectrum. For this reason, it is sometimes approximated by more regular functions [144].

10.4 Surface roughness and specular reflection

Assume a plane wave is incident on a rough surface. The Rayleigh roughness parameter γ is given by

$$\gamma = 2kh \sin\theta \tag{10.5}$$

where k is the wavenumber of the incident wave, and θ is the grazing angle made by the incident direction with the local tangent to the surface. The Rayleigh roughness parameter expresses the phase difference between two rays with grazing angle θ reflecting from portions of the surface separated by the rms height deviation h, as shown in Figure 10.1.

When γ is very small, all rays reflect approximatively in phase and the surface can be treated as smooth, producing specular reflection only. As γ increases, the surface gets rougher and specular reflection decreases in intensity,

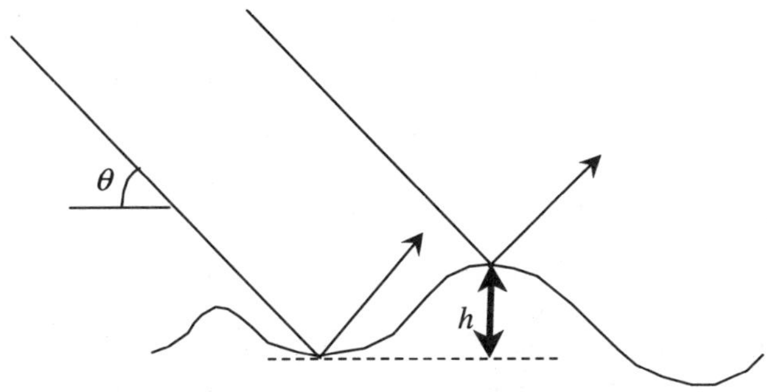

Figure 10.1 *Rayleigh roughness parameter*

as more and more energy gets scattered in other directions than the specular direction [61].

10.5 HF propagation over the rough sea

10.5.1 Small perturbation method

When the Rayleigh roughness parameter and the surface slopes are much smaller than unity, the small perturbation method is applicable to computation of the scattered field [61]. The first order term of the specular field expansion with the small perturbation method is zero [61]. However even second order variations can have a substantial effect on long range HF propagation over the sea surface, and a more accurate estimate of the specular field is then required. In [7, 8], Barrick combines the second order small perturbation method with the Leontovich approximation to compute the effective surface impedance $\overline{\Delta}$ of the rough sea surface at HF and VHF frequencies, for a vertically polarized incident wave propagating horizontally. The resulting formula is

$$\overline{\Delta} = \Delta + \frac{1}{4}\int_{-\infty}^{+\infty}\int_{-\infty}^{+\infty} f(p,q)W(p,q)\,dp\,dq \tag{10.6}$$

where

$$f(p,q) = \frac{k^2p^2 + ik\Delta(p^2+q^2+2kp)^{1/2}(p^2+q^2-kp)}{ik(p^2+q^2+2kp)^{1/2} + \Delta(k^2-p^2-q^2-2kp)} \tag{10.7}$$

Figure 10.2 shows the real and imaginary parts of effective surface impedance as a function of frequency, for sea states 0 and 5. The electrical constants of sea water are $\epsilon_r = 80$, $\sigma = 4$ S/m. The Phillips spectrum was used for

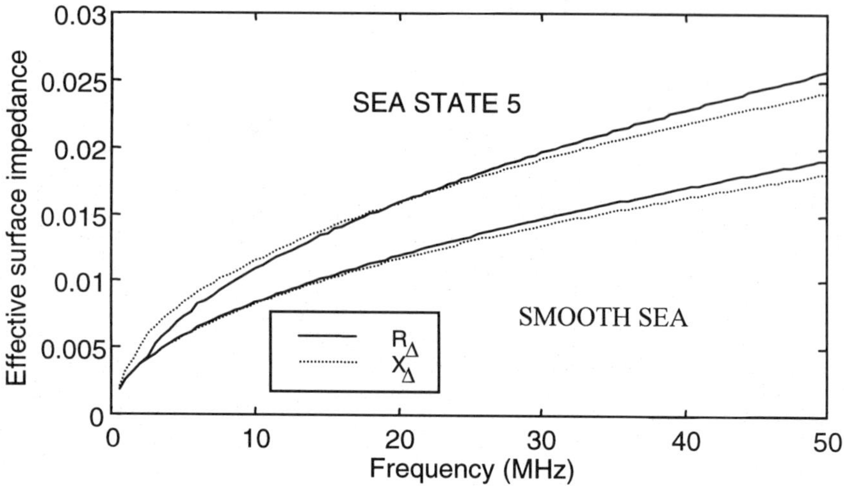

Figure 10.2 *Effective surface impedance* $\overline{\Delta} = R_\Delta - iX_\Delta$ *as a function of frequency*

the rough sea calculation, assuming downwind propagation. We note that the surface impedance remains small even for the rough sea case, which means that a strong surface wave should still be sustained. This will be confirmed by the results in the next section.

10.5.2 Propagation at 10 MHz

In [46], Dockery and Kuttler applied the DMFT method to compute ground wave propagation over the sea surface at 10 MHz, using the Barrick model to compute effective impedance at different wind speeds, based on the semi-isotropic Phillips spectrum. Table 10.2 gives the effective surface impedance at 10 MHz based on these values for different sea states.

The transmitter is vertically polarized and located at mean sea level. A standard linear atmosphere is used, with a modified refractivity gradient of 0.118 M-units/km. Figure 10.3 shows the surface wave results obtained by Dockery and Kuttler with their parabolic equation model TEMPER, together with those obtained by Barrick [8] using a mode theory surface wave program developed by Berry and Christman [19]. Results are presented in terms of excess path loss relative to the smooth sea case, for a receiver located at mean sea level. The parabolic equation and mode theory models are in excellent agreement. We see that, as announced in the previous section, a strong surface wave is present for all sea states, degrading of course as the sea roughness increases.

Table 10.2 Effective surface impedance at 10 MHz

Sea state	Effective surface impedance
0	$0.0084 - 0.0083i$
2	$0.0086 - 0.0100i$
3	$0.0100 - 0.0109i$
4	$0.0109 - 0.0115i$
5	$0.0118 - 0.0122i$

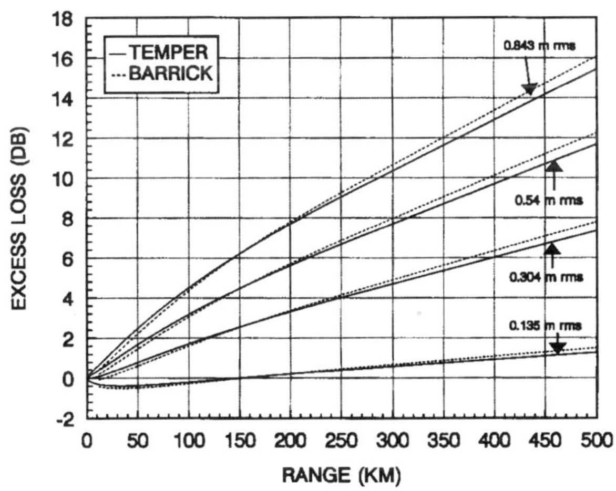

Figure 10.3 Rough surface effects on surface wave propagation at 10 MHz ©IEEE [46]

10.6 Ducting propagation over the rough sea surface

10.6.1 Roughness reduction factor

At microwave frequencies, the small perturbation approach of [7] is no longer feasible, since the height deviations of the sea surface are usually large compared to the wavelength of the incident electromagnetic field. The approach taken instead by many authors is based on the Kirchhoff approximation [61].

The assumption here is that the height $\zeta(x, y)$ of the rough surface is a slowly varying function with respect to the incident wavelength, in the sense that the radius of curvature at each point of the surface is very large compared to the wavelength. The Kirchhoff approximation treats the surface as locally plane, which results in a simplified expression for the field scattered in any direction. In particular, the average specularly reflected field $\overline{\phi}$ is given by

$$\overline{\phi} = \phi_0 \int_{-\infty}^{\infty} \exp(2ik\zeta \sin\theta) P(\zeta) d\zeta \tag{10.8}$$

where ϕ_0 is the specularly reflected field for a flat surface and $P(\zeta)$ is the probability density function of surface heights. The derivation of this formula can be found in [61]. We can restate this by writing the effective reflection coefficient R for the rough surface as

$$R = \rho R_0 \tag{10.9}$$

where R_0 is the reflection coefficient for a flat surface and the roughness reduction factor ρ is given by

$$\rho = \int_{-\infty}^{\infty} \exp(2ik\zeta \sin\theta) P(\zeta) d\zeta \tag{10.10}$$

This result was obtained by Ament [1] for the particular case of a Gaussian surface height distribution, of the form

$$P_1(\zeta) = \frac{1}{\sqrt{2\pi}h} \exp\left(-\frac{\zeta^2}{2h^2}\right) \tag{10.11}$$

where h is the random mean square (rms) deviation of the surface height. In that case, the roughness reduction factor ρ_1 is given by

$$\rho_1 = \exp\left(-\frac{1}{2}\gamma^2\right) \tag{10.12}$$

where γ is defined by eqn. 10.5. Some years later, Beard [17] noticed that experimentally measured reflection coefficients tended to be larger than predicted by eqn. 10.12 for large values of γ. Miller *et al.* [111] then used a more realistic model for the sea surface, given by

$$P_2(\zeta) = \frac{1}{\pi^{3/2}h} \exp\left(-\frac{\zeta^2}{8h^2}\right) K_0\left(\frac{\zeta^2}{8h^2}\right) \tag{10.13}$$

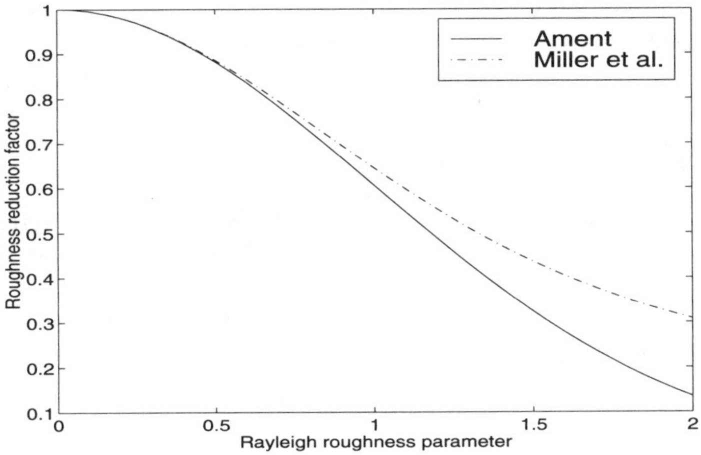

Figure 10.4 *Roughness reduction factor as a function of Rayleigh roughness parameter*

where K_0 is the modified Bessel function of the second kind of order 0. This probability density function represents a random surface height distribution of the form $\zeta = u \sin \alpha$, where u is Gaussian with mean zero and standard deviation $h\sqrt{2}$, α is uniformly distributed on $[-\pi/2, \pi/2]$, and u and α are mutually independent [111]. The variable α represents sea swell. This model yields the roughness reduction factor ρ_2 given by

$$\rho_2 = \exp\left(-\frac{1}{2}\gamma^2\right) I_0\left(\frac{1}{2}\gamma^2\right) \tag{10.14}$$

where I_0 is the modified Bessel function of order 0. This expression gives a much better fit to the experimental curves [111], and is the one generally used by radiowave propagation modellers [63]. Figure 10.4 shows the behaviour of ρ_1 and ρ_2: the fall-off is much slower for ρ_2.

The Kirchhoff approximation is problematic in several respects. Naturally depolarization effects are neglected since the framework is scalar, but this should have negligible consequences for forward propagation. A more serious difficulty is that shadowing effects are neglected: reflection in the specular direction is computed assuming that all the specularly reflecting elements of the rough surface are equally illuminated. At very low grazing angles, shadowing effects can substantially affect the results. Another potential source of error is that the propagation medium above the rough surface is assumed to be homogeneous, which is of course not the case for ducting applications. We should also bear in mind that eqn. 10.8 is a single scatter approximation which does not model multiple scatter. Many authors have analysed the drawbacks of the Kirchhoff approximation [116, 160]. However it has the merit of simplicity, and there is currently no alternative for long-range microwave propagation over the sea.

10.6.2 Estimation of grazing angles

For a smooth reasonably conducting surface, the Leontovich approximation yields an impedance boundary condition which is independent of incidence angle. This is no longer the case for the rough sea surface models considered in Section 10.6.1. The surface impedance δ for a plane wave with grazing angle θ is related to the reflection coefficient R by

$$\delta = \sin\theta \, \frac{1-R}{1+R} \tag{10.15}$$

Denoting the smooth surface impedance by δ_0, we have

$$\delta = \sin\theta \, \frac{(1+\rho)\,\delta_0 + (1-\rho)\,\sin\theta}{(1-\rho)\,\delta_0 + (1+\rho)\,\sin\theta} \tag{10.16}$$

where ρ is the roughness reduction factor. Here we take the definition of eqn. 10.14. At microwave frequencies and moderate wind speeds, the rough surface impedance obtained from eqn. 10.15 is strongly dependent on grazing angle. This is illustrated in Figure 10.5, which plots the real part of δ for a frequency of 10 GHz for different roughness conditions. The sea surface electric constants are $\epsilon_r = 80$ and $\sigma = 4$ S/m. The real part of the rough surface impedance is shown for wind speeds of 5 and 10 m/s, corresponding to rms wave heights of 0.13 m and 0.51 m, for the Phillips ocean wave spectrum model. We note that the imaginary part of the rough surface impedance remains quite small for these parameters. Impedance values for the transparent case ($R = 0$) are shown too.

After a steep decrease at small grazing angles, the rough surface impedance curves start increasing again, and eventually behave like the generalized impedance function for a transparent boundary. This is to be expected since the reflection coefficient decays to zero with increasing grazing angle. The curves intersect at the Brewster angle point $(\theta_B, \sin\theta_B)$. The Brewster angle for this case is 6.4°.

It is clear from Figure 10.5 that a propagation model using local boundary conditions requires a good estimate of grazing angle at each point of the surface. One difficulty is that the concept of local grazing angle may not always be applicable: for example certain types of ducting conditions can give rise to several intersecting families of rays, and in diffraction or skip zones, no rays at all might exist.

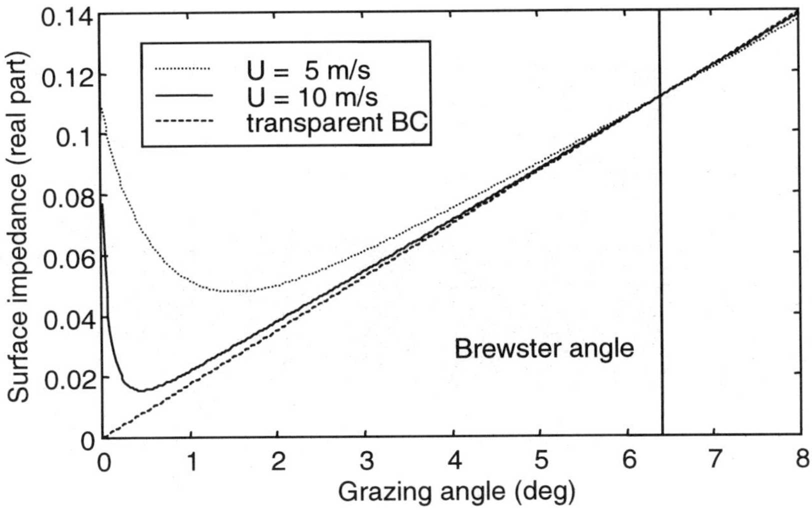

Figure 10.5 Rough sea surface impedance at 10 GHz

Assuming a local grazing angle can be defined, grazing angle estimation has been studied extensively by Dockery and Kuttler [46]. They consider the relative merits of spectral estimation and geometrical optics. The spectral estimation method consists of two stages. First, the field is propagated from range x to range $x+\Delta x$ using a Dirichlet boundary condition at the surface. A spectral estimator (for example the MUSIC algorithm [140]) is then applied to find the dominant spectral component at the advanced range. The solution is then recomputed, using the DMFT algorithm with the impedance corresponding to the estimated grazing angle. This method has the merit of providing a smoothly varying grazing angle as a function of range. It has been shown in [46] that it works well when surface refractivity gradients are not too strong, but that wavefront distortions can cause severe underestimation of the grazing angles when large surface refractivity gradients are present. Alternatively, the grazing angle can be computed from a ray-trace. Of course the usual restrictions of geometrical optics apply, so that at some ranges no rays at all may be found although the field is non-zero.

10.6.3 Surface duct propagation

We apply the DMFT method to X-band propagation in a strong surface duct, using ray-optics to compute local grazing angles. The refractivity profile is given in Table 10.3. The source is a vertically polarized 10 GHz antenna located at 25 m above mean sea level. Figure 10.6 shows path loss contours computed with the DMFT parabolic equation algorithm for propagation over a smooth

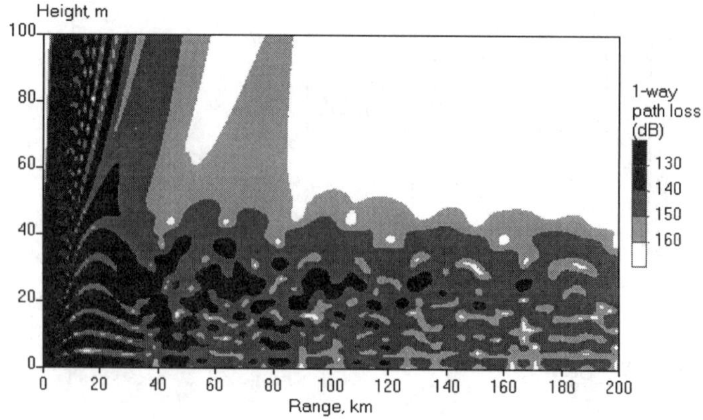

Figure 10.6 Propagation at 10 GHz in surface duct over smooth sea surface

sea. Since the amplitude of the reflection coefficient is very close to unity for grazing angles less than a few degrees, very little energy is lost by successive bounces over the sea surface and the signal propagates to larges distances in the surface duct with weak attenuation. Surface grazing angles computed from geometrical optics are shown in Figure 10.7. Successive bounces are clearly seen.

Table 10.3 Surface duct profile for rough sea simulation

Height (m)	M (M-units)
0	330
50	320
1050	438

Figure 10.8 shows path loss contours computed with the DMFT parabolic equation algorithm for propagation over a rough sea with an rms wave height of 1 metre, corresponding to a wind speed of 14 m/s for the Phillips model. Effective impedance at each range is calculated from the geometrical optics grazing angles of Figure 10.7. In contrast with the smooth sea case, significant energy is lost at each bounce on the sea surface and ducting is not supported.

Figure 10.9 shows a comparison of PE results obtained with the TERPEM model [93] with mode theory results obtained with the MLAYER code [15].

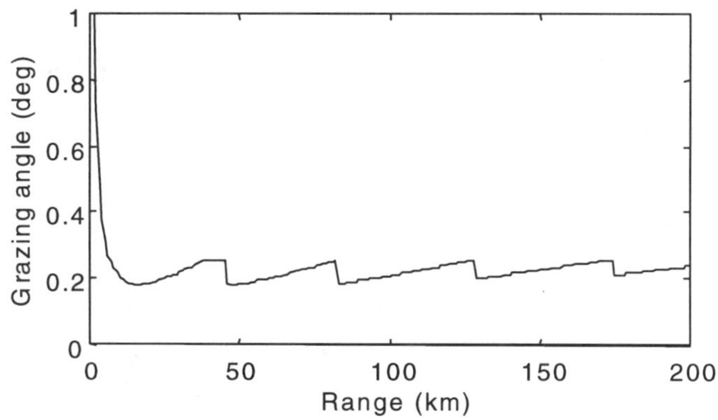

Figure 10.7 *Grazing angles computed from geometrical optics*

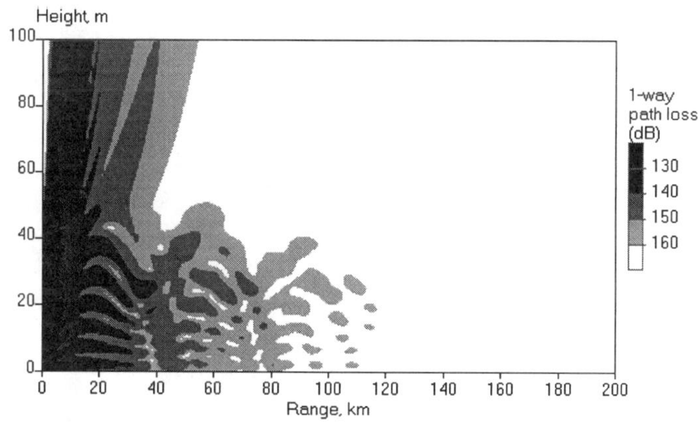

Figure 10.8 *Propagation at 10 GHz in surface duct over rough sea surface, rms waveheight 1 m*

Within the mode theory framework, rough surfaces are modelled by applying the impedance boundary condition separately to each modal function. Agreement between the PE and mode theory results is excellent, showing that for this case geometrical optics grazing angles provide a good approximation to the spectral behaviour of the wavefront at the surface.

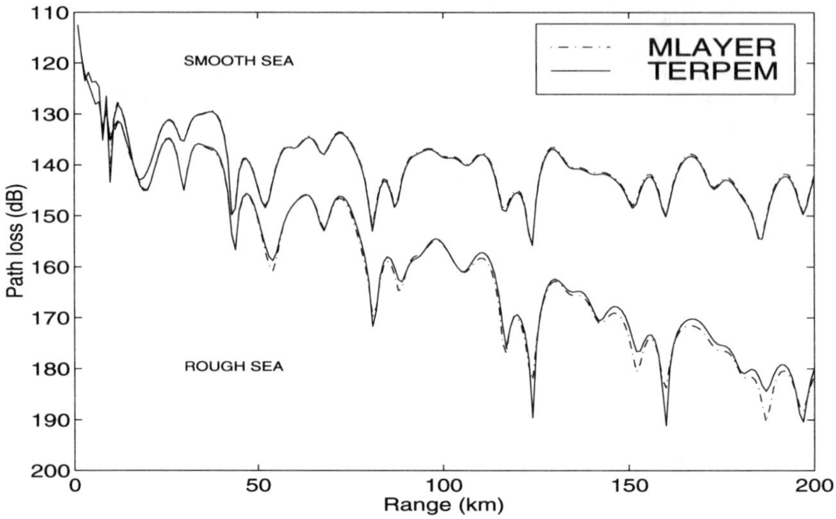

Figure 10.9 *Comparison of mode theory and PE results for rough and smooth sea surfaces*

Chapter 11
Hybrid methods

11.1 Introduction

Some radar applications involve calculations of the electromagnetic field in very large domains, up to several hundreds of km in range and several km in height. Even larger sizes may be required for the modelling of refractive effects on Earth/space paths. As PE integration times depend on frequency, propagation angles and domain size, calculations become prohibitively expensive for such large domains.

Fortunately two crucial properties hold for most tropospheric problems. Firstly, strong variations of the atmospheric refractive index seldom occur above a kilometre or so, and terrain features are often limited to low altitudes. This means that perturbations of the propagation medium are usually confined in height. Secondly, since the refractive index of the air remains very close to unity, complex atmospheric refraction effects are limited to small propagation angles from the horizontal. It should then be possible to reserve the powerful, but relatively slow PE techniques for field calculations at low heights and propagation angles, and to use faster methods to produce a solution at all angles and heights of interest. Indeed extremely efficient hybrid PE/ray-trace techniques were developed by Hitney [58] to solve large oversea propagation problems for low emitters. The Radio Physical Optics model is presented in Section 11.2.

In Chapter 8, Laplace transform methods were used to derive perfectly transparent boundary conditions. In fact they also yield a convenient way to extend PE solutions to large heights. This is the horizontal PE method (HPE) [90], which is described in Section 11.3. With hybrid techniques merging vertical PE (VPE), HPE and ray-tracing algorithms, quite general propagation problems involving terrain and atmospheric refraction can be solved quickly. In Section 11.4, we show how HPE methods can be combined with ray-trace techniques to solve high antenna problems very efficiently, using the ideas of

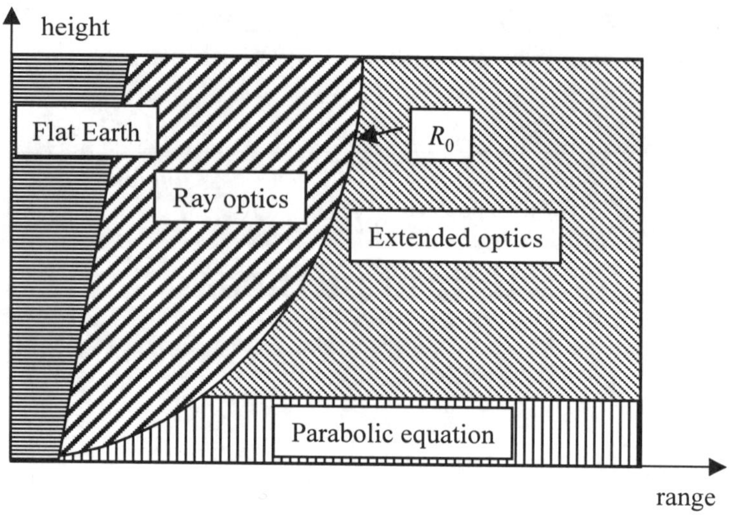

Figure 11.1 RPO regions

Chapter 8 to model sources above the vertical PE domain boundary. Finally we apply hybrid techniques to Earth-space propagation problems in Section 11.5.

11.2 Radio Physical Optics

The Radio Physical Optics (RPO) model [58] uses a combination of ray optics and parabolic equation methods to speed up computations for propagation over flat terrain. The domain is decomposed into four regions, as shown in Figure 11.1. In the flat Earth region, a simple two-ray model is used, assuming the rays propagate in straight lines. The idea here is that refractive effects are negligible for rays launched at angles of more than a few degrees from the horizontal, and there is hence no point in using more expensive ray-tracing algorithms. A threshold value of 5° is proposed in [58].

The ray optics (RO) region is determined by a limiting reflected ray R_0 with grazing angle θ_0 at the ground. Our aim is to ensure that field strength can easily be computed at any point in the RO region from the combination of a direct and reflected ray. We must determine the threshold value θ_0 in such a way that the ray R_0 can be traced without difficulty. In order to achieve this, we choose the grazing angle θ_0 large enough to avoid spherical diffraction problems at the Earth's surface and any ducting effects. In the range-independent case, the value suggested in [58] is

$$\theta_0 = \max\left(0.002, 2.5 \times \left[\frac{0.01772}{f^{\frac{1}{3}}}\right]\right) + \Delta\theta \tag{11.1}$$

where f is the frequency in MHz and $\Delta\theta$ accounts for possible ducting effects. The expression in square brackets is the diffraction region limit angle given by Reed and Russell in [133]. The quantity $\Delta\theta$ is given by

$$\Delta\theta = \sqrt{m^2(0) - m_{\min}^2} \tag{11.2}$$

where $m(0)$ is the modified refractive index at the surface and $m_{\min}$ is the minimum value of the refractive index over all heights of interest.

We can carry out a similar procedure for direct rays, finding limiting launch angles α_1 and α_2 such that there is no difficulty with the tracing of direct rays launched with angles greater than α_1 or smaller than α_2. If the transmitter is located is a strong ducting layer, α_1 will be strictly larger than α_2. This case is illustrated in Figure 11.2: we can find direct rays at any point of regions A_1 and A_2, but the direct ray-trace may be in difficulty for points in region A_0. If the antenna is low, region A_0 is usually limited to very small ranges and this is of little consequence. A serious problem could arise however for high antennas. For this reason the RPO model is normally limited to antennas with heights less than 100 m. In this case the RO region is determined by the reflected part of the limiting ray R_0.

In the ray optics region, the simple ray-trace model described in Chapter 5 can be used to calculate path loss since we have ensured that no trapping or diffraction effects are present. At each point of the RO region, the complex propagation factors of the direct and reflected rays are combined, taking the reflection coefficient at the surface into account.

The parabolic equation region is determined by the limiting ray R_0 and a maximum altitude beyond which the extended optics model is used. The extended optics model [58] extends the field computed with the parabolic equation model at a given height z_b to higher altitudes, using geometrical optics, as shown in Figure 11.3. The idea is that if z_b is sufficiently large, rays above that height do not undergo significant focusing or defocusing. Then to a good approximation, the field amplitude relative to free space can be assumed constant along a ray. This constant value is taken to be that computed with the PE model at height z_b. In this way one avoids using differential geometrical optics techniques to compute field strength along the ray, which could yield quite inaccurate results in the low altitude part of the calculation. Instead, the high accuracy PE value is used. Of course this technique is extremely fast since only elementary ray-trace calculations are required.

The main difficulty with this approach is that it is not usually possible to find rays going through all points of the horizontal $z - z_b$. The extended

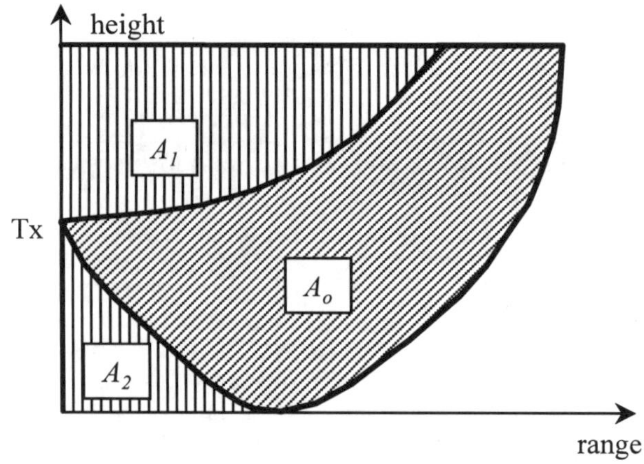

Figure 11.2 Determination of RO region for antenna in duct

optics model gets around this problem by assuming the existence of a "horizon ray" at height z_b. In other words there exists a limiting range x_0 such that at all ranges $x \leq x_0$ we can find a reflected ray going through (x, z_b), whereas no such rays exist for ranges $x > x_0$. Let β_0 be the elevation angle of the limiting ray at (x_0, z_b). Then for ranges beyond the optical limit x_0, we also use β_0 as the initial elevation angle. In other words, beyond the optical limit, we create artificial rays parallel to the horizon ray, as shown in Figure 11.3. There is so far no theoretical justification for this approach, which relies on the assumption that the tangent to wavefronts beyond the horizon does not vary with range. Of course no "secondary sources" should exist, since that would result in multiple horizon rays. Such a situation could certainly occur in the presence of irregular terrain, where energy could be scattered off a number of terrain facets which would then act as secondary sources.

The RPO model is extremely efficient for problems involving low antennas and propagation over flat terrain. We show an application to propagation in the elevated duct described in Table 11.1. The results have been obtained with the Advanced Refractive Effects Prediction System (AREPS) [125].

Figure 11.4 shows the coverage diagram of a horizontally polarized 10 GHz source located 75 ft above mean sea level. The antenna beam pattern is Gaussian, with a half-power beamwidth of 1°. For this case, the extended optics region starts at an altitude of 2160 feet and a range of 42.7 nautical miles. The different models merge very smoothly, providing seamless path loss contours. In particular the extended optics model correctly extends the field leaking from the duct. Run times were of just a few seconds on a desktop computer. For this case, most of the calculation time is spent on the PE algorithm, which

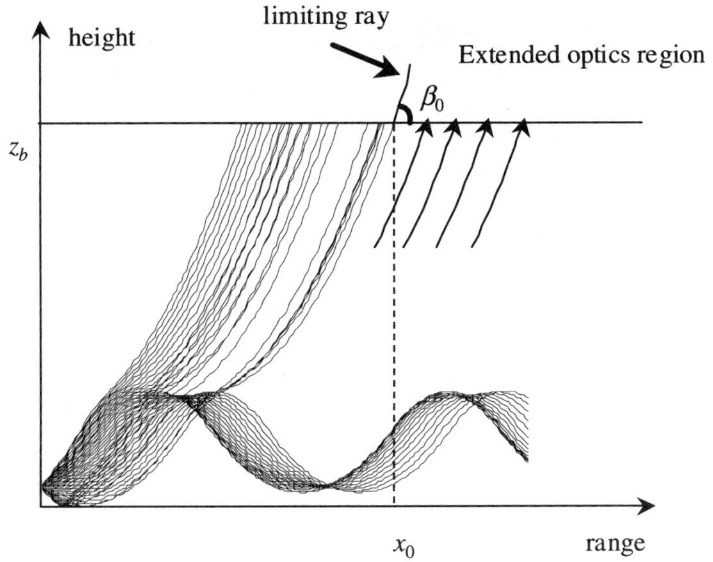

Figure 11.3 Extended optics method, showing limiting ray

Table 11.1 M-profile for RPO example

Height (m)	M (M-units)
0.0	330.0
50.0	340.0
100.0	335.0
200.0	346.8

used a 1024 transform size. Figure 11.5 shows a comparison with a pure PE model at a range of 200 nautical miles. Agreement is excellent. We note that the pure PE model was limited to angles of 1 degree at most from the horizontal. This already necessitated a sine transform of size 16,384 because of the maximum height of 5000 ft. A larger transform still would have been required to model the higher launch angles in the ray optics and flat Earth regions.

11.3 Horizontal PE

We now turn to the horizontal PE method [90], which provides a more general solution to the problem solved by the extended optics method. Assume that

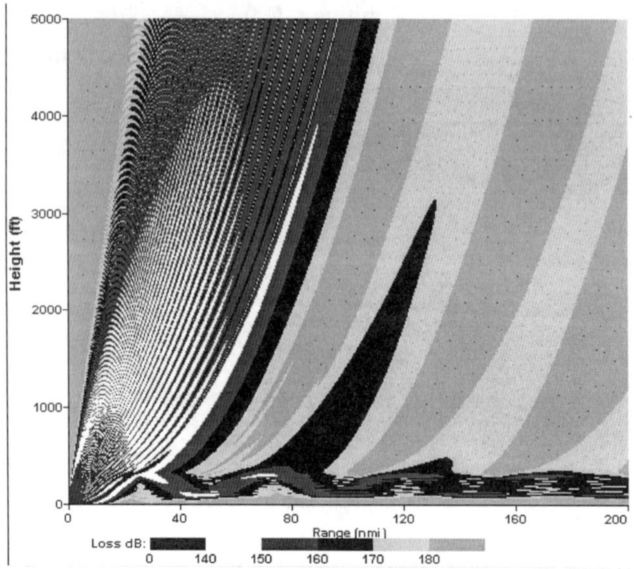

Figure 11.4 Coverage diagram obtained with AREPS hybrid model [125]

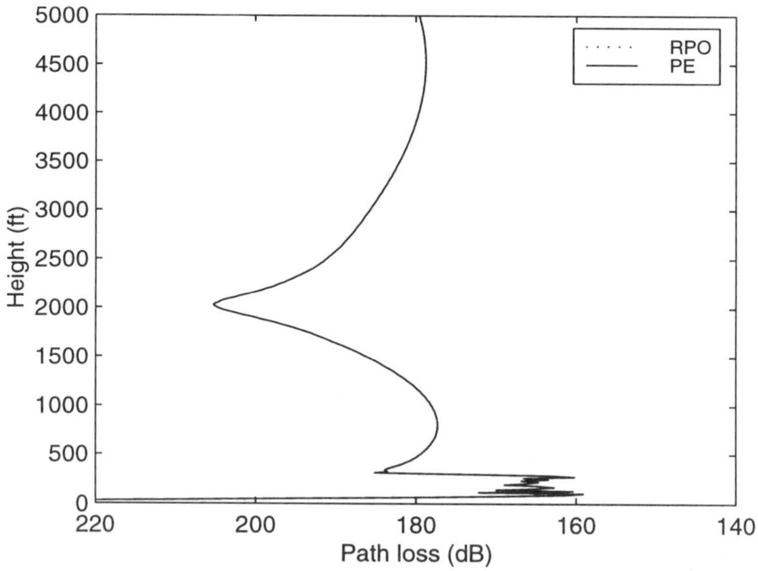

Figure 11.5 Comparison of RPO hybrid model and PE model at 200 nmi

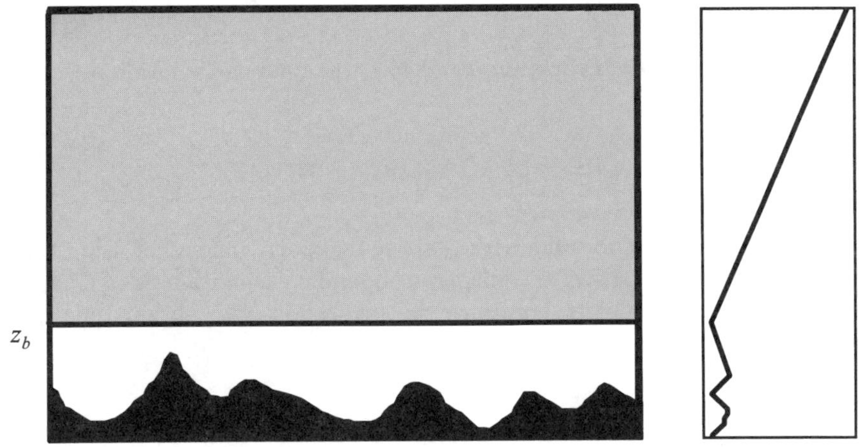

Figure 11.6 Framework for HPE calculations

the medium above the boundary z_b is upwards transmitting, in the sense that no energy in reflected back below z_b by refractive index variations above z_b. This is the case in particular for an exponential atmosphere above z_b. We suppose that the solution on the initial horizontal at height z_b is already known, and we wish to extend that solution to heights above z_b, as illustrated in Figure 11.6. Below z_b, any perturbations of the medium are allowed, as well as terrain variations: the horizontal PE method automatically calculates the effect of these perturbations above height z_b. We assume that above z_b, the refractive index is independent of range; in fact as mentioned in Chapter 8, a slightly weaker assumption is enough, and we can add a range-dependent offset to the refractive index. In this section, we also assume that the initial field is zero above z_b. This restriction will be lifted in Section 11.4, allowing us to model sources located above the boundary.

We start with the Laplace transform in range of the SPE above z_b. Since the initial field vanishes for $z \geq z_b$, this is given by

$$\frac{\partial^2 U}{\partial z^2}(s,z) + \Big(2iks + k^2(n^2-1)\Big)\,U(s,z) = 0 \tag{11.3}$$

For the well-behaved case of an upward-transmitting medium, the solution is bounded and uniquely determined. The cases of a constant and of a linear atmosphere, which we have already touched on in Chapter 8, provide the building blocks for practical application of the method.

11.3.1 Homogeneous medium

If the refractive index is constant equal to 1, the bounded solution of eqn. 11.3 is given by

$$U(s,z) = U(s,z_b)\exp(-i\Omega z\sqrt{s}) \tag{11.4}$$

where $\Omega = (1+i)\sqrt{k}$, for all s with positive real part. Instead of using this expression to derive a perfectly transparent boundary condition as in Chapter 8, we can invert the Laplace transform at any height z to obtain the solution above z_b directly as

$$u(x,z) = \int_0^x u(x-\xi,0)g(\xi,z)\,d\xi \tag{11.5}$$

where the singular kernel g satisfies

$$\mathcal{L}g = \exp\left(-i\Omega z\sqrt{k}\right) \tag{11.6}$$

We have [53]

$$g(x,z) = -\frac{i\Omega z}{2\sqrt{\pi}x^{\frac{3}{2}}}\exp\left(\frac{\Omega^2 z^2}{4x}\right) \tag{11.7}$$

The convolution formula of eqn. 11.5 shows that the solution at a given range x only involves ranges less than x, which is consistent with the paraxial approximation. Eqn. 11.5 is not appropriate for numerical implementation. Instead we use the relationship between Laplace and Fourier transforms to invert eqn. 11.4. We shall need the Fourier transform $\mathcal{F}_x$ with respect to range, defined by

$$\mathcal{F}_x\{u(x,z)\} = V(q,z) = \int_{-\infty}^{+\infty} u(x,z)\mathrm{e}^{-2i\pi qx}dx \tag{11.8}$$

where we have assumed that for fixed z, $u(x,z)$ is square-integrable in range. If u is sufficiently regular, we have

$$\mathcal{L}U(2i\pi q,z) = \mathcal{F}_x(q,z) \tag{11.9}$$

By analytic continuation of eqn. 11.4, we obtain the field in terms of the Fourier transform in range

$$u(x,z) = \mathcal{F}_x^{-1}\left\{A(q,z-z_b)\mathcal{F}_x\{u(x,z_b)\}\right\} \tag{11.10}$$

where the multiplier A is defined by

$$A(q,z) = \begin{cases} \exp\left(-2z\sqrt{\pi k q}\right) & \text{for } q \geq 0 \\ \exp\left(2iz\sqrt{-\pi k q}\right) & \text{for } q \leq 0 \end{cases} \tag{11.11}$$

We see that since the multiplier A is bounded, if $u(x, z_b)$ is square-integrable as a function of range, so is $u(x, z)$ at any height above z_b. We also note that although the direct Fourier transform is effectively one-sided since $u(x, z)$ is zero for negative x, the inverse Fourier transform is two-sided. We shall look further into this in Section 11.3.3.

Computer implementation of eqn. 11.10 with Fast Fourier Transforms requires one FFT for the initial calculation of the Fourier transform of $u(x, z_b)$, followed by one inverse FFT for each horizontal of interest. Since eqn. 11.10 is exact for any height above z_b, the solution at a given height may be obtained directly without the need to compute the field at intermediate heights.

11.3.2 Linear medium

We now consider a linearly increasing squared refractive index profile of the form

$$n^2(x,z) = 1 + \alpha(z - z_b) + \beta\ ,\ \ z \geq z_b \tag{11.12}$$

where the slope α is positive (the constant case has been treated above and ducting is excluded from this analysis). For this case eqn. 11.3 is a variant of the Airy equation (see Appendix A) and the problem is similar to the non-local boundary condition problem for a linear medium treated in Section 8.5.2. The bounded solution of eqn. 11.3 is given by

$$U(s,z) = U(s,z_b)\frac{w_1\left\{\zeta_s(z - z_b)\right\}}{w_1\left\{\zeta_z(0)\right\}} \tag{11.13}$$

where

$$\zeta_s(z) = -\left(\alpha k^2\right)^{1/3} z - \frac{k^2\beta + 2iks}{(\alpha k^2)^{2/3}} \tag{11.14}$$

and w_1 is the "outgoing" Airy function $\mathrm{Ai}\{z\exp(2i\pi/3)\}$. As before we assume that u is square-integrable on each horizontal, in order to express the solution in terms of Fourier transforms as

$$u(x,z) = \mathcal{F}_x^{-1}\left\{B(q,z)\mathcal{F}_x\{u(x,z_b)\}\right\} \tag{11.15}$$

where the multiplier B is a ratio of Airy functions

$$B(q, z) = \frac{w_1\left\{\zeta_{2i\pi q}(z - z_b)\right\}}{w_1\left\{\zeta_{2i\pi q}(0)\right\}} \tag{11.16}$$

Using asymptotic expansions [117], one can show that B is bounded in q for fixed z. It follows that u is square-integrable at any height above z_b provided it is square-integrable on the initial horizontal.

More general upward transmitting profiles can be treated with the horizontal PE method. We approximate the squared refractive index profile with a piecewise linear function. Then in each horizontal slice where the refractivity slope is constant, the horizontal PE method is applied with the appropriate exponential or Airy kernel. An additional Airy function multiplication is needed at each slope breakpoint, although no additional FFTs or inverse FFTs are required.

11.3.3 Spectral decomposition and numerical performance

The Fourier transform variable q is related to the horizontal plane wave decomposition of u: for negative values of q, we have

$$q = -\frac{1 - \cos\theta}{\lambda} \sim -\frac{\theta^2}{2\lambda} \tag{11.17}$$

where θ is the angle of propagation from the horizontal. This can be seen by considering the reduced function u corresponding to a plane wave of unit amplitude propagating at angle θ_0 from the horizontal

$$u(x, z) = \exp ik\left\{x(\cos\theta_0 - 1) + z\sin\theta_0\right\} \tag{11.18}$$

The Fourier transform of u with respect to x at height zero is the Dirac function $\delta\left(q + (1 - \cos\theta_0)/\lambda\right)$, which proves eqn. 11.17. Positive values of q correspond to evanescent waves. From eqns. 11.11 and 11.16, it can be seen that for a fixed value of q, the multiplier tends to zero as the height z increases. It is tempting to neglect the contribution of positive values of q and to use the one-sided inverse Fourier transform. This is indeed possible provided the horizontal spectral components of the field above z_b vanish at small grazing angles.

In order to determine optimal sampling in range, we apply the Nyquist criterion, which relates the range step Δx to the maximum propagation angle of interest θ_m at height z_b by

$$\Delta x \leq \frac{\lambda}{\theta_m^2} \tag{11.19}$$

Correct reconstruction of the signal is then automatically obtained at all heights, since no further direct Fourier transforms are required. This is quite striking in a linear or exponential atmosphere, where the antenna beam will curve up as it propagates: the Airy kernels implement the appropriate shift to higher propagation angles. There is no resolution problem for the inverse Fourier transforms, provided of course that the maximum range of interest at the higher altitudes is included in the domain.

The HPE starter at the initial height z_b may be computed with any available method, provided both phase and amplitude are calculated. Of course, the usual vertical PE method is particularly appropriate, since it calculates directly the reduced field u, but other methods or combinations of methods may be used if necessary. There are no constraints on terrain and refractivity structures below the boundary z_b. In particular there are no special difficulties if multiple reflections from terrain are present.

The range resolution required by HPE determines that of the algorithm used to calculate the field on the initial horizontal. If the vertical PE is used as a starter, HPE constraints often mean that the VPE range step must become smaller than in independent use. Even so, there is still a large overall computational gain in most cases. Consider a domain of 300 km by 5 km. The terrain is flat, and we assume that no anomalous refractivity features are present above 0.1 km. We consider frequencies of 3 and 10 GHz, with maximum propagation angles of 1.5° and 0.5° respectively.

Table 11.2 gives computing requirements when the vertical PE algorithm (split-step/FFT) is used in the entire domain, versus those for the hybrid vertical/horizontal PE model. For the latter, the vertical PE is calculated up to a height of 0.1 km and extended above this height with the HPE method. Due to upwards refraction, propagation angles increase as the beam propagates. If we assume a standard well-mixed atmosphere above 0.1 km, a ray making an angle of 1.5° with the horizontal at 0.1 km will reach an angle of 2.9° at 5 km. The horizontal PE method automatically accounts for this, but FFT sizes for the vertical PE have to be increased substantially if it is to be used in the entire domain, in order to cope with the larger propagation angles.

For a fixed frequency, computation times increase roughly as the square of θ_m. Large computing gains are obtained if the angular sector over which the PE calculations are made can be reduced. As with the Radio Physical Optics method, it makes sense then to restrict the use of the PE to a reduced angular sector, and to compute the field at higher angles with faster methods, for instance a simplified ray-trace technique. The best opportunities for this arise for propagation over flat terrain. For propagation over irregular terrain, high angle propagation may occur due to terrain diffraction or reflection, and this type of phenomenon cannot be modelled adequately with ray optics.

An interesting consequence of eqn. 11.19 is that performance improves with increasing frequency for propagation over a flat surface, because the propagation angle above which ray optics can safely be used increases with wavelength [58]. This is reflected by the choice of maximum propagation angles in the table: the use of a maximum angle of 0.5° at 10 GHz results in very substantial computing gains.

Table 11.2 Computing requirements for a domain of 300 km by 5 km

Frequency		3 GHz	10 GHz
Maximum propagation angle		1.5°	0.5°
Vertical PE in whole domain	Δx	0.5 km	0.5 km
	FFT size	32768	65536
Vertical PE up to 0.1 km	Δx	0.125 km	0.250 km
	FFT size	512	512
Horizontal PE above 0.1 km	Δz	10 m	10 m
	FFT size	4096	2048
VPE/HPE hybrid to VPE time ratio		12 %	3%
VPE/HPE hybrid to VPE memory ratio		12%	3%

11.3.4 Examples

We start with a surface duct case. The refractivity profile is bilinear, with a slope of −20 M-units/km between 0 and 100 m and a slope of 118 M-units/km above 100 m. Figure 11.7 shows the coverage diagram of a horizontally polarized 3 GHz source located at a height of 50 m. A hybrid model has been used, with vertical PE calculations below 100 m, extended above 100 m with the horizontal PE method. Computation parameters correspond to the first column of Table 11.2. Figure 11.8 shows path loss against height at a range of 200 km, obtained with the hybrid model and with a pure PE model for comparison. The pure PE results have been offset by 20 dB for clarity. Agreement is excellent between the two methods.

We now apply the horizontal PE method to a case involving a combination of irregular terrain and tropospheric ducting. The source is horizontally polarized and located 50 m above ground, and the frequency is 2 GHz. The refractivity profile is trilinear, with a 2 M-units ducting layer between heights of 180 and 200 m above mean sea level. Figure 11.9 shows the resulting cover-

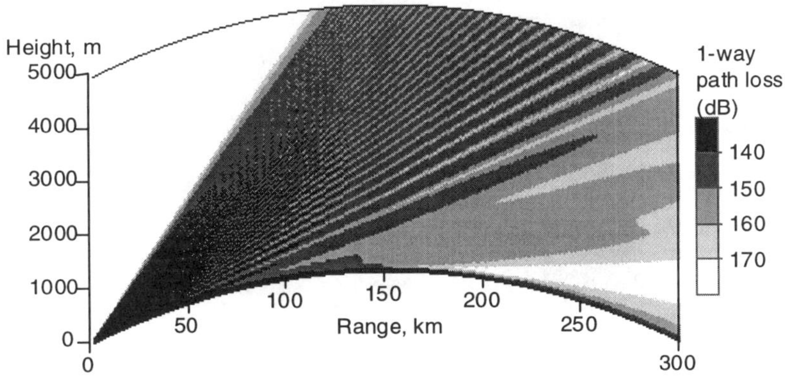

Figure 11.7 *3 GHz antenna in ducting environment; HPE has been used above 100 m*

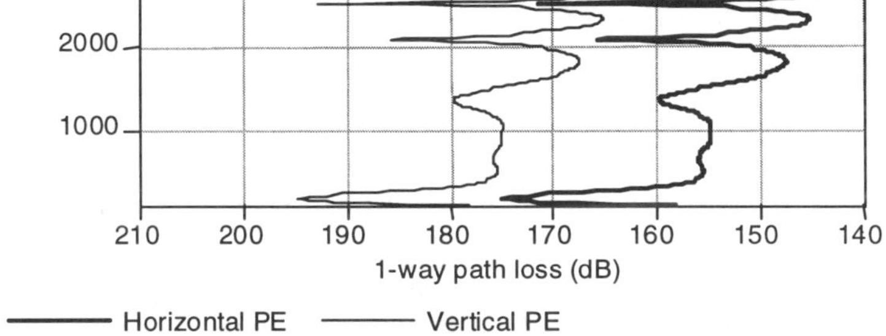

Figure 11.8 *Comparison of vertical and horizontal PE results at range 200 km for environment of Figure 11.7*

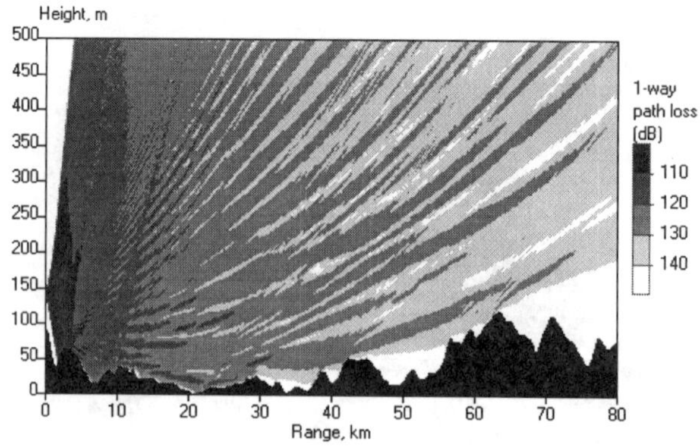

Figure 11.9 3 GHz source in environment with terrain and ducting; HPE has been used above 200 m

age diagram. Results below 200 m have been obtained with a finite-difference implementation of the PE method, using ground electrical constants $\epsilon_r = 15$ and $\sigma = 0.01$ S/m. The horizontal PE has been used above 200 m. Propagation angles were modelled up to 7° from the horizontal, requiring a range step size of 10 m. Figure 11.10 shows path loss against range at a height of 400 m for the hybrid and pure PE models. Pure vertical PE results have been offset by 20 dB. Agreement between the two methods is excellent, down to the fine detail of the lobing due to multiple terrain reflections. We note that RPO would not be suitable for this case, since the ray-trace cannot be used over irregular terrain.

11.4 High antennas

By using the same ideas as in Chapter 8, the horizontal PE method can be generalized to the case of a source above the boundary. Assume that a solution $u_0(x, z)$ with the same initial field as u above z_b is known in the domain above z_b. Then u can be extended above z_b with the following steps: first we extend $v = u - u_0$ to the domain above z_b with the horizontal PE method, using as values on the initial horizontal the difference field $v(x, z_b)$. This is possible since $v(x, z)$ is zero for $z \geq z_b$. Then we add the known solution u_0 to the extended v to retrieve the wanted field u above z_b.

We have seen in Chapter 8 that for Gaussian sources, the solution is available in closed form when the medium is linear above z_b. We apply this to the case of a bilinear modified refractivity profile, with a slope of -100 M-

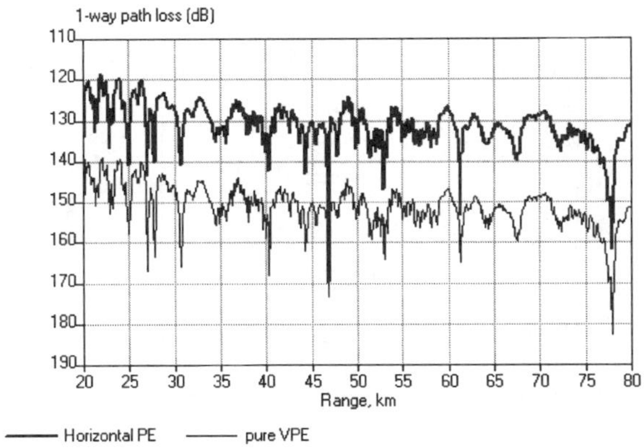

Figure 11.10 *Comparison between hybrid and pure PE results for terrain and ducting example, at height 400 m*

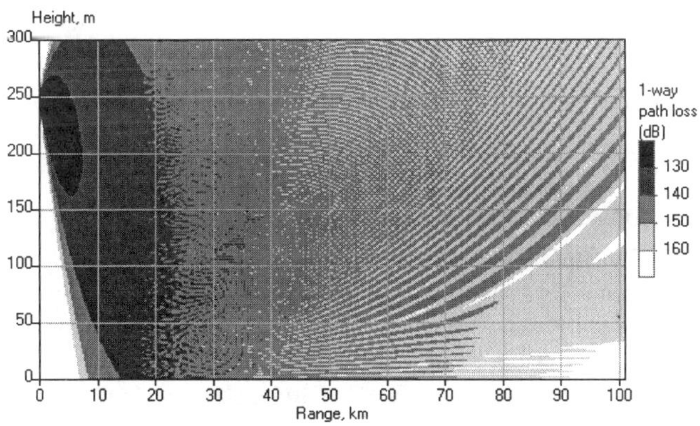

Figure 11.11 *Path loss contours of 10 GHz source in surface duct using analytic auxiliary function, with HPE solution above 50 m*

units/km between 0 and 50 m and of 118 M-units/km above 50 m. Figure 11.11 shows path loss contours for a horizontally polarized 10 GHz source located at a height of 250 m. The antenna is pointing down with an elevation angle of $-0.32°$. The beam pattern is Gaussian, with half-power beamwidth $1°$. The field above 50 m has been obtained with the generalized HPE method, using the analytical form given by eqn 8.66 for the auxiliary solution u_0. Below 50 m, a finite-difference code using the incoming energy Airy NLBC has been used.

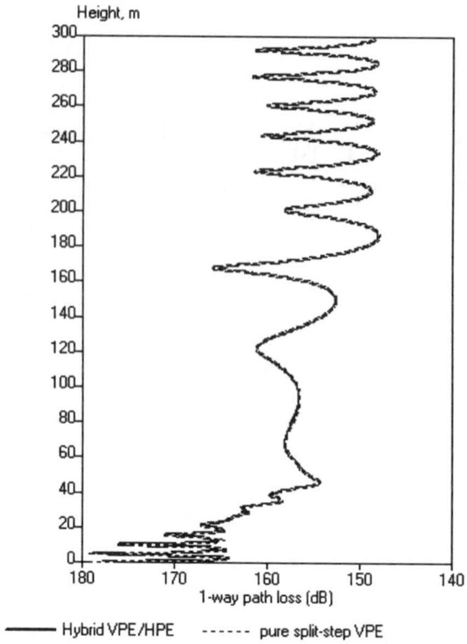

Figure 11.12 *Comparison of hybrid HPE/VPE using analytic auxiliary function and results obtained with pure PE algorithm*

The field above 50 m is accurately calculated with the generalized HPE method, as demonstrated in Figure 11.12, which shows a comparison with results obtained with a split-step PE algorithm applied in the whole domain.

In practice the medium above z_b is often more complicated, and there is no closed-form solution for u_0. For point-sources, a ray-tracing algorithm may then be used to compute u_0 numerically. Since the medium above z_b is upward-transmitting, there are no difficulties there with the ray-trace. Below z_b, the choice of medium is arbitrary. In Chapter 8, we placed a perfectly reflecting mirror at height z_b to derive incoming energy boundary conditions for the split-step/Fourier PE. Here the requirements are very different: we do not need the auxiliary solution to vanish below z_b, but we do need to be able to compute it easily at any point above z_b.

The optimal choice is a linear medium extending down to $-\infty$, with a slope matching the slope at z_b, and chosen in order to have continuity at z_b. In other words if

$$\frac{\partial n^2}{\partial z}(x, z_b) = \alpha \tag{11.20}$$

we take

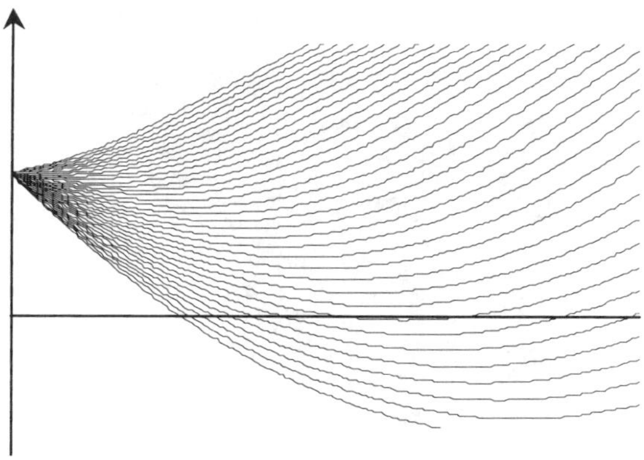

Figure 11.13 Ray-tracing in linearly extended medium

$$n^2(x, z) = n^2(x, z_b) + \alpha(z - z_b) \tag{11.21}$$

for $z \le z_b$. In this unbounded medium, there are no horizon or focusing problems. Figure 11.13 shows the type of ray-trace results obtained for a medium matched to the standard atmosphere, with $\alpha/2 = 118$ M-units/km.

The infinite linear medium below z_b curves the rays back into the higher part of the domain. This upgoing energy cancels out when applying the generalized HPE procedure, leaving only the desired upgoing energy which comes from field u below z_b.

We apply the infinite linear medium technique to the case of a 3 GHz antenna at a height of 1000 m, in the environment described by Table 11.3. A surface duct is present, together with a subrefractive layer between 100 and 200 m.

Table 11.3 M-profile values for generalized HPE with ray-trace

Height (m)	M (M-units)
0	305
100	300
200	320
300	332

The antenna is tilted downwards with an elevation angle of $-0.35°$. The

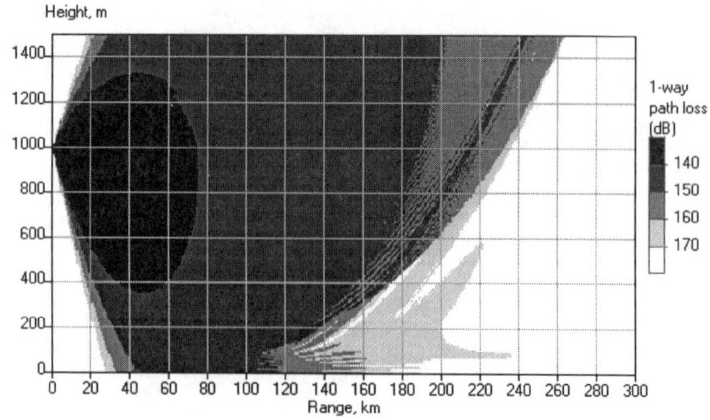

Figure 11.14 Hybrid HPE/VPE results using a ray-trace calculation for the auxiliary field

path loss contours shown in Figure 11.14 have been obtained using the generalized HPE above 128 m and the split-step Fourier algorithm with incoming energy below that height. As a result of the low heights required, the sine transform size for the vertical PE algorithm is only 512. By contrast the transform size required for a pure VPE computation in the whole domain is 8192. We note the effect of the surface duct on the coverage of this high antenna: a "radar hole" appears above the duct, due to the fact that energy arriving from above will be deflected downwards in the ducting layer and reflect at a shorter range than in a standard atmosphere. The results of the generalized HPE method are in excellent agreement with those of the pure vertical PE. This is illustrated in Figure 11.15, which shows path loss against height at a distance of 300 km for the two techniques.

11.5 Earth-space paths

Low elevation Earth-space radio links are affected by refractive conditions in the lower troposphere. Anomalous propagation effects can distort path delays, degrading the accuracy of GPS calculations if low elevation satellites are used. There has also been some interest in the possibility of using inverse techniques to infer tropospheric refractivity features from their effects on low elevation satellite signals [2, 59]. This is a situation where perturbations are located in the first one or two kilometres above the Earth's surface, whereas the satellite altitudes are several hundreds or even thousands of kilometres. It is hence essential to use a fast method to extend low altitude calculations to satellite

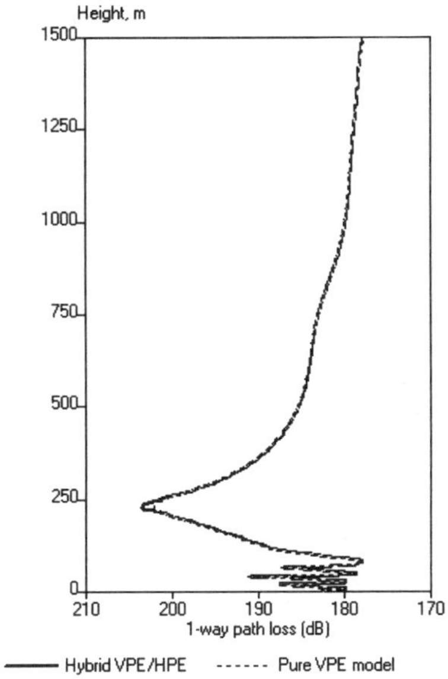

Figure 11.15 *Comparison of hybrid HPE/VPE using ray-trace auxiliary function and results obtained with pure PE algorithm*

heights.

In [59], the extended optics technique was adapted to deal with very large heights. The main difference with the low altitude case is that spherical geometry is used for the ray-trace above a few kilometres. The horizontal PE method can also be adapted to deal with this situation [87]. The main difficulty is that in view of the large altitudes involved, the low altitude Earth flattening transformation which gives rise to a linear modified refractive index is not valid. One could in principle get around this problem by using the logarithmic Earth flattening transformation of eqn. 4.38, which is valid at all heights. This leads to a differential equation in the Laplace transform domain which has a unique bounded solution. Unfortunately numerical evaluation of the resulting kernel via Taylor and asymptotic series is a somewhat daunting task. Instead we use successive flattening transformations as shown in Figure 11.16, each leading to a linear modified refractive index. With slices of a few hundreds of km, the error due to the use of a linear refractive index remains small. We note however that even for a ray launched horizontally, the propagation angle relative to the Earth's surface becomes large at large ground ranges. This means that the SPE is no longer sufficiently accurate for very large satellite heights, and hence neither is the horizontal PE method based on it.

With this simple approach, one Airy kernel multiplication is required for

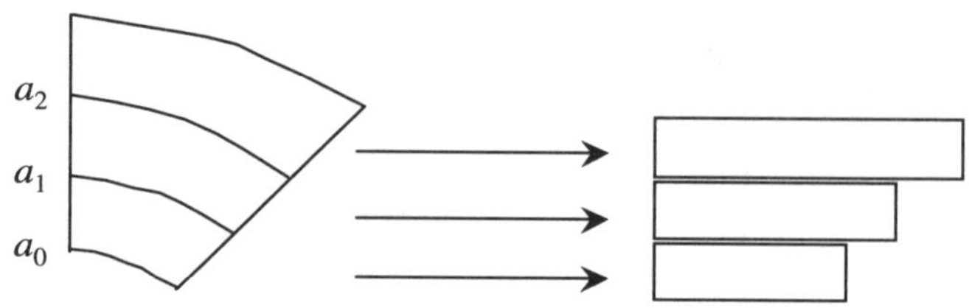

Figure 11.16 *Successive changes of coordinates for Earth-space applications*

each slice. This technique was applied to satellite measurements at 1239 MHz [2, 59]. The P76-5 wideband satellite was orbiting 1000 km above the horizon. Recordings were made at Point Loma on the Pacific coast, San Diego, California, using a horizontally polarized receiver located 34 m above mean sea level. Refractivity profiles were obtained from radiosonde ascents at the receiver site. Figures 11.17 and 11.18 show two examples of measured refractivity, respectively for 20 July 1978 at 19.45 local time and 28 July 1978 at 20.00 local time. The numerical values can be found in Tables 11.4 and 11.5.

Table 11.4 *M-profile for 20 July 1978 at 19.45 LT*

Height (m)	M (M-units)
43.0	364.8
204.1	380.0
349.5	396.4
359.2	323.8
386.8	326.4
617.3	355.5
1111.6	419.9
1540.1	481.8
1907.7	528.5
2749.4	642.9
3219.8	703.2
3545.5	748.0
4592.7	891.4
5430.8	1007.5

Figures 11.19 and 11.20 show the coverage of the receiver up to 3000 m altitude for these two cases. The very strong surface duct present on 20 July causes substantial trapping even at the comparatively low frequency considered. On 28 July, a large elevated duct was present, sufficiently strong to bend

Table 11.5 M-profile for 28 July 1978 at 20.00 LT

Height (m)	M (M-units)
43.0	359.0
133.0	373.8
465.3	410.8
487.8	396.5
593.2	391.3
787.0	388.7
1004.4	413.5
1543.1	486.3
1935.1	529.5
2517.5	644.1
2688.6	648.9
3217.4	723.2
3815.6	807.6
3925.5	829.1

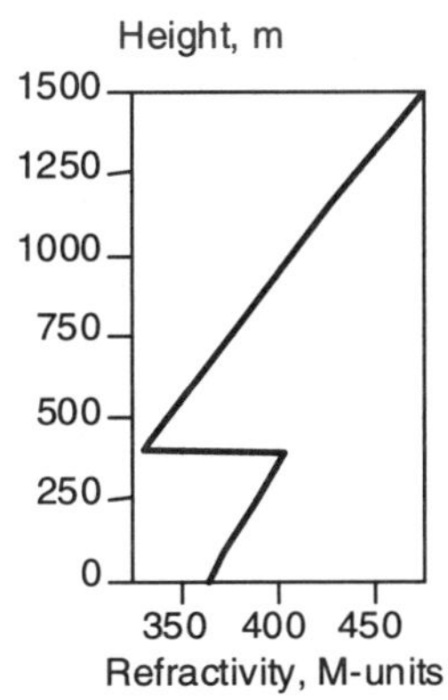

Figure 11.17 Refractivity profile measured at Point Loma, 20 July 1978, local time 19.45

the beam, although not achieving trapping.

Extended optics [59] and the horizontal PE method [87] were used to extend the field upwards. As in [2, 59], the refractivity model consists of a linear region, an exponential region and a constant region, defined as a function of height above the Earth's surface. In the linear region, which extends from 2.5 km to 9 km, the refractivity N decreases to a value of 105 N-units. In

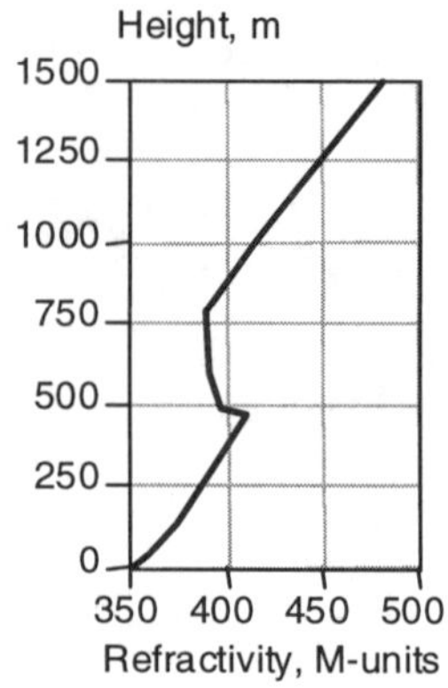

Figure 11.18 *Refractivity profile measured at Point Loma, 28 July 1978, local time 20.00*

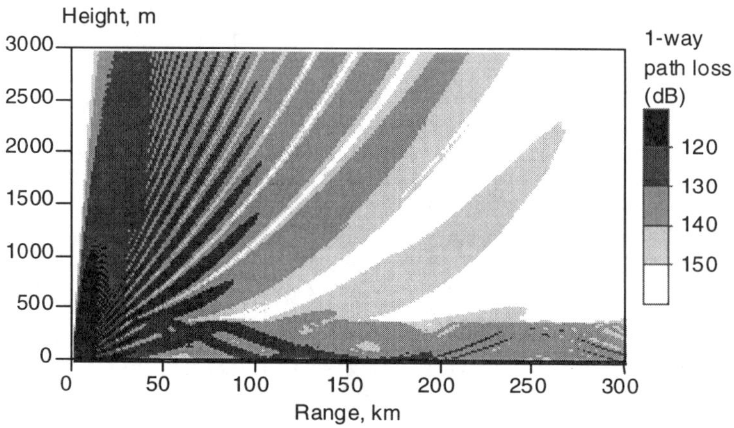

Figure 11.19 *Low altitude coverage of L-band receiver, 20 July 1978, local time 19.45*

the exponential region, which extends from 9 km to 50 km, the refractivity is given by

$$N(z) = 105\,\mathrm{e}^{-0.1424(z-9)} \tag{11.22}$$

where z is height in kilometres. From 50 km to satellite height, the refractivity is taken to be zero.

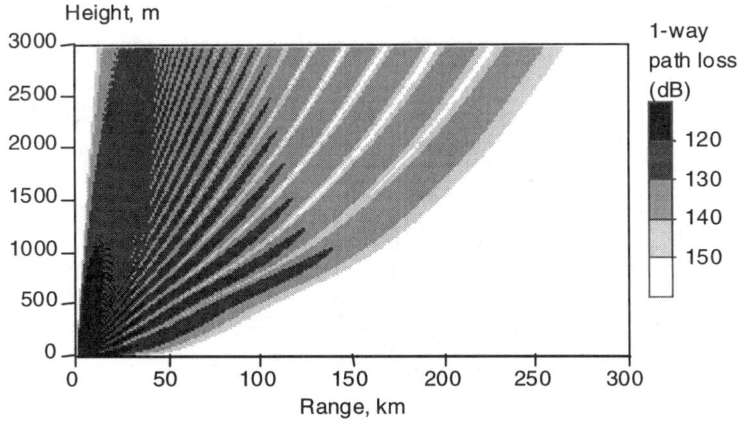

Figure 11.20 *Low altitude coverage of L-band receiver, 28 July 1978, local time 20.00*

Since the satellite is moving relative to the Earth, it is more economical for this application to invoke reciprocity, treating the receiver as the source. The field at 1000 km altitude is then obtained directly for all ground ranges. We note that at the large ground ranges of interest, the propagation angle relative to the Earth's surface goes up to 28° or so, so that the accuracy of the narrow-angle horizontal PE begins to deteriorate.

Figures 11.21 and 11.22 show the satellite signal plotted as a function of ground range. Measured results are given, together with results obtained with the horizontal PE method and the extended optics method [59]. Since only relative values of the measured signal were available, an arbitrary path loss offset was added to the observed data. The two models give almost identical results. Agreement with experimental results is not as good; this could well be due to range-dependent refractive effects, which could not be modelled in the absence of appropriate meteorological data. However the qualitative behaviour of the measured signal is in reasonable agreement with the modelled results: for 20 July, the signal remains high at extended ranges, due to leakage from the surface duct, while for 28 July a sharp decrease occurs beyond the radio horizon, which is extended but not suppressed by the elevated duct.

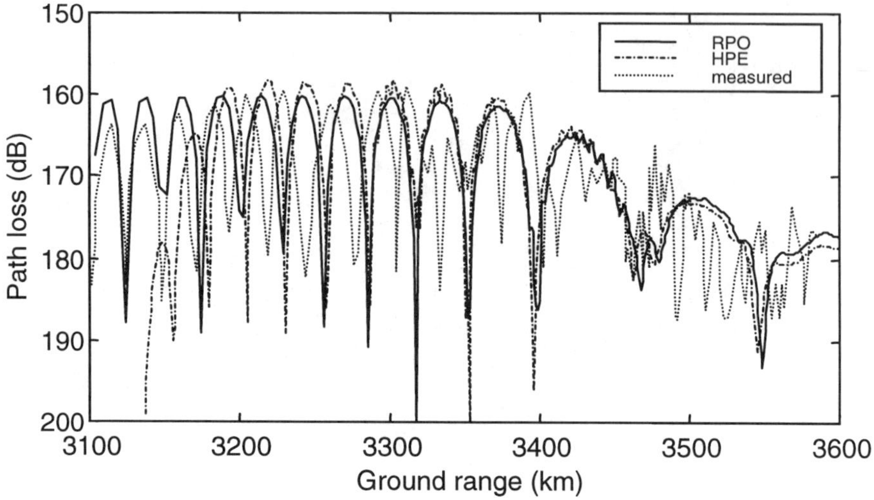

Figure 11.21 *Comparison of measured and modelled results for P76-5 satellite, altitude 1000 km, 20 July 1978 [59, 87]*

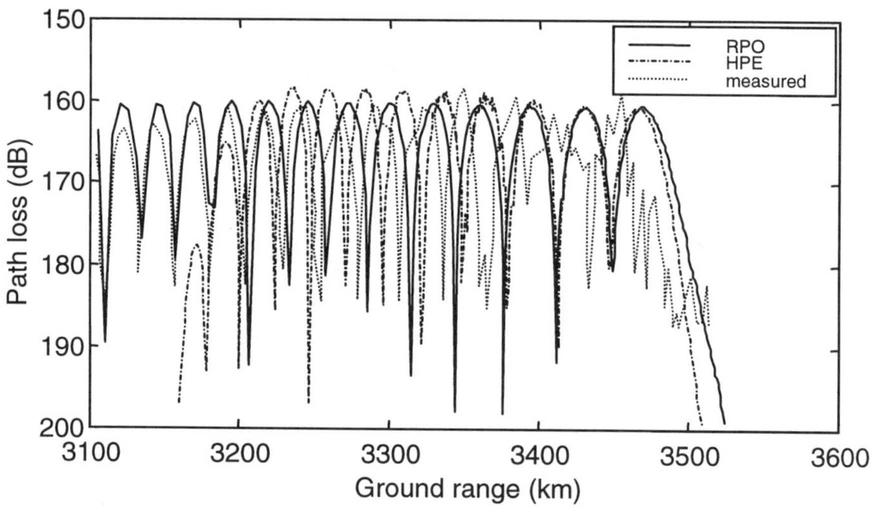

Figure 11.22 *Comparison of measured and modelled results for P76-5 satellite, altitude 1000 km, 28 July 1978 [59, 87]*

Chapter 12

Two-dimensional scattering

12.1 Introduction

Ideally a PE algorithm should model wide-angle scatter accurately and be numerically efficient while having good stability properties. For scattering applications, it should also have enough flexibility to represent complex boundary conditions. Unfortunately these properties tend to be mutually exclusive, and we shall see that in general a compromise has to be made depending on the application. We have already looked at numerical schemes for long-range applications. There the emphasis was on speed, since integration domains were extremely large relative to the wavelength, and on the modelling of refractive index variations in the background propagation medium. For this type of application, split-step Fourier methods provide a very powerful tool.

This chapter is concerned with scattering by two-dimensional objects. Integration domains are generally far smaller than for long-range applications, perhaps a few hundred wavelengths rather than thousands or millions, and it is modelling of object boundaries that becomes crucial, rather than that of refractive index variations, as shown in Figure 12.1. Assuming a wavelength of 10 cm, corresponding to a frequency of 3 GHz, a typical tropospheric propagation domain of 100 km in range by 1 km in height represents 10^6 by 10^4 wavelengths as shown in Figure 12.1(a). By contrast, a typical scattering problem like that of Figure 12.1(b) requires a domain of 5 by 4 m, or 50 by 40 wavelengths.

Hence requirements on numerical efficiency become less stringent for object scattering applications and we can accept far smaller range steps and grid sizes. With this in mind, finite-difference methods now become quite attractive. In fact Fourier methods are not an option in general, since they do not have the necessary flexibility for boundary modelling: for example although suitable Fourier transforms exist for modelling an impedance boundary on one side of the domain, no Fourier method can cope with a double-sided problem like that

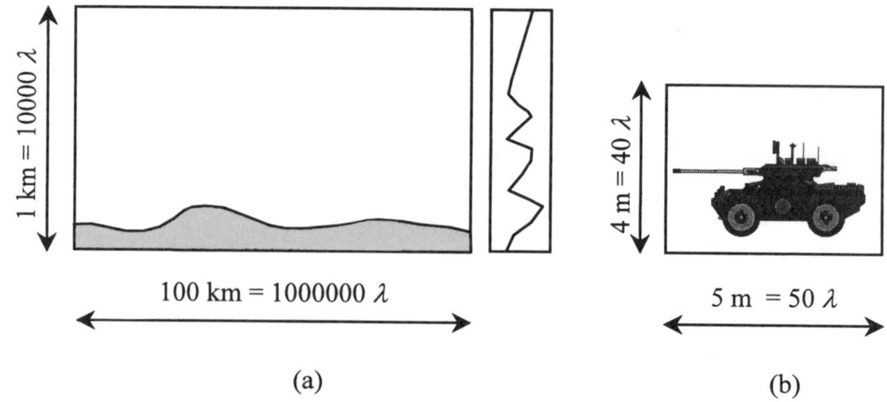

Figure 12.1 *PE modelling applications assuming a 3 GHz source, showing (a) long-range propagation, involving very large domain, non-homogeneous medium and relatively simple boundaries and (b) object scattering, involving small domain and complex boundaries*

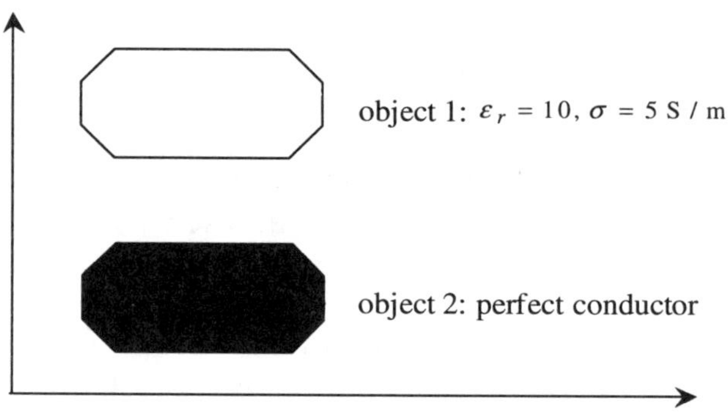

Figure 12.2 *Problems involving two scattering objects*

shown in Figure 12.2, where different types of boundary conditions have to be enforced on each object. This kind of difficulty becomes even more intractable for three-dimensional problems. There Fourier methods are applicable only to very limited object geometries. In this chapter we concentrate on finite-difference schemes. The two-dimensional schemes described here will later be used as building blocks for three-dimensional scattering algorithms.

Sections 12.2 and 12.3 are rather technical, giving the derivation of the different schemes and their numerical properties. In Section 12.2, we look at

Padé schemes for solving both the narrow-angle and the wide-angle parabolic equations [172]. The numerical properties of the lower order Padé schemes are examined in detail. Section 12.3 describes the very wide-angle split-step Padé schemes introduced by Collins [34, 35] and explains how they relate to higher order Padé approximants. Section 12.4 deals with perfectly transparent boundary conditions suitable for wide-angle schemes. The work of [121, 129] is generalized to split-step Padé schemes with plane wave incidence [23, 175].

In Section 12.5, we look at different ways of modelling the scattered field with parabolic equation techniques. The reflecting facet approach [92] is a natural extension of the ideas used for modelling terrain in radiowave propagation applications. The rotating PE approach [94, 174] solves for the scattered field rather than the total field. This allows the paraxial direction to be decoupled from the incident direction: by rotating the paraxial direction, the scattered field may be computed in any angular sector of interest. The full 360° bistatic radar cross-section of an object can be computed in a small number of rotating PE runs. Section 12.6 gives examples of scattering calculations for two-dimensional objects.

12.2 Padé schemes

A Padé-(p_1, p_2) approximant of a function $f(t)$ is a rational function F of the form

$$F(t) = \frac{P_2(t)}{P_1(t)} \tag{12.1}$$

where P_1 and P_2 are polynomials of degrees p_1 and p_2 respectively. The number of degrees of freedom in the determination of P_2/P_1 is $p_1 + p_2 + 1$, so that it is possible to match the Taylor expansion of f up to order $p_1 + p_2$. When solving the parabolic equation

$$\frac{\partial u}{\partial x} = ik\left(-1 + \sqrt{1+Z}\right)u = 0 \tag{12.2}$$

where the differential operator Z is defined by

$$Z = \frac{1}{k^2}\frac{\partial^2}{\partial z^2} + n^2(x,z) - 1 \tag{12.3}$$

two approaches are possible: we can use a Taylor expansion of the square-root and solve the resulting approximate equation, or we can write directly the formal solution

$$u(x+\Delta x, z) = \mathrm{e}^{ik\Delta x\left(\sqrt{1+Z}-1\right)} \cdot u(x,z) \tag{12.4}$$

With a first-order expansion of the square-root, the former approach leads to the narrow-angle PE. The formal solution of the narrow-angle PE can be expressed in terms of the narrow-angle exponential operator

$$u(x+\Delta x, z) = \mathrm{e}^{ik\Delta x Z/2} \cdot u(x,z) \tag{12.5}$$

It has been shown in [4, 5] that higher order Padé approximations of the square-root operator $\sqrt{(1+Z)}$ result in numerical instabilities. Instead, we use the second approach and look for Padé approximants P_2/P_1 of the function

$$f(t) = \exp\left(ik\Delta x(\sqrt{1+t}-1)\right) \tag{12.6}$$

We note that discretization in x has already been implicitly carried out since the range step Δx is used to define the function f. We then replace eqn. 12.4 with

$$P_1(Z) \cdot u(x+\Delta x, z) = P_2(Z) \cdot u(x,z) \tag{12.7}$$

This is already a very useful simplification since we have replaced the unwieldy exponential pseudo-differential operator with polynomials in Z, which are just ordinary partial derivatives. The next stage is to discretize in z. This is done with the usual central difference expression for the second-order derivative. If we denote by A_1, A_2 the matrices corresponding to the action of the discretized versions of $P_1(Z), P_2(Z)$, we end up with a matrix expression of the form

$$A_1 U_{m+1} = A_2(U_m + B_m) \tag{12.8}$$

where U_m represents the solution at range $m\Delta x$ and heights $\Delta z, \ldots, N\Delta z$ and B_m is a vector incorporating the non-homogeneous part of the boundary conditions. We see that if $p_1 > 0$, A_1 is non-trivial and the scheme requires inversion of a linear system to obtain U_{m+1}. The scheme is then classified as implicit. When $p_1 = 0$, no inversion is required and the scheme is explicit.

In order to reach optimal choices of PE algorithms, we shall need a few notions commonly used in numerical analysis. More information on finite-difference schemes can be found in [135, 145]. The finite-difference scheme is said to be convergent if

$$U_m(j) \to u(m\Delta x, j\Delta z) \tag{12.9}$$

as the mesh sizes $\Delta x, \Delta z$ tend to zero. This guarantees that the vector U_m is a good approximation to the true solution if the grid is sufficiently fine. Here the situation is complicated by the fact that we are dealing with pseudo-differential operators. It is clear that a given Padé scheme can only be a reasonable approximation of the square-root operator for certain directions of propagation, and hence the usual notion of convergence does not apply. The best we can hope for is that the error can be made small in a given angular sector of propagation directions.

We look at stability and consistency properties of the scheme. To simplify notations, we recast eqn. 12.8 in the form

$$\mathbf{U}_{m+1} = A\mathbf{U}_m + \mathbf{B}_m \tag{12.10}$$

where $A = A_1^{-1}A_2$. The finite-difference scheme is said to be consistent if the local truncation error

$$T(m\Delta x, j\Delta z) = \tilde{u}_{m+1}(j) - A\tilde{u}_m(j) - B_m \tag{12.11}$$

tends to zero as the mesh sizes $\Delta x, \Delta z$ tend to zero, where

$$\tilde{u}_m(j) = u(m\Delta x, j\Delta z) \tag{12.12}$$

Again the usual notion of consistency is not appropriate since we are dealing with pseudo-differential operators: we can only hope that the truncation error will tend to zero as the order of the Padé scheme gets larger. We shall assume that the Taylor expansion machinery is applicable to our exponential pseudo-differential operator. If we match the Padé-(p_1, p_2) approximant to the Taylor expansion of order $p_1 + p_2$, the leading error term in Z is of order $p_1 + p_2 + 1$, which indeed tells us that we can get as close as we wish to the exponential pseudo-differential operator by using high order Padé approximants.

The error in Z depends on the direction of propagation: for a plane wave propagating at angle θ from the horizontal, the operator Z corresponds to multiplication by $\sin^2\theta + n^2(x, z) - 1$. Hence the error due to replacing the exponential operator by the Padé-(p_1, p_2) approximant matching the Taylor expansion of order $p_1 + p_2$ is proportional to

$$\max\left\{\sin^2\theta + n^2(x, z) - 1\right\}^{p_1+p_2+1} \tag{12.13}$$

where the maximum is taken over the spatial and angular domains of interest. Provided that refractive index variations are not too large and that we only consider solutions with angular components in a given sector, the term in curly brackets is less than 1, and we can get arbitrarily small errors by taking p_1+p_2 large enough.

The next stage is to consider the error due to the discretization of Z. With the three-point central difference approximation of Z, the leading error term in z is of order 2, so any Padé method for the PE is at best second-order accurate in z. The leading error term is also of order $p_1 + p_2 + 1$ in Δx. We note that eqn. 12.4 is actually the integrated formal solution of the original PE given by eqn. 12.2. In numerical analysis terminology, the numerical scheme corresponding to our Padé approximation is $(p_1 + p_2)$th order accurate in x. This determines the numerical efficiency of the method, since the range steps have to be small enough to preserve accuracy.

The use of higher order Padé approximants improves accuracy, but makes boundary modelling very difficult to implement. Finite-difference approximations of higher order derivatives in z require more grid points, and alternatives have to be found for points close to or on the boundaries. In practice, this limits p_1 and p_2 to be at most 1 if object boundaries are present.

We now look at the very important concept of stability, which deals with the propagation of errors. A finite-difference scheme necessarily causes truncation errors, and computer implementations always imply the presence of rounding errors. We need some guarantee that such errors will not blow up as the scheme propagates. We use the following definition: a finite-difference scheme is unconditionally stable if any initial error remains bounded as it propagates, with a bound that does not depend on the grid sizes Δx and Δz.

We use the matrix form of eqn. 12.10 to state this property in terms of the matrix A of the scheme. In this analysis, the term B_m which represents the non-homogeneous part of the boundary conditions is considered as known. Then if an error V_0 is made on the initial values vector U_0, both U_m and the perturbed solution $U_m + V_m$ propagate according to eqn. 12.10, and the error V_m satisfies the homogeneous equation

$$V_{m+1} = AV_m \tag{12.14}$$

Let N be the order of the matrix A. If A can be diagonalized, unconditional stability can be formulated in terms of its eigenvalues $\lambda_1, \ldots, \lambda_N$ as

$$|\lambda_j| \leq 1,\ j = 1, \ldots, N \tag{12.15}$$

for any choice of $\Delta x, \Delta z$. In order to look at stability properties of Padé schemes, we define the matrix A of the scheme in terms of the matrix D which corresponds to the discretization of Z. We assume for simplicity that the field is zero at the top and bottom of the domain. The matrix D is then of the form

$$D = \frac{1}{k^2 \Delta z^2} \begin{bmatrix} -2+b_1 & 1 & & & \\ 1 & -2+b_2 & 1 & & \\ & & \cdot & & \\ & & & \cdot & \\ & & & -2+b_{N-1} & 1 \\ & & & 1 & -2+b_N \end{bmatrix} \tag{12.16}$$

where

$$b_j = k^2 \Delta z^2 \left(n^2(j\Delta z) - 1 \right), \; j = 1, \ldots, N \tag{12.17}$$

We limit the analysis to the case where the refractive index is real. Then the b_js are real, and the matrix D is real and symmetric. It follows that D is diagonalizable, and that all its eigenvalues $\mu_j, j = 1, \ldots, N$, are real. It is straightforward to show that A is also diagonalizable, with eigenvalues

$$\lambda_j = \frac{P_2(\mu_j)}{P_1(\mu_j)} \tag{12.18}$$

Since A is diagonalizable, the Padé scheme corresponding to A is unconditionally stable if and only if all the eigenvalues are of modulus at most 1. We now look in more detail at the properties of Padé-(1,1), Padé-(0,1) and Padé-(2,1) schemes, which are the most commonly used.

12.2.1 Padé-(1,1) scheme

The Padé-(1,1) approximant which has the same Taylor expansion of order 2 as the function f is given by

$$R_1(t) = \frac{1 + \frac{1}{4}(1 - ik\Delta x)t}{1 + \frac{1}{4}(1 + ik\Delta x)t} \tag{12.19}$$

This corresponds to the Claerbout approximation of eqn. 2.25. The Crank-Nicolson scheme we derived in Chapter 3 for the narrow-angle PE can also be viewed as a Padé approximant. If we denote by g the function corresponding to the narrow-angle exponential operator

$$g(t) = \exp\left(\frac{ik\Delta xt}{2}\right) \tag{12.20}$$

the Padé-(1,1) approximant with the same Taylor expansion of order 2 as g is given by

$$S_1(t) = \frac{1 + \frac{1}{4}(ik\Delta x)t}{1 - \frac{1}{4}(ik\Delta x)t} \tag{12.21}$$

which corresponds to the narrow-angle Crank-Nicolson scheme. Both the narrow-angle and wide-angle Crank-Nicolson schemes are second-order accurate in x, which allows relatively large range steps. We now show that they are also unconditionally stable. For the wide-angle Padé-(1,1) scheme, we have

$$\lambda_j = \frac{1 + \frac{1}{4}(1 - ik\Delta x)\mu_j}{1 + \frac{1}{4}(1 + ik\Delta x)\mu_j} \tag{12.22}$$

Since the μ_js are real, all the eigenvalues are of modulus 1, which shows that the wide-angle Crank-Nicolson scheme is unconditionally stable. The same applies to the narrow-angle Crank-Nicolson scheme. In fact we can show that Crank-Nicolson schemes preserve the complex inner product norm

$$||\mathbf{u}|| = \left\{\sum_{j=1}^{N} |u_j|^2\right\}^{\frac{1}{2}} \tag{12.23}$$

Since matrix D is real and symmetric, we can find a basis of eigenvectors $\mathbf{e}_1, \ldots, \mathbf{e}_N$ corresponding to $\lambda_1, \ldots, \lambda_N$, which is orthonormal with respect to the complex inner product. Now if we take any vector $\mathbf{u}$, we can write $\mathbf{u}$ as

$$\mathbf{u} = \sum_{j=1}^{N} u'_j \mathbf{e}_j \tag{12.24}$$

Since the basis $\mathbf{e}_1, \ldots, \mathbf{e}_N$ is orthonormal, we have

$$||\mathbf{u}|| = \left\{\sum_{j=1}^{N} |u'_j|^2\right\}^{\frac{1}{2}} \tag{12.25}$$

Using the fact that $\mathbf{e}_1, \ldots, \mathbf{e}_N$ are eigenvectors for A, we have

$$A\mathbf{u} = \sum_{j=1}^{N} \mu_j u'_j \mathbf{e}_j \tag{12.26}$$

and since the eigenvalues are of modulus 1, we get

$$||A\mathbf{u}|| = ||\mathbf{u}|| \tag{12.27}$$

So A preserves the inner-product norm as announced. In general, the matrix A depends on range: this happens of course if the refractive index structure is range-dependent, but also in the presence of irregular terrain or obstacles. The analysis above also works for the range-dependent case: at each range, the matrix A for that range corresponds to an isometry, and hence errors remain bounded as they propagate.

More generally for any positive integer p, the Padé-(p,p) approximant results in a norm-preserving scheme. To see this, we note that if $P_2(t)/P_1(t)$ is the Padé-(p,p) approximant for $f(t)$, then $P_1(t)/P_2(t)$ is the Padé-(p,p) approximant for $1/f(t)$. If t is real and greater than -1, we have $1/f(t) = \overline{f(t)}$. This implies $P_1 = \overline{P_2}$. Then for any real t, we have

$$|P_1(t)| = |P_2(t)| \tag{12.28}$$

so that all the eigenvalues have modulus 1. The norm preserving property follows, using the orthonormal basis of eigenvectors argument as before.

Although Crank-Nicolson schemes are unconditionally stable, we see that since the eigenvalues are of modulus 1, errors do not decay as they propagate. More generally, if oscillations are introduced in the solution, due for example to the presence of discontinuities in the boundary data, they will propagate indefinitely, resulting in a noisy solution. Clearly a scheme with eigenvalues of modulus strictly less than 1 will ensure that errors decay as they propagate. This is the reason for introducing the Padé-(1,0) scheme.

12.2.2 Padé-(1,0) scheme

The Padé-(1,0) approximant is given by

$$R_0(t) = S_0(t) = \frac{1}{1 - \dfrac{ik\Delta x}{2}t} \tag{12.29}$$

For this case the same result is obtained for both the narrow-angle and the wide-angle exponential operators, since the Padé-(1,0) scheme cannot do better than matching the first order expansion of the square-root operator. Despite its narrow-angle limitations, this scheme is extremely useful in view of its computational simplicity, particularly for extension to three-dimensional problems. The Padé-(1,0) scheme is the classical implicit scheme [145], which has better stability properties than the Crank-Nicolson scheme, while keeping the implementation of boundary conditions simple. Basically the second order derivative in z is evaluated at the forward range only, rather than by averaging between previous and forward ranges as in the Crank-Nicolson scheme. This scheme is only first-order accurate since $N_1 + N_2 = 1$, hence it requires small range steps. However it has very good stability properties. The eigenvalues of the Padé-(1,0) scheme are given by

$$\lambda_j = \frac{1}{1 - \frac{ik\Delta x}{2}\mu_j} \tag{12.30}$$

Hence all eigenvalues have modulus strictly less than 1, and we can repeat the analysis of the Padé-(1,1) scheme to get

$$||A\mathbf{u}|| < ||\mathbf{u}|| \tag{12.31}$$

For the general case of propagation in a non-homogeneous medium, one cannot get much further in the absence of estimates on the eigenvalues. However in the case where the b_js are zero (propagation in vacuum), which is the most important for object scattering applications, the eigenvalues are given explicitly by [145]

$$\mu_j^0 = -\frac{4}{k^2\Delta z^2}\sin^2\left(\frac{j\pi}{2(N+1)}\right),\ j = 1,\ldots,N \tag{12.32}$$

with corresponding eigenvectors $\mathbf{e}_j$ given by

$$\mathbf{e}_j(l) = \sqrt{\frac{2}{N}}\sin\left(\frac{jl\pi}{N+1}\right),\ l = 1,\ldots,N \tag{12.33}$$

We note that high frequency components of the solution correspond to high values of j. Putting

$$r = \frac{2\Delta x}{k\Delta z^2} \tag{12.34}$$

we have

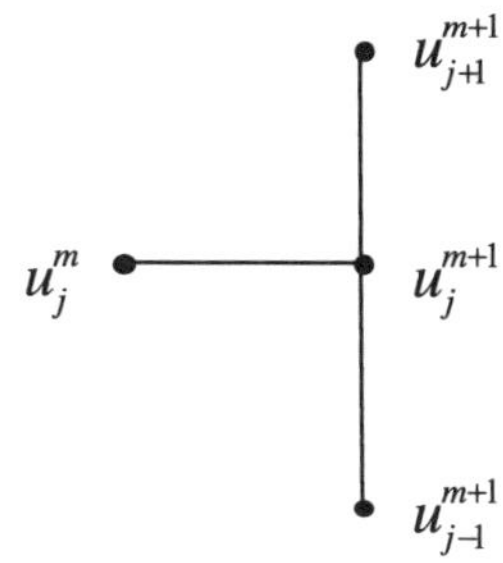

Figure 12.3 Structure of Padé-(1,0) scheme

$$|\lambda_j| = \frac{1}{\sqrt{1 + r^2 \sin^4\left(\frac{j\pi}{2(N+1)}\right)}} \tag{12.35}$$

For a fixed value of the ratio r, we see that the higher frequency component are attenuated very rapidly, providing an automatic "clean-up" of the solution from noise due to boundary discontinuities. This corresponds to the L_0-stability described in [145]. The wave equation itself has this "cleaning-up" property: high frequency components attenuate with range. It is of course desirable to use a numerical scheme which also has this property, which is why the Padé-(1,0) scheme is very useful despite its accuracy limitations. By contrast the Crank-Nicolson schemes propagate all components of the field without attenuation, thus artificially maintaining high frequency components which should in fact decay with range. For this reason the Crank-Nicolson schemes are most useful for long-range problems with smooth boundaries. L_0-stable schemes are more suitable for object scattering applications, which often involve boundary discontinuities.

The discretized equations for the Padé-(1,0) scheme are given by

$$\frac{u_j^{m+1} - u_j^m}{\Delta x} = \frac{i}{2k\Delta z^2}\left(u_{j-1}^{m+1} + u_{j+1}^{m+1} - 2u_j^{m+1}\right) \tag{12.36}$$

This is a 4-point finite-difference scheme, as shown in Figure 12.3.

Eqn. 12.36 is valid for grid points which are not on scattering object boundaries. The resulting system has to be supplemented by equations at boundary points. Figure 12.4 schematically describes the grid points close to the object for sloping boundaries, assuming the boundary of the object can be represented on a rectangular grid. An equation for point A_1 is obtained from eqn. 12.36.

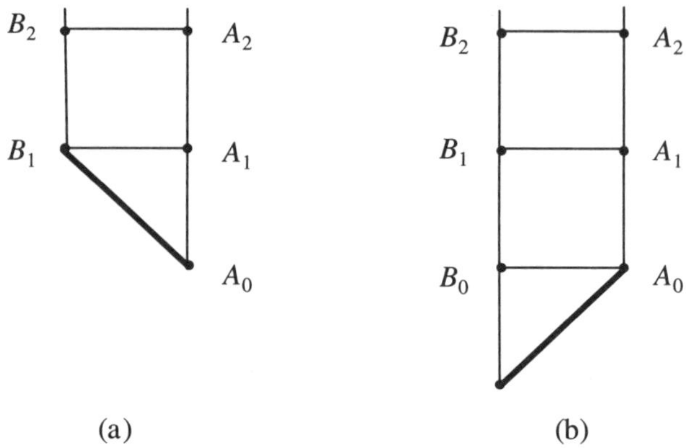

Figure 12.4 *Boundary modelling with Padé-(1,0) scheme for (a) ascending and (b) descending slopes*

For point A_0 on the boundary of the object, we need to enforce the appropriate boundary condition. We denote by a_i the value of u at point A_i. For Dirichlet boundary conditions, the supplementary equation for A_0 is of the form

$$a_0 = \alpha \tag{12.37}$$

where α is the value to be enforced at point A_0. For Neumann boundary conditions, estimates of the normal derivative at point A_0 are needed, and hence we require finite-difference expressions for the partial x and z derivatives. The partial z-derivative is obtained directly using a single-sided finite-difference expression, for example the 3-point expression

$$\frac{\partial u}{\partial z}(A_0) \sim \frac{2a_1 + a_2 - 3a_0}{4\Delta z} \tag{12.38}$$

Direct evaluation of the partial x-derivative is not recommended. Instead we use the PE to replace the partial range derivative with the second order z-derivative. This can be evaluated at point A_0 with a single-sided finite-difference expression. The simplest one, based on three points only, amounts to using the same finite-difference expression for the second order z-derivative at A_0 and A_1, writing

$$\frac{\partial^2 u}{\partial z^2}(A_0) \sim \frac{a_0 + a_2 - 2a_1}{\Delta z^2} \tag{12.39}$$

We note that this method has no difficulty modelling large ascending slopes: the fact that point B_0 could then be located quite far from the boundary has no bearing on the algorithm. However for descending slopes, we see that the range step (which is also the distance between B_1 and A_1) will have to be small in order to maintain the fixed z-grid.

For more accurate representation of the scattering object, an adaptive scheme can be used where the distance between A_0 and A_1 is adjusted so that A_0 is exactly on the scatterer. It is straightforward to modify the finite-difference expressions for the first z-derivative at A_0 and the second order z-derivative at A_1 accordingly.

12.2.3 Padé-(2,1) scheme

The classical implicit scheme suppresses numerical high frequency oscillations that might appear at boundary discontinuities, for example at building edges, but that scheme is unsuitable for large domains in view of the small range step it requires. Then it is desirable to use a higher order scheme which suppresses unwanted oscillations. This can be done with a Padé-(2,1) scheme. For the narrow-angle exponential operator, the Padé-(2,1) approximant which matches the Taylor expansion to order 3 is [173]

$$B_2(t) = \frac{1 + \frac{i\delta}{6}t}{1 - \frac{i\delta}{3}t - \frac{\delta^2}{24}t^2} \tag{12.40}$$

where

$$\delta = k\Delta x \tag{12.41}$$

The corresponding Padé-(2,1) approximant for the wide-angle exponential operator is

$$A_2(t) = \frac{1 + at}{1 + bt + ct^2} \tag{12.42}$$

with

$$a = \frac{-\delta^2 + 3i\delta + 3}{6(i\delta + 1)}, b = \frac{2\delta^2 + 3}{6(i\delta + 1)}, c = -\frac{i\delta(\delta^2 + 3)}{24(i\delta + 1)} \tag{12.43}$$

These schemes are third-order accurate in range and thus allow quite large range-steps. The eigenvalues for the narrow-angle Padé-(2,1) scheme are given by

$$\lambda_j = \frac{1 + \frac{i\delta}{6}\mu_j}{1 - \frac{\delta^2}{24}\mu_j^2 - \frac{i\delta}{3}\mu_j} \tag{12.44}$$

Hence the modulus is given by

$$|\lambda_j|^2 = \frac{1 + \frac{\delta^2}{36}\mu_j^2}{1 + \frac{\delta^2}{36}\mu_j^2 + \frac{\delta^4}{576}\mu_j^4} \tag{12.45}$$

showing that the scheme is unconditionally stable. As for the Padé-(1,0) scheme, all the eigenvalues have modulus strictly less than 1, and high frequency components attenuate rapidly whatever the value of the ratio r, so the scheme is L_0-stable. The same properties hold for the wide-angle Padé-(2,1), for which the moduli of the eigenvalues are given by

$$|\lambda_j|^2 = \frac{\alpha_j}{\alpha_j + \beta_j^2} \tag{12.46}$$

with

$$\alpha_j = 576(1 + \delta^2 + \mu_j) + 384\delta^2\mu_j + (16\delta^4 + 48\delta^2 + 144)\mu_j^2 \tag{12.47}$$

and

$$\beta_j = (3 + \delta^2)\delta^2\mu_j^2 \tag{12.48}$$

To illustrate the difference in stability properties between Padé-(1,1) and Padé-(2,1) schemes, we consider the aperture shown in Figure 12.5. The initial field is one on the aperture and zero outside it, approximating a plane wave normally incident on a slit. Figure 12.6 shows the field as a function of height 30 wavelengths away from the aperture, computed with both the Padé-(1,1) and Padé-(2,1) schemes. The discontinuities in the initial field at the aperture edges become persisting disturbances in the Padé-(1,1) (or Crank-Nicolson) scheme. By contrast the disturbances quickly vanish with the Padé-(2,1) scheme.

Figure 12.7 shows the filtering effect of the Padé-(2,1) scheme: starting with the noisy Padé-(1,1) results at a distance of 30λ from the aperture, the Padé-(2,1) scheme is applied with a range step of 0.1λ. The figure shows

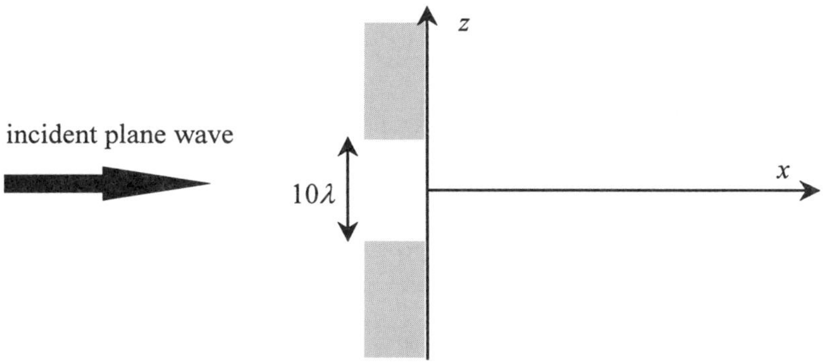

Figure 12.5 Geometry for aperture simulation

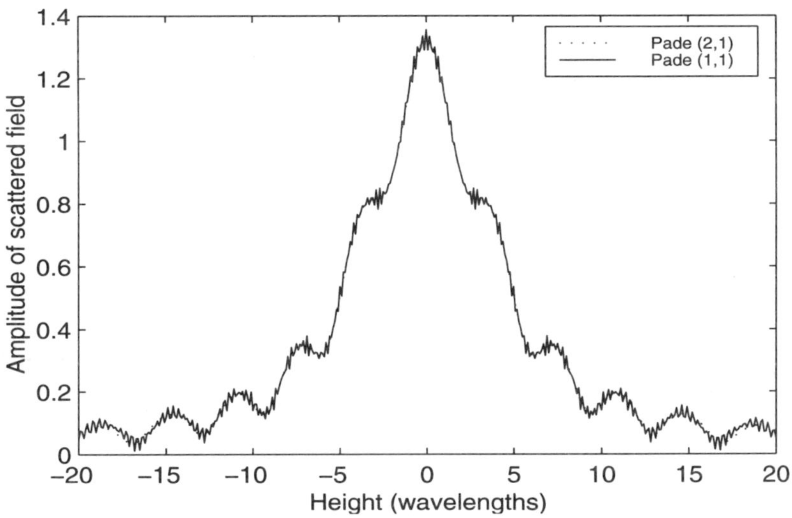

Figure 12.6 Comparison of Padé-(1-1) and Padé-(2-1) schemes 30 wavelengths from aperture

results after one, two and three steps. The Padé-(2,1) scheme is extremely powerful as a regularizing algorithm: in less than half a wavelength, it has completely eliminated the boundary disturbances.

With the combination of L_0-stability and third-order accuracy in range, the Padé-(2,1) schemes are very efficient numerically. Their main drawback is the difficulty of modelling boundaries because more grid points are involved in the discretization. We also note that the Padé-(2,1) matrices are pentadiago-

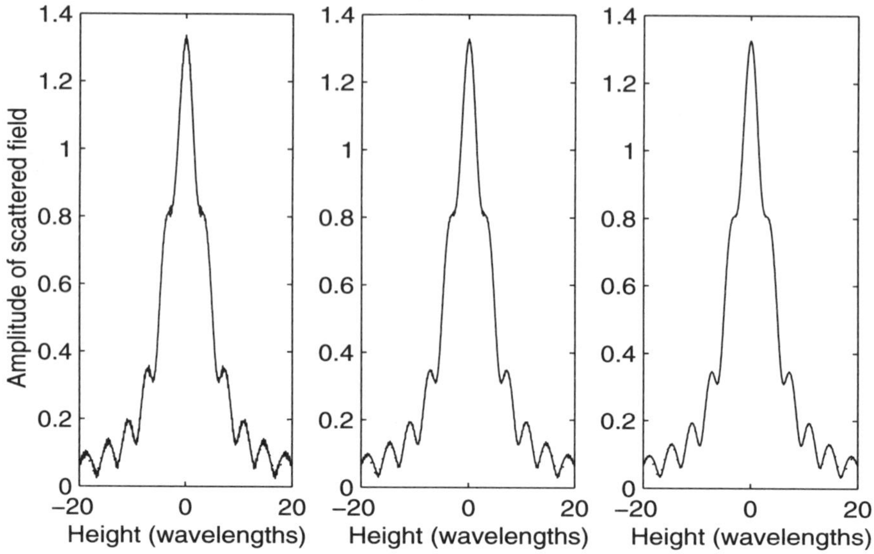

Figure 12.7 *Filtering effect of Padé-(2-1) scheme: results (a)* 30.1, *(b)* 30.2, *(c)* 30.3 *wavelengths from aperture. Dotted curves: results obtained with Padé-(2,1) scheme only. Continuous curves: results obtained with Padé-(1,1) scheme up to distance* 30λ *from aperture*

nal instead of tridiagonal, which results in more complex inversion algorithms. Padé-(2,1) schemes are extremely useful for eliminating unwanted oscillations once the field has been marched beyond the the scattering objects, for example for building scatter applications [172]. An example is given in Section 13.6. They can also be used with double pass methods for radar-cross-section calculations, as described in Section 13.4.2.

12.3 Split-step Padé method

The split-step Padé method was introduced by Collins in [34] to solve problems involving very large propagation angles. We have seen in the previous section that with one Padé approximant, better accuracy at large propagation angles requires high order polynomials. The resulting linear systems become quite cumbersome, as the matrices involve more and more non-zero entries. The split-step method gets around that problem by decomposing the Padé approximant into partial fractions.

The original derivation given by Collins [34] goes as follows: starting with the formal solution of the PE given in eqn. 12.4, we approximate directly the wide-angle exponential pseudo-differential operator with a Padé-(1,1) sum of the form

$$e^{ik\Delta x\left(\sqrt{1+Z}-1\right)} \sim 1 + \sum_{l=1}^{L} \frac{a_l Z}{1 + b_l Z} \tag{12.49}$$

where the $2L$ coefficients a_l, b_l will be determined at a later stage. Substituting this expression in eqn. 12.4, we get

$$u\left(x + \Delta x, z\right) = u\left(x, z\right) + \sum_{l=1}^{L} \left(1 + b_l Z\right)^{-1} a_l Z u\left(x, z\right) \tag{12.50}$$

The sum on the RHS can be computed easily by solving for each term separately. If we put

$$v_l\left(x + \Delta x, z\right) = \left(1 + b_l Z\right)^{-1} a_l Z u\left(x, z\right) \tag{12.51}$$

we have

$$u\left(x + \Delta x, z\right) = u(x, z) + \sum_{l=1}^{L} v_l\left(x + \Delta x, z\right) \tag{12.52}$$

and we can solve independently for each v_l. In this sense, we have a split-step solution. However we do not make any further approximations at this stage, This makes this method very different from the split-step Fourier solutions which involve the splitting of non-commuting operators.

We can relate this to the previous section by thinking of the Padé-(1,1) expansion of eqn. 12.49 as the decomposition of a high order Padé approximant of the function $f(t)$ into a sum of the constant term 1 and of L simple fractions, writing

$$F(t) = 1 + \sum_{l=1}^{L} \frac{a_l t}{1 + b_l t} \tag{12.53}$$

By summing the right hand side, we can also express F as

$$F(t) = \frac{P_2(t)}{P_1(t)} \tag{12.54}$$

where P_1 and P_2 are polynomials of degrees p_1 and p_2 respectively. Assuming the b_ls are distinct, we have $p_1 = L$ and $p_2 \leq L$. We now turn to the determination of the coefficients. For optimal accuracy, it is a natural idea to ask for identical Taylor expansions on both sides up to order $2L$. Then F is the Padé-(L, L) approximant of f, and we already know that the resulting scheme is unconditionally stable, but not L_0-stable. This means that in practice the scheme will be unusable for wide-angle applications. In particular waves with an imaginary angle, which should propagate as evanescent waves, are instead sustained indefinitely. We think in terms of the vertical plane wave spectrum, and consider a reduced plane wave of the form

$$u(x, z) = \exp(ik(\cos\alpha - 1)x + ik\sin\alpha z) \tag{12.55}$$

When $|\sin\alpha| \leq 1$, this corresponds to a plane wave propagating at the physical angle α from the horizontal. For outgoing waves, $\cos\alpha$ must be positive. When $|\sin\alpha| > 1$, we must choose $\cos\alpha$ in such a way that the wave is evanescent, hence $\cos\alpha$ must have a positive imaginary part, which means that

$$\cos\alpha = i\sqrt{\sin^2\alpha - 1} \tag{12.56}$$

We can relate this to the function $f(t)$ of eqn. 12.6, by noting that the variable t corresponds in spectral terms to $-\sin^2\alpha$. When we were only looking at relatively narrow-angle schemes, there was no contribution from evanescent waves, and it was quite safe to use all the degrees of freedom on our Padé approximants to matching Taylor expansions. With very-wide-angle schemes, there is no longer a cut-off far away from components with imaginary angles. If we do not take any precautions, these components are almost guaranteed to cause disastrous errors in the numerical solution. The answer is to ensure L_0-stability, which will follow automatically if the degree p_1 of the numerator in eqn. 12.54 is strictly less than L. This is equivalent to

$$\lim_{|t|\mapsto\infty} F(t) = 0 \tag{12.57}$$

which corresponds for the coefficients to the condition

$$\sum_{l=1}^{l=N} \frac{a_l}{b_l} = -1 \tag{12.58}$$

In particular, we see that the approximant tend to zero as t tends to $-\infty$, giving the evanescence condition at $-\infty$. We still have $2L - 1$ constraints available for Taylor expansion matching

$$F^{(m)}(0) = f^{(m)}(0), \quad m = 1, \ldots, 2L - 1 \tag{12.59}$$

Eqn. 12.49 automatically ensures $f(0) = F(0)$. The constraints yield a system of $2L$ non-linear equations in $\mathbf{A} = (a_1, \ldots, a_L, b_1, \ldots, b_L)$. This can be solved with a modified Newton method as in [33]. Let Γ be the vector function defining the constraints. We want to find a solution of the vector equation

$$\Gamma(\mathbf{A}) = 0 \tag{12.60}$$

We denote by J the Jacobian matrix of Γ, given by

$$J_{k,l} = \frac{\partial \Gamma_k}{\partial a_l}, \quad J_{k,L+l} = \frac{\partial \Gamma_k}{\partial b_l}, \quad k = 1, \ldots, 2L, \quad l = 1, \ldots, L \tag{12.61}$$

The modified Newton method consists in constructing a sequence $\mathbf{A}_m$ starting from a first guess $\mathbf{A}_0$ and with

$$\mathbf{A}_{m+1} = \mathbf{A}_m - \epsilon J^{-1} \Gamma(\mathbf{A}_m) \tag{12.62}$$

Iteration is stopped once the required error threshold is attained. The use of the relaxation parameter $\epsilon < 1$ can improve convergence initially if the first guess is not sufficiently close to the solution.

We now solve for the auxiliary functions v_ls. These satisfy

$$(1 + b_l Z)\, v_l(x + \Delta x, z) = a_l Z u(x, z) \tag{12.63}$$

Discretizing in z, we obtain finite-difference equations for the v_l's in the form

$$\begin{aligned} &v_l^{j+1} + v_l^{j-1} + \left(\frac{s^2}{b_l} + s^2(n^2 - 1) - 2\right) v_l^j \\ &= \frac{a_l}{b_l} \left\{u_l^{j+1} + u_l^{j-1} + \left(s^2(n^2 - 1) - 2\right) v_l^j\right\} \end{aligned} \tag{12.64}$$

where

$$v_l^j = v(x + \Delta x, j\Delta z), \quad s = k\Delta z \tag{12.65}$$

As usual, boundary conditions at the top and bottom of the domain need to be added in order to obtain a well-determined system of equations. For example if the terrain boundary is flat and located at $z = 0$, the Dirichlet boundary condition $u(x, 0) = 0$ is satisfied provided each v_l also satisfies it.

Local impedance boundary conditions are treated in a similar way. More complicated techniques are necessary for non-local boundary conditions, as will be shown in Section 12.4. In any case each v_l is the solution of a tridiagonal system, which is solved by Gaussian elimination as in the previous section.

Using the fact that

$$\sum_{l=1}^{N} \frac{a_l}{b_l} = -1 \tag{12.66}$$

we can rewrite the SSP in the simpler form

$$u(x + \Delta x, z) = -\sum_{l=1}^{l=N} (1 + b_l Z)^{-1} a_l u(x, z) \tag{12.67}$$

which is a Padé-(1,0) expansion. This is simpler to implement than the Padé-(1,1) sum, since the second order partial z-derivatives at the forward range are not required. We have seen the advantages of this type of scheme earlier with the classical implicit (or Padé-(1,0)) scheme versus the Crank-Nicolson (or Padé-(1,1)) scheme. The crucial difference is that using a Padé-(1,0) sum does not degrade the numerical properties of the split-step Padé scheme: we are using a simpler formulation, but the sum is still the same high order Padé approximant as before.

Since the split-step Padé (SSP) method with L terms corresponds to the Padé-$(L, L-1)$ approximant, it is accurate to $(2L-1)$th-order in x, so that quite large range steps can be used. Examples of the numerical efficiency of the method can be found in [34, 35]. The SSP method can treat very wide-angle propagation if enough terms are included. It follows from eqn. 12.13 that the number of terms required for a given accuracy in a given angular sector depends on refractive index variations.

With eight terms, propagation up to 70° can be modelled accurately. Consider the case of a plane wave propagating in vacuum at 70° from the horizontal. To fix ideas we take a frequency of 3 GHz, corresponding to a wavelength of 10 cm. Figure 12.8 shows the real part of the reduced field u as a function of range, computed respectively with the SPE and 8-term SSP algorithms. The theoretical values are shown for comparison. Whereas the narrow-angle results show a very large phase error, the SSP results are in excellent agreement with the theory.

12.4 Wide-angle NLBCs

The non-local boundary conditions of Chapter 8 can be extended to wide-angle parabolic equation schemes [121, 129, 175]. The simplest way to do this is to

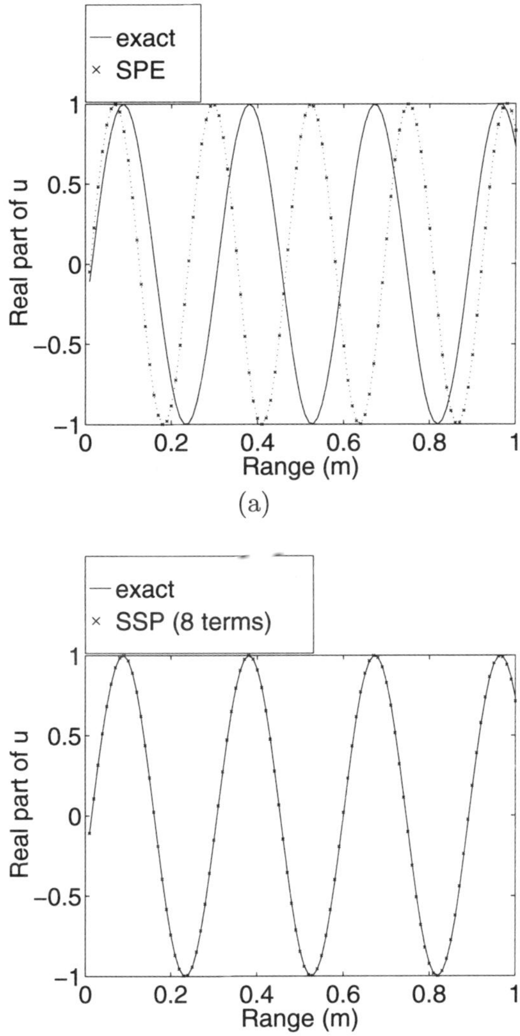

Figure 12.8 *Performance of (a) narrow-angle and (b) 8-term split-step Padé PE codes for 3 GHz plane wave propagating* 70° *from the horizontal [175]*

consider the one-way wave equation with the exact square-root operator Q

$$\frac{\partial u}{\partial x} = -ik(1-Q)u \tag{12.68}$$

where $Q = \sqrt{1+Z}$. We treat the case of a homogeneous medium above height z_b. Assume for simplicity that $n(x,z) = 1$ for $z \geq z_b$.

We treat first the diffractive case, assuming that the initial field is zero for $z \geq z_b$. Taking the Laplace transform in range, we get for $z \geq z_b$

$$pU(p,z) = -ik(1-Q)U(p,z) \tag{12.69}$$

which can also be written as

$$ik\,QU(p,z) = (p+ik)\,U(p,z) \tag{12.70}$$

We now apply ikQ to both sides, getting

$$-k^2\left(\frac{1}{k^2}\frac{\partial^2 U}{\partial z^2}(p,z) + U(p,z)\right) = (p+ik)^2 U(p,z) \tag{12.71}$$

which can be rewritten as

$$\frac{\partial^2 U}{\partial z^2}(p,z) + (p^2 + 2ikp)U(p,z) - \frac{\partial u}{\partial x}(0,z) = 0 \tag{12.72}$$

We notice that this is also the Laplace transform in range of the reduced wave equation in vacuum

$$\frac{\partial^2 u}{\partial x^2} + 2ik\frac{\partial u}{\partial x} + \frac{\partial^2 u}{\partial z^2} = 0 \tag{12.73}$$

provided we have for $z \geq z_b$

$$\frac{\partial u}{\partial x}(0,z) = 0 \tag{12.74}$$

In fact the derivation of eqn. 12.72 proves that this last equation indeed holds for solutions of the outgoing wave equation in the diffractive case, a property which is not obvious a priori.

Now since u satisfies the outgoing wave equation, eqn. 12.68, U must be bounded, and wanted solutions of eqn. 12.72 are of the form

$$U(p,z) = U(p,0)\,\exp\left(iz\sqrt{p^2+2ikp}\right) \tag{12.75}$$

Differentiating with respect to z, we obtain for $z \geq z_b$

$$\frac{\partial U}{\partial z}(p,z_b) = i\sqrt{p^2+2ikp}\,U(p,z_b) \tag{12.76}$$

We get the wide-angle non-local boundary condition by taking the inverse Laplace transform of this expression. We write

$$\sqrt{p^2+2ikp} = p\left(\frac{\sqrt{p^2+2ikp}}{p} - 1\right) + p \tag{12.77}$$

From [53, 75], we have

$$\mathcal{L}w(p) = i\left(\frac{\sqrt{p^2+2ikp}}{p} - 1\right) \tag{12.78}$$

where

$$w(x) = -k\mathrm{e}^{-ikx}\left(J_0(-kx) - iJ_1(-kx)\right) \tag{12.79}$$

and J_0 and J_1 are the Bessel functions of order 0 and 1. This yields the wide-angle non-local boundary condition

$$\frac{\partial u}{\partial z}(x,z_b) = i\frac{\partial u}{\partial x}(x,z_b) + \int_0^x \frac{\partial u}{\partial x}(\xi,z_b)w(x-\xi)d\xi \tag{12.80}$$

In the transmitting case, this must be modified by adding the appropriate incoming energy term

$$I(x) = \frac{\partial v}{\partial z}(x,z_b) - i\frac{\partial v}{\partial x}(x,z_b) - \int_0^x \frac{\partial v}{\partial x}(\xi,z_b)w(x-\xi)d\xi \tag{12.81}$$

where v is any solution above z_b with the same initial field as u above z_b, as described in Section 8.6. The case of plane-wave incidence is the most important for wide-angle scattering applications. For plane-wave incidence, the incoming energy term can be given in closed form [175]. We take for v the reduced plane wave

$$v(x,z) = \exp\left(ik\left[x\sin^2\frac{\theta}{2} + (z - z_b)\sin\theta\right]\right) \tag{12.82}$$

Then we have

$$\begin{aligned} e^{ikx} I(x) &= ik\left(\sin\theta + 2i\sin^2\frac{\theta}{2}\right) e^{ikx\cos\theta} \\ &-2ik^2\sin^2\frac{\theta}{2}\int_0^x e^{ik\xi\cos\theta}\{J_0(-k[x-\xi]) - iJ_1(-k[x-\xi])\}d\xi \end{aligned} \tag{12.83}$$

The integral can be evaluated using the expansion [99]

$$\int_0^x e^{-it\cos\theta} J_\nu(t)dt = 2e^{-ix\cos\theta}\sum_{k=0}^{\infty} i^k \frac{\sin([k+1]\theta)}{\sin\theta} J_{k+1+\nu}(x) \tag{12.84}$$

The resulting expression for the incoming energy term is

$$\begin{aligned} e^{ikx} I(x) &= ik\left(\sin\theta + 2i\sin^2\frac{\theta}{2}\right) e^{ikx\cos\theta} \\ &+2ik^2\sin\frac{\theta}{2}\sum_{k=0}^{\infty} i^k \sin\left(\frac{2k+1}{2}\theta\right) J_{k+1}(kr) \end{aligned} \tag{12.85}$$

12.4.1 Split-step Padé implementation

The main use for the non-local boundary condition we have just derived is with split-step Padé codes which model propagation at very large angles. Using the notation of Section 12.3, we write the split-step Padé solution as

$$u(x + \Delta x, z) = u(x,z) + \sum_{l=1}^{L} v_l(x + \Delta x, z) \tag{12.86}$$

where the v_ls are given by

$$v_l(x + \Delta x, z) = (1 + b_l Z)^{-1} a_l Z u(x,z) \tag{12.87}$$

Our problem is to find boundary conditions for the v_ls such that the non-local boundary condition given by eqn. 12.80 is satisfied by u. The main difficulty is that the v_ls at the advanced range $x + \Delta x$ are defined as a function of u

at range x. In order to write a non-local boundary condition for the v_ls, we need to define them at $x = 0$. For the case of plane-wave incidence given by eqn. 12.82, we can assume that the medium is a vacuum at ranges $x \leq 0$ and use eqn. 12.82 to define the incident field at negative ranges. We can then extend the definition of v_l as well. Now in a vacuum eqn. 12.87 reduces to the ordinary differential equation

$$\frac{\partial^2 v_l}{\partial z^2}(x + \Delta x, z) + \frac{k^2}{b_l} v_l(x + \Delta x, z) = k^2 \frac{a_l}{b_l} \frac{\partial^2 u}{\partial z^2}(x, z) \tag{12.88}$$

For plane wave incidence, we can compute the RHS explicitly and the differential equation satisfied by v_l on the initial vertical becomes

$$\frac{\partial^2 v_l}{\partial z^2}(0, z) + \frac{k^2}{b_l} v_l(0, z) = -k^2 \frac{a_l}{b_l} \sin^2\theta \, \mathrm{e}^{-2ik\Delta x \sin^2\theta + ikz\sin\theta} \tag{12.89}$$

The solution which satisfies the original eqn. 12.87 is

$$v_l(0, z) = \frac{a_l \sin^2\theta}{b_l \sin^2\theta - 1} \mathrm{e}^{-2ik\Delta x \sin^2\theta - ikz\sin\theta} \tag{12.90}$$

In order to derive the incoming energy term $I_l(x)$ for v_l, we note that the solution in a vacuum corresponding to v_l is simply

$$v_l^0(x, z) = \frac{a_l \sin^2\theta}{b_l \sin^2\theta - 1} \mathrm{e}^{-2ik\Delta x \sin^2\theta} v(x, z) \tag{12.91}$$

where v is the solution in a vacuum corresponding to u. By linearity, we have

$$I_l(x) = \frac{a_l \sin^2\theta}{b_l \sin^2\theta - 1} \mathrm{e}^{-2ik\Delta x \sin^2\theta} I(x) \tag{12.92}$$

If we enforce the non-local boundary condition

$$\frac{\partial v_l}{\partial z}(x, z_b) = i\frac{\partial v_l}{\partial x}(x, z_b) + \int_0^x \frac{\partial v_l}{\partial x}(\xi, z_b) w(x - \xi) d\xi + I_l(x) \tag{12.93}$$

on each v_l, then by linearity, u satisfies the correct non-local boundary condition

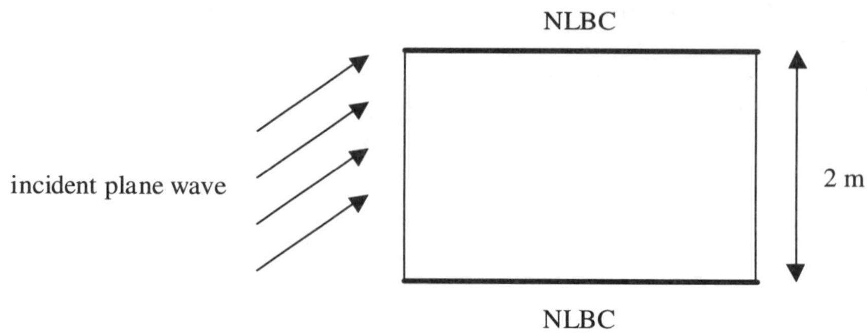

Figure 12.9 Integration domain for split-step/Padé simulations

$$\frac{\partial u}{\partial z}(x, z_b) = i\frac{\partial u}{\partial x}(x, z_b) + \int_0^x \frac{\partial u}{\partial x}(\xi, z_b)w(x-\xi)d\xi + I(x) \tag{12.94}$$

This provides very accurate domain truncation, as illustrated by a comparison of split-step Padé results and the analytical solution for a plane wave propagating in a vacuum [175]. The frequency is 3 GHz. The propagation domain is shown in Figure 12.9. The non-local boundary condition is applied both at the top and bottom of the domain, which spans 2 m or 20 wavelengths. Figure 12.10 shows the real part of the field on a vertical cut at a distance of 4 m for a plane wave propagating at an angle of 80° from the horizontal. Split-step Padé results for $L = 10$ are shown. Results are in excellent agreement with the reference values, showing at the same time that the split-step Padé solution is very accurate up to very large propagation angles and that the non-local wide-angle boundary condition provides extremely clean truncation.

12.5 Modelling the scattered field

12.5.1 Reflecting facet model

In all that follows, we assume that the scatterers are non-penetrable, and that we can solve for the field outside the scattering objects by using suitable boundary conditions on the object surfaces. A first approach for scattering applications is to treat forward scattering by objects in exactly the same way as propagation over irregular terrain, as in Chapter 7. The boundary of the scatterer is discretized on a rectangular grid. The PE is used to propagate the forward scattered field along the positive x-direction, applying suitable

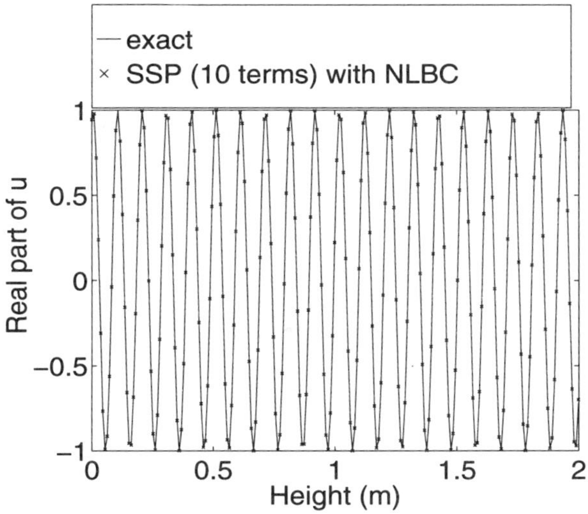

Figure 12.10 *Plane wave propagation at 80° from the horizontal, showing field a range 4 m computed with 10-term split-step/Padé method and wide-angle NLBC and exact solution; the frequency is 3 GHz*

boundary conditions on the scatterer. The initial field is the known incident field, a plane wave for most applications. The backward propagated field is obtained with a second run, this time marching in the negative x-direction. The object is treated as a sequence of reflecting facets which act as sources for the backward propagating energy [92, 95, 23]. In this case the simulation starts at some vertical beyond the scatterer, setting the initial field to zero. Boundary conditions on each facet are given by the appropriate polarization-dependent reflection coefficients, which may vary along the scattering object. The technique is illustrated schematically in Figure 12.11.

With the facet approach, the scattered field is computed in a forward paraxial cone and a backward paraxial cone. Note that with this approach, we compute the forward total field, which is the sum of the incident field and the forward scattered field, and the backward scattered field directly. The bistatic radar-cross-section can be computed from near-field results using near-field/far-field transformations (see Appendix B).

12.5.2 Rotating PE method

The approach of the previous section for the backscattered field can be generalized to arbitrary scattering directions with the rotating PE method [94, 174]. We start with a field component ψ and write

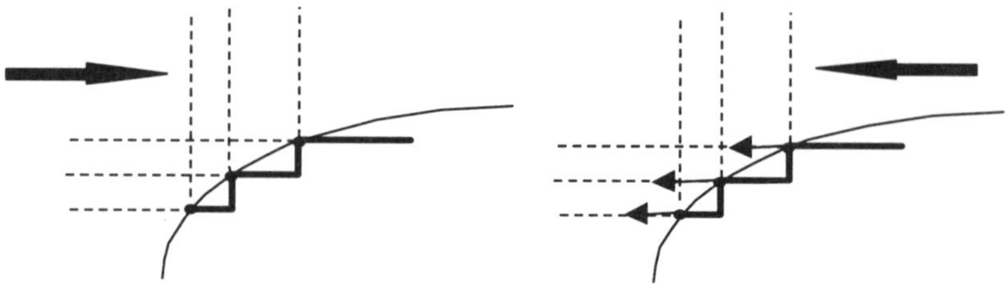

(a) (b)

Figure 12.11 *Facet model for (a) forward and (b) backward scattering calculations*

$$\psi = \psi^i + \psi^s \tag{12.95}$$

where ψ^i is the incident field, or the field that would exist in the absence of the scattering objects, and ψ^s is the scattered field, which is the perturbation caused by the scatterers. Then ψ^i satisfies the wave equation

$$\Delta\psi^i(x,z) + k^2 n^2(x,z)\psi^i(x,z) = 0 \tag{12.96}$$

for all (x, z), where $n(x, z)$ is the refractive index of the background medium, while ψ and ψ^s satisfy the wave equation outside the scattering objects only. We choose the positive x-direction as the paraxial direction and denote by u^s the reduced PE field corresponding to ψ^s

$$u^s(x,z) = \mathrm{e}^{-ikx}\psi^s(x,z) \tag{12.97}$$

Note that the paraxial direction is completely arbitrary and is not linked to the incident direction. In what follows, we are interested in the energy scattered in a cone centred along the paraxial direction. In that cone, the reduced scattered field u^s satisfies approximately the parabolic equation

$$\frac{\partial u^s}{\partial x} + ik(1-Q)u^s = 0 \tag{12.98}$$

outside the scattering objects. Instead of solving for the total field, we solve for the scattered field, extracting the part that propagates in the chosen paraxial cone, as shown in Figure 12.12.

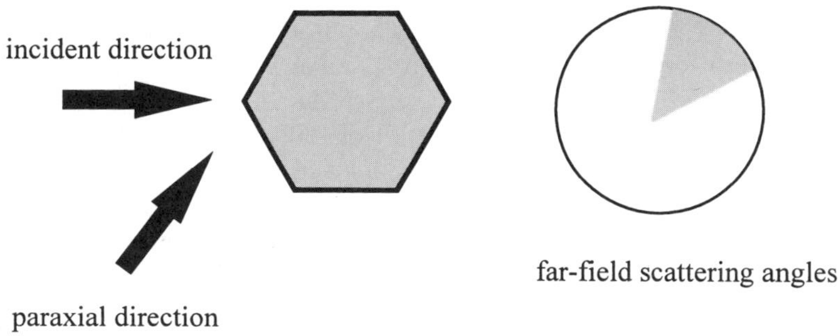

Figure 12.12 Rotating PE method

For many applications, the background propagation medium is a vacuum, and the unperturbed field can be obtained analytically. We note however that the rotating PE method is applicable to far more general situations, where the background medium may exhibit refractive index variations and contain interfaces. Then the unperturbed field must be obtained numerically, for example with a long-range PE code.

12.5.3 Boundary conditions

On the scatterers, u^s satisfies non-homogeneous boundary conditions since the boundary conditions involve the incident field. For example for a perfectly conducting object and horizontal polarization, ψ is the transverse electric field E_y which must be zero on the scatterers, so the boundary condition is

$$u^s(P) = -e^{-ikx}\psi^i(P) \tag{12.99}$$

for a point P on the boundary of the object. For vertical polarization, ψ is the transverse magnetic field H_y. Its normal derivative must be zero on the scatterer

$$\frac{\partial \psi}{\partial \mathbf{n}}(P) = 0 \tag{12.100}$$

where P is a point on the scatterer and $\mathbf{n}(P)$ is the outer normal to the surface of the scatterer at P. Using the PE to evaluate the range derivative, we get the non-homogeneous boundary condition for u^s in the form

$$\left\{ikn_x Q u^s + n_z \frac{\partial u}{\partial z}\right\}(P) = 0 \tag{12.101}$$

where n_x and n_z are the coordinates of the outer normal. Surface impedance boundary conditions can be modelled in a similar way with non-homogeneous mixed boundary conditions. It is easy to see that in the backscattering case, when the paraxial direction makes an angle of 180° with the incident direction, solving for u^s while enforcing these non-homogeneous boundary conditions on the scatterer is just the same as the facet treatment. However if we replace the physical argument about "sources" by the above mathematical treatment, we can generalize the method to any choice of paraxial direction. For the forward propagation case, the two formulations are equivalent, differing only by the choice of boundary condition (homogeneous for total field integration, non-homogeneous for scattered field integration).

12.5.4 Initial field

We assume the scatterers are located in the half-space $x > 0$ and start integration at range zero. With the rotating PE method, the initial condition is simply that the field should be zero on the initial vertical. This somewhat surprising property is due to the fact that we are solving for the scattered field in a paraxial cone centred along the positive x-direction. Since all the scattering sources are located in the half-space $x > 0$, the initial scattered field for the rotating PE is zero. There is no paradox here: because of the paraxial approximation, the initial PE field is not the actual physical scattered field, which is of course not zero in front of the object. Rather, the initial PE field represents the scattered energy which propagates towards the object, in the paraxial direction. That this energy should be zero follows from the fact that the scattered field always propagates away from the object.

12.5.5 Numerical implementation

Eqn. 12.98 can be solved with a variety of schemes. The simplest is of course to use the narrow-angle approximation of Q, together with a suitable Padé scheme. The most flexible for this application is the Padé-(1,0) scheme, which makes it very simple to implement arbitrary boundary conditions while suppressing ripples caused by sharp changes in impedance or slope. As it is only first order accurate, this scheme requires quite small range steps. Unlike long-range propagation problems, object-scattering problems are usually limited to a few hundreds of wavelengths at most, and this is not normally a drawback. It should be pointed out that the non-homogeneous boundary conditions include a right-hand-side term with a phase variation in range which depends on the angle between incident and paraxial directions. The worst case is that of backscattering, which involves an exponential term in $\exp(-2ikx)$ on the RHS of the boundary condition. Because of this fast phase variation in range, small range steps are required anyway for paraxial directions making a non-negligible angle with the incident direction.

If very-wide-angle results are required, the split-step Padé method can be adapted to solve for the scattered field. For this, it is better to use the simpler form of the SSP solution given by eqn. 12.67. We write the SSP solution as

$$u^s(x + \Delta x, z) = -\sum_{l=1}^{L} w_l^s(x + \Delta x, z) \tag{12.102}$$

where the functions w_l^s are defined by

$$w_l^s(x + \Delta x, z) = -\frac{a_l}{1 + b_l Z} u^s(x, z) \tag{12.103}$$

The coefficients (a_l, b_l) correspond to the Padé expansion of Section 12.3. It is straightforward to write a finite-difference expression for w_j inside the integration domain. For points on the boundary of a scatterer, we must express the w_js in such a way that u^s will satisfy the required non-homogeneous boundary condition. Consider for example the Dirichlet case. Eqn. 12.99 will be approximately satisfied if we have

$$w_l^s(x + \Delta x, z_b) = -w_l^i(x + \Delta x, z_b) \tag{12.104}$$

for a boundary point $P = (x + \Delta x, z_b)$, where w_l^i corresponds to a Padé term for the incident field. Assume the background medium is a vacuum. Then w_l^i solves the differential equation

$$\left(1 + b_l \frac{1}{k^2} \frac{\partial^2}{\partial z^2}\right) w_l^i = a_l u^i \tag{12.105}$$

For plane-wave incidence the solution can be obtained explicitly, but in general numerical integration is required.

12.5.6 Near-field/far-field considerations

Near-field calculations are normally carried out in a small domain containing the object, and the bistatic radar cross-section is obtained via the near-field/far-field transformations derived in Appendix B. Far-field bistatic scattering results can be obtained at all scattering angles by rotating the paraxial direction. With a narrow-angle code, giving good results in a 30° angular sector, a dozen runs suffice to reconstruct the whole scattering pattern.

Near-field results obtained with the rotating PE method are not necessarily accurate. For example with a narrow-angle code, the near-field results do not include wider-angle contributions and are hence incorrect. However this does not affect far-field results in the paraxial cone of interest, provided

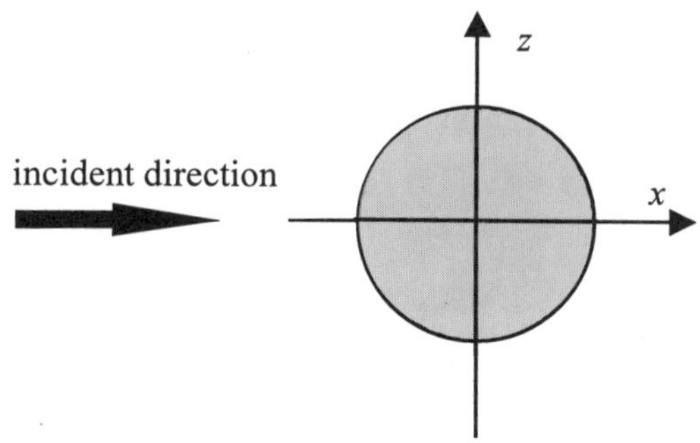

Figure 12.13 Geometry for scattering by circular cylinder

energy scattered by the object does not undergo large changes of direction. An example where the narrow-angle algorithm is inadequate is given in Section 12.6.2. It is certainly the case that split-step Padé techniques can alleviate this type of problem for Dirichlet boundary conditions. At this stage it is not known whether more demanding boundary conditions can be treated with SSP methods, nor whether they can be generalized to three-dimensional problems.

12.6 Examples

In all the examples that follow, the background medium is assumed to be a vacuum.

12.6.1 Cylinders

For plane or cylindrical wave incidence, the field scattered by perfectly conducting circular cylinders can be expressed as a series of Hankel functions [24]. We consider a cylinder of radius a, and a plane wave of unit amplitude incident in the positive x-direction as shown in Figure 12.13.

Figure 12.14 shows the real part of the scattered field computed with the split-step Padé method for a perfectly conducting cylinder of normalized radius $ka = 10$. The incident plane wave is horizontally polarized and propagating from the left. Eight terms have been used for the SSP. The perfectly transparent boundary condition described in Section 12.4.1 has been used to truncate the domain. Wide-angle scatter is modelled, and the non-local truncation performs very well. However it is clear that side scatter is not represented: this is a consequence of the paraxial approximation, since energy propagating at 90° from the marching direction is outside the paraxial cone.

Figure 12.15 shows bistatic RCS results in the forward and backward direction calculated with both the narrow-angle and the split-step Padé PE

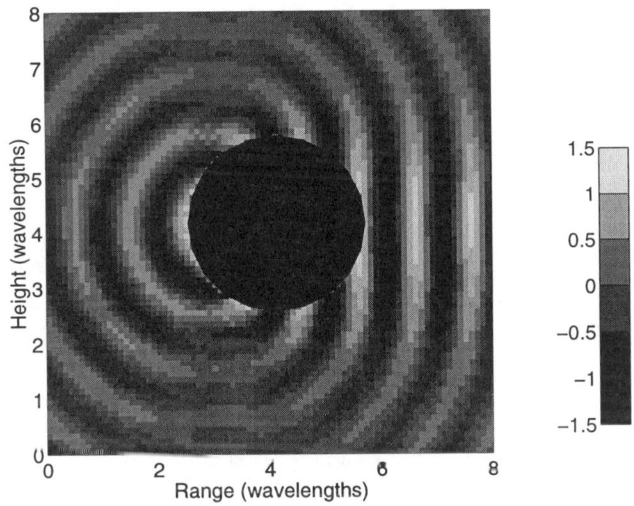

Figure 12.14 *Contours of real part of the field scattered by cylinder of normalized radius $ka = 10$*

algorithms [23]. The Crank-Nicolson (or Padé-(1,1)) scheme has been used to compute each term of the SSP sum. Theoretical results are shown for comparison. The narrow-angle version is only accurate up to 15° or so as expected, while the 8-term split-step Padé PE is very accurate up to 40° or so. From the results in Section 12.3, one would expect the 8-term SSP algorithm to be accurate up to angles of 70°. The loss of accuracy is probably due to numerical difficulties with the boundary conditions on the object, caused by the lack of L_0-stability of the Crank-Nicolson scheme.

We now look at calculations made with the rotating PE method, using the Padé-(1,0) scheme. Figure 12.16 shows results of single rotating PE runs for a perfectly conducting cylinder of radius 5λ under horizontally polarized plane wave incidence. Theoretical results are also shown. For both forward scatter (paraxial direction 0°) and side scatter (paraxial direction 90°), agreement with the theory is excellent up to angles of 15° from the paraxial direction. In Figure 12.17, seven runs of the rotated PE have been used to compute the all-angle bistatic RCS of the cylinder. Results are shown for both polarizations. Agreement is excellent at all angles for horizontal polarization, and slightly less good for vertical polarization for which the boundary condition on the object is troublesome, even with the L_0 stable Padé-(1,0) scheme.

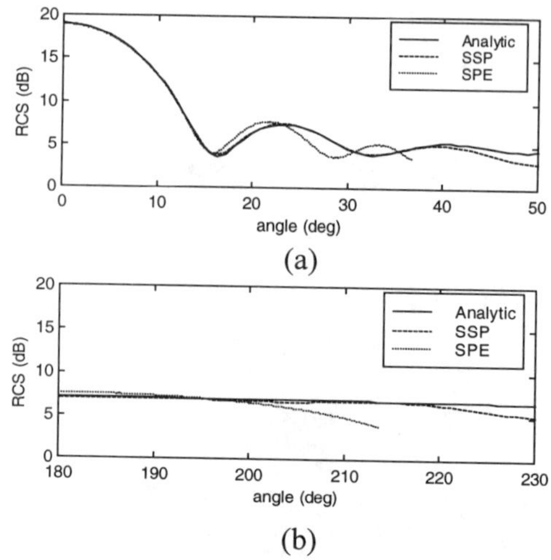

Figure 12.15 *Bistatic RCS results for horizontally polarized plane incident on perfectly conducting cylinder of normalized radius* $ka = 10$, *showing (a) forward scatter and (b) backscatter*

12.6.2 L-shaped object

We have already mentioned the problem of surface waves, which cannot be modelled accurately with PE techniques when objects are sufficiently small to support waves travelling all around the object. A related difficulty is that for non-convex objects, energy scattered by one part of the object in a given direction can sometimes contribute to scatter from another part of the object in a completely different direction. This is illustrated by the L-shaped object shown in Figure 12.18. Energy scattered from the top of the vertical branch of the L (ray 1) can hit the horizontal branch and be reflected away (ray 2). Although the energy along ray 2 is travelling in the paraxial direction, it cannot be modelled by the narrow-angle rotated PE because ray 1 is not in the paraxial cone.

This is demonstrated by the following numerical example: we choose dimensions $a = 10\lambda$, $b = 3\lambda$, $c = d = 0.5\lambda$, and a paraxial direction at $20°$ from the horizontal. Then ray 1 makes an angle of $40°$ with the paraxial direction and is not part of the narrow-angle cone. We model the object with the Dirichlet boundary condition, i.e. as a perfect conductor for horizontally polarized incidence. Figure 12.19 shows bistatic RCS results obtained with the narrow-angle PE rotated by $20°$, and with a split-step Padé method using 8 terms. The latter is accurate up to $70°$ and provides a reference solution. The narrow-angle rotated PE solution differs substantially from the reference, showing that the

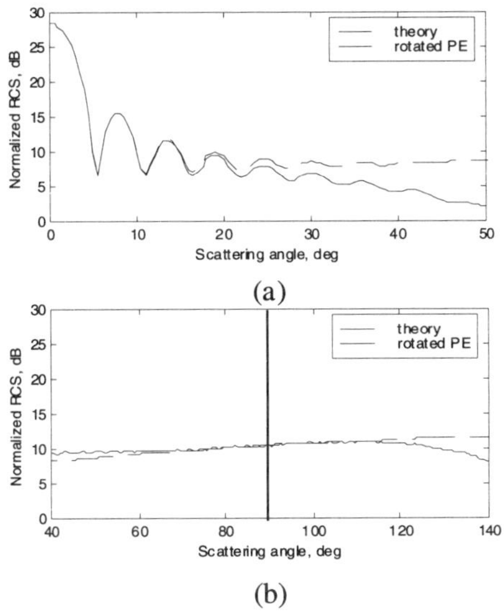

Figure 12.16 *Bistatic RCS of perfectly conducting cylinder of radius* 5λ *computed with a single rotating PE run for paraxial directions of (a)* $0°$ *and (b)* $90°$ *(horizontal polarization)*

far-field results are affected significantly by this problem. Figure 12.20 shows near-field results obtained with the 8-term split-step Padé method for this case, confirming that energy diffracted by the top of the L-shape is accurately taken into account by the wider-angle scheme.

12.6.3 2D aircraft shape

Figure 12.21 shows the field scattered by an idealized two-dimensional aircraft shape. A horizontally polarized plane wave of frequency 57.2 MHz is incident on the object from below. The maximum dimensions of the aircraft are 19 m in range and 4.2 m in height. A Crank-Nicolson implementation of the 8-term split-step Padé method has been used. The distortions due to the paraxial approximations are clearly visible in scattering directions close to the horizontal. This case has also been treated with a finite element method [20]. Figure 12.22 displays bistatic RCS results computed with the split-step Padé PE and finite element methods, showing rather good agreement. This case has also be treated at microwave frequencies [92].

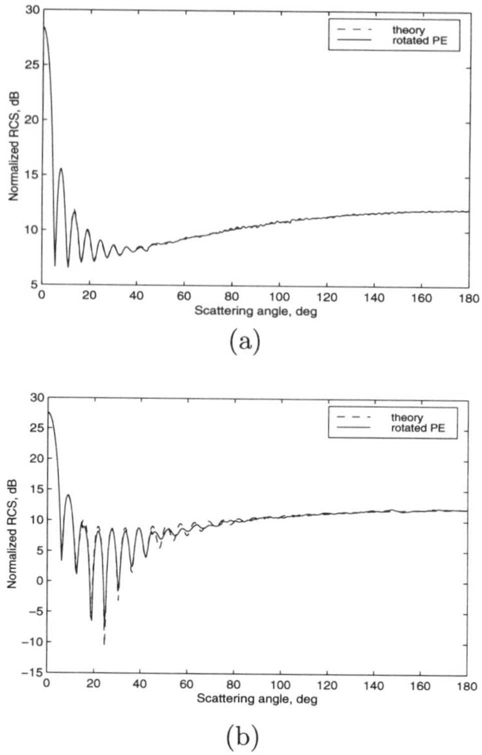

Figure 12.17 Bistatic RCS of cylinder of radius 5λ calculated with 7 rotating PE runs for (a) horizontal and (b) vertical polarizations

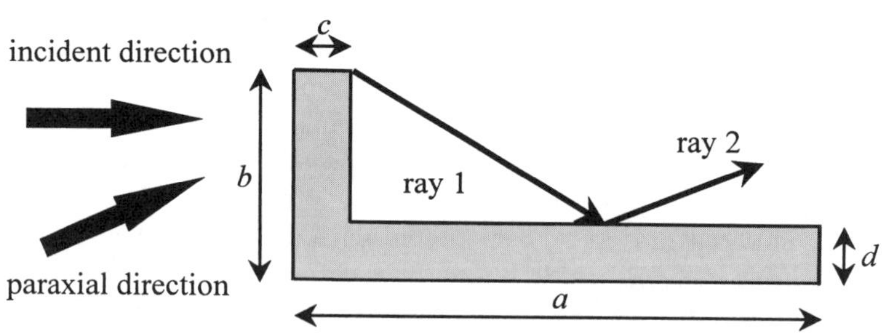

Figure 12.18 Rotating PE difficulties for L-shaped object

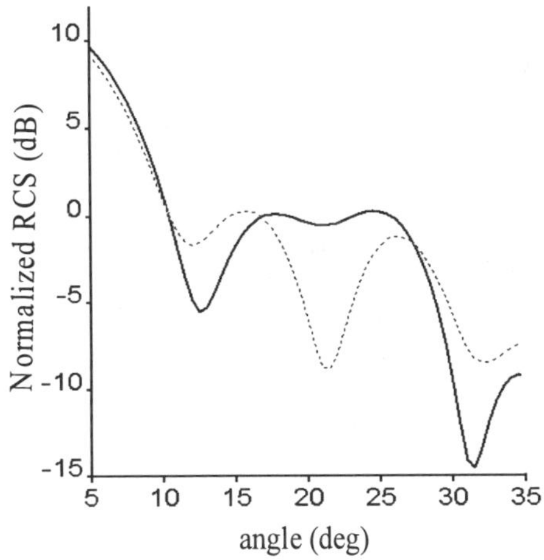

Figure 12.19 *RCS results for L-shaped object with rotated narrow-angle PE code (continuous line) and wide-angle PE code(dotted line)*

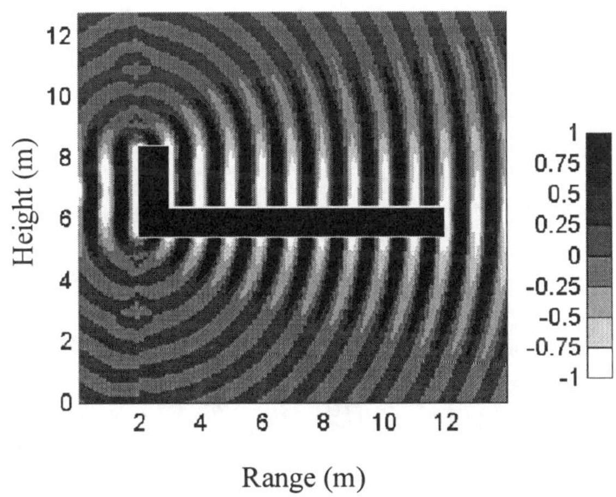

Figure 12.20 *Real part of scattered field computed with split-step Padé PE code for L-shaped object*

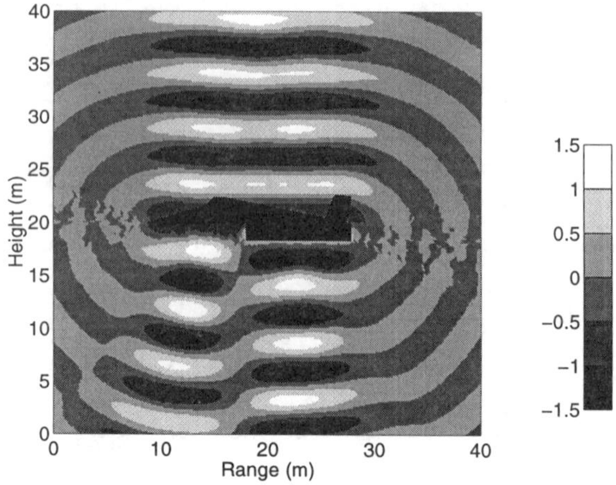

Figure 12.21 *Contours of real part of scattered field, 57.2 MHz plane wave incident from below*

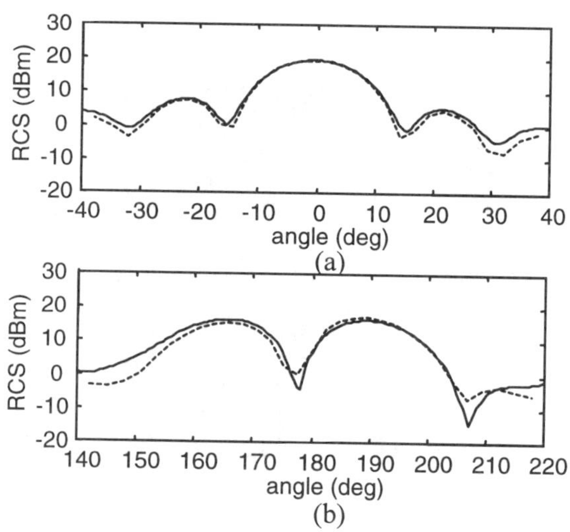

Figure 12.22 *Comparison of bistatic RCS results for idealized aircraft shape for (a) forward and (b) backward scatter, finite element solution [20] (continuous line) and split-step Padé PE (dotted line)*

Chapter 13

Three-dimensional scattering for the scalar wave equation

13.1 Introduction

In this chapter we consider three-dimensional scattering problems which can be modelled with the scalar wave equation. Acoustic scattering by non-elastic objects is included in this category, but of course most electromagnetic scattering problems involve coupling between scalar components of the field through the boundary conditions on the object and need vector solutions. Some important electromagnetic applications can however be treated with the scalar wave equation.

We first look at axially symmetric problems, which in fact reduce to two dimensions with a modified version of the parabolic equation [74, 129]. The framework for the axial symmetry PE is given in Section 13.2. The application of axial symmetry PE techniques to diffractive X-ray optics was pioneered by Kopylov *et al.* in [74]. Their work on the modelling of thin and thick Fresnel zone plates is presented in Section 13.3.

We then consider more general problems which require genuine three-dimensional PE algorithms. These require more sophisticated numerical techniques than two-dimensional problems, as there is usually no simple way to invert the linear systems involved [172, 170, 94]. Three-dimensional PE algorithms suitable for scattering applications are described in Section 13.4. The main difficulty resides in the modelling of boundary conditions on the scattering objects, which requires the inversion of large sparse matrices. The use of a double-pass method, minimizing the size of the domain where the sparse matrix system must be solved, substantially reduces computation costs. The robustness of the method allows its use for quite irregular surfaces: an application to rough surface modelling is presented in Section 13.5. Scalar three-dimensional PE techniques can be applied to a number of urban propagation problems. Examples of building scatter calculations are presented in Section 13.6.

13.2 Axial symmetry parabolic equation

We use cylindrical coordinates (r, θ, x) as shown in Figure 13.1. We assume axial symmetry around the x-axis and choose the positive x-axis as the paraxial direction. Following [74, 129], we write the narrow-angle PE for the reduced field u as

$$\Delta_\perp u(x,r) + 2ik\frac{\partial u}{\partial x} + k^2(n^2(x,r) - 1)u = 0 \tag{13.1}$$

Here $\Delta_\perp$ is the transverse Laplacian, which in the case of axial symmetry is expressed as

$$\Delta_\perp u(x,r) = \frac{1}{r}\frac{\partial}{\partial r}\left(r\frac{\partial u}{\partial r}(x,r)\right) \tag{13.2}$$

for $r > 0$. Because of axial symmetry, the radial derivative vanishes at zero and we have

$$\Delta_\perp u(x,0) = 2\frac{\partial^2 u}{\partial r^2}(x,0) \tag{13.3}$$

Eqn. 13.1 is not amenable to Fourier transform solutions because of the $1/r$ factor. However there is no difficulty of implementation with finite-difference schemes. The Crank-Nicolson scheme for example is given by [74]

$$\begin{aligned} 2ik\frac{u_j^{m+1} - u_j^m}{\Delta x} &+ \frac{u_{j+1}^{m+1} - 2u_j^{m+1} + u_{j-1}^{m+1} + u_{j+1}^m - 2u_j^m + u_{j-1}^m}{2\Delta r^2} \\ &+ \frac{u_{j+1}^{m+1} - u_{j-1}^{m+1} + u_{j+1}^m - u_{j-1}^m}{4\Delta x \Delta r^2} + k^2 b_j^m \frac{u_j^{m+1} + u_j^m}{2} = 0 \end{aligned} \tag{13.4}$$

where

$$u_j^m = u(x_m, r_j) = u(m\Delta x, j\Delta r) \tag{13.5}$$

and

$$b_j^m = n^2\left(\frac{x_m + x_{m+1}}{2}, r_j\right) \tag{13.6}$$

For applications involving composite materials with refractive index close to unity, it is not appropriate to model interfaces with surface impedance boundary conditions. Instead propagation inside the materials should be modelled,

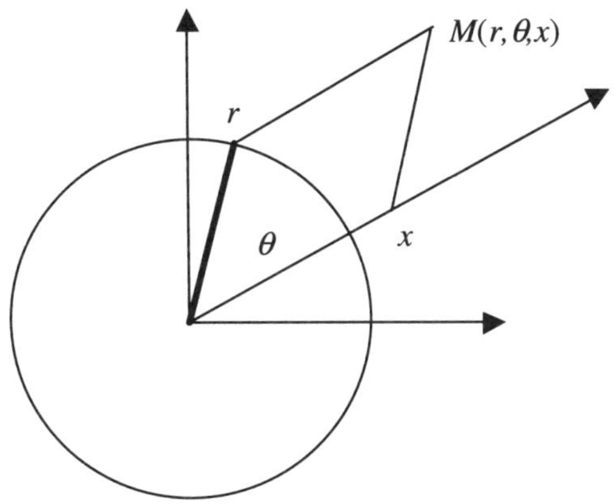

Figure 13.1 Cylindrical coordinates for axially symmetric problems

and this involves taking into account the discontinuities at interfaces. As suggested in [74], the integration scheme has better numerical properties if the discrete average in eqn. 13.6 is replaced by the integral

$$\overline{b_j^m} = \frac{1}{j\Delta r^2 \Delta x} \int_{m\Delta x}^{(m+1)\Delta x} dx \int_{(j-1/2)\Delta r}^{(j+1/2)\Delta r} n^2(x,r) r dr \tag{13.7}$$

which provides a more accurate representation of refractive index variations.

13.2.1 Domain truncation

Non-local boundary conditions can be derived for axially symmetric problems in much the same way as for the two-dimensional scalar case [74, 129]. We suppose that all the perturbations are located inside a cylinder C of radius r_0, with

$$n(r) = 1, \quad r \geq r_0 \tag{13.8}$$

We derive a non-local boundary condition on C. Then we can limit integration to the interior of the cylinder, as shown in Figure 13.2.

Eqn. 13.1 can be solved in closed form outside C. The appropriate Laplace transform is

$$U(s,r) = \int_0^{+\infty} u(x,r) e^{-sx} dx \tag{13.9}$$

We assume that the initial field is zero outside C

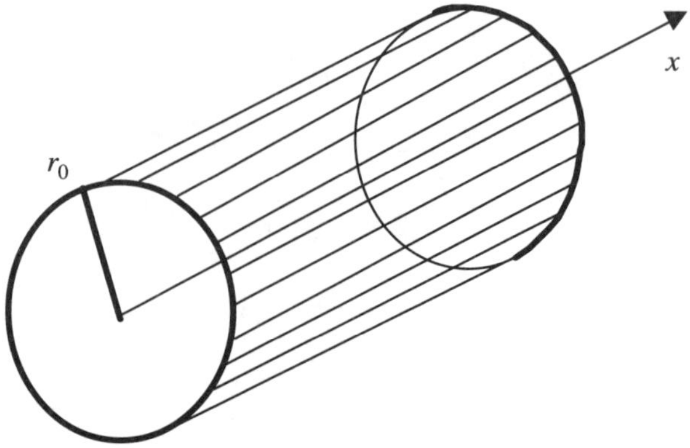

Figure 13.2 Domain truncation for axially symmetric problems

$$u(x,r) = 0, \quad r \geq r_0 \tag{13.10}$$

The outgoing solutions of the Laplace transform of eqn. 13.1 are of the form

$$U(s,r) = A(s)H_0^{(1)}(r\sqrt{2iks}) \tag{13.11}$$

where $H_0^{(1)}$ is the Hankel function of order 0 of the first kind. Differentiating with respect to r, we get the boundary condition

$$\frac{\partial U}{\partial r}(s,r_0) = \sqrt{2iks}\frac{H_1^{(1)}(r_0\sqrt{2iks})}{H_0^{(1)}(r_0\sqrt{2iks})}U(s,r_0) \tag{13.12}$$

which leads to a non-local equation of the form

$$\frac{\partial u}{\partial r}(x,r_0) = \int_0^x \frac{\partial u}{\partial x}(\xi,r_0)g_0(x-\xi)d\xi \tag{13.13}$$

where the convolution kernel g_0 is the inverse Laplace transform of

$$G_0(s) = \sqrt{\frac{2ik}{s}}\frac{H_0^{(1)\prime}(r_0\sqrt{iks})}{H_0^{(1)}(r_0\sqrt{2iks})} \tag{13.14}$$

Using contour integrals to invert the Laplace transform, one can express the convolution kernel in terms of the zeros of the Hankel function $H_0^{(1)}$. Another option which is simpler to implement is to use asymptotic expressions, which are accurate provided the radius of the cylinder is sufficiently large compared to the wavelength [74, 129]. We have the asymptotic formula [117]

$$H_\nu^{(1)}(z) \sim \sqrt{\frac{2}{\pi z}} \exp\left(i\left[z - \frac{\nu}{2} - \frac{\pi}{4}\right]\right) \tag{13.15}$$

as z tends to infinity. Now we can write the inverse Laplace transform f of a function F as the integral

$$f(x) = \frac{1}{2\pi i}\int_{\beta-i\infty}^{\beta+i\infty} F(s)e^{sx}ds \tag{13.16}$$

where β can take any real value larger than a threshold value β_0 [57]. From eqn. 13.15, we get if $kr_0^2 \gg 1$

$$G_0(s) \sim -i\sqrt{\frac{2ik}{s}},\, s = \beta + i\gamma \tag{13.17}$$

uniformly in γ, which gives the approximation of the inverse Laplace transform as

$$g_0(x) \sim \sqrt{\frac{2k}{\pi x}} \exp\left(-i\frac{\pi}{4}\right) \tag{13.18}$$

This kernel is the one we obtained for the 2D transparent boundary condition in Chapter 8. In fact the condition $kr_0^2 \gg 1$ allows us to treat the radius of curvature of the cylinder as infinite, which amounts to the 2D problem of Chapter 8.

If the incident field does not vanish outside C, an incoming energy term must be introduced as in Chapter 8. Let u_0 be any function satisfying eqn. 13.1 for $r \geq r_0$, and such that $u_0(0,r) = u(0,r)$ for $r \geq r_0$. Then the non-local boundary condition for u can be written as

$$\frac{\partial u}{\partial r}(x, r_0) = \int_0^x \frac{\partial u}{\partial x}(\xi, r_0)g_0(x-\xi)d\xi + I(x) \tag{13.19}$$

where the incoming energy term $I(x)$ is given by

$$I(x) = \frac{\partial u_0}{\partial r}(x, r_0) - \int_0^x \frac{\partial u_0}{\partial x}(\xi, r_0)g_0(x-\xi)d\xi \tag{13.20}$$

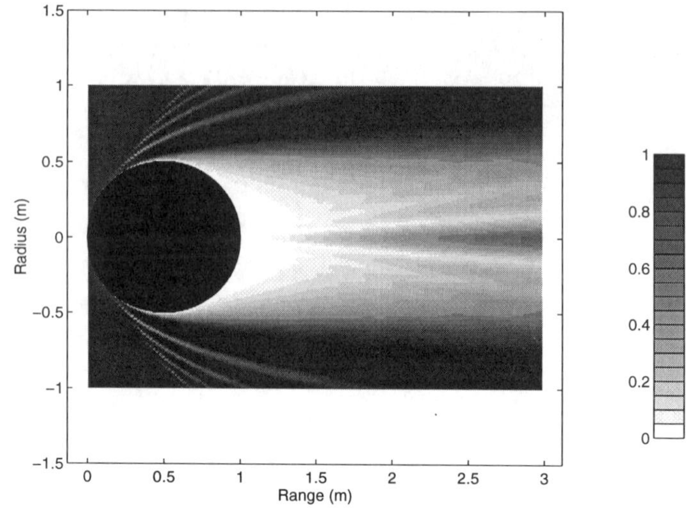

Figure 13.3 Real part of field scattered by acoustically soft sphere of radius $5\lambda = 0.5$ m, computed with axial symmetry PE

In practice the most frequent case is that of plane wave incidence, with the incident direction of propagation along the axis of symmetry. In that case we can take u_0 constant in the whole domain, and the incoming energy term is identically zero. Hence eqn. 13.13 is also valid for the case of normal plane wave incidence. The non-local truncation technique for cylindrical coordinates can be generalized to the case when there is no axial symmetry [74, 129], although it is often more convenient then to use Cartesian coordinates and PML truncation, as described in Section 13.4.

Non-local truncation for the axially symmetric case has been applied to acoustic scattering by an acoustically soft object for which the Dirichlet boundary condition holds [95]. Figure 13.3 shows the real part of the field scattered by an acoustically soft sphere of radius 0.5 m, under plane wave incidence coming from the left. The incident wavelength is 0.1 m. The non-local truncation works very well at all angles of incidence on the boundaries. Comparisons with 3D results which do not exploit the axial symmetry are given in Section 13.4.

13.3 Application to X-ray optics

Soft X-rays are low-energy X-rays which can be used to probe organic or ionorganic matter [110]. Efficient diffractive optical systems use Fresnel zone plates, which consist of partial screens with alternating opaque and transparent "zones", constructed to improve focusing efficiency. Fresnel zone plates are

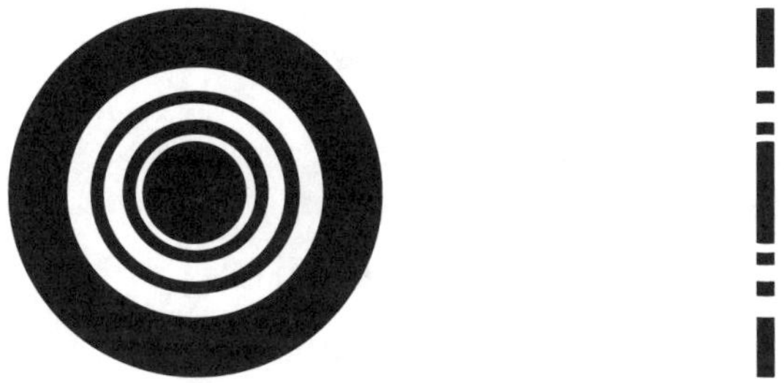

Figure 13.4 Transverse and side views of Fresnel zone plate

usually axially symmetric, as shown in Figure 13.4. The number of zones for soft X-ray optics is large, typically of the order of several hundreds.

Although closed-form solutions exist for ideal Fresnel lenses, this is not the case for more realistic thick zone plates. Since the dimensions of the zone plates are very large relative to the wavelength, finite element techniques are out of the question: for typical lenses, the maximum diameter of the zone plate is of the order of several thousand wavelengths, while the focus can be located several millions of wavelengths away from the lens, giving a domain comparable in size to that of long-range radar propagation problems. Simplified solutions are available for certain types of zone plates, but they only give limited information on the behaviour of the field [109, 138]. Of course it is very tempting to apply parabolic equation techniques to this problem. Fortunately at X-ray frequencies most materials are almost transparent, so that it is possible to model X-ray propagation through the optical elements of thick Fresnel zone plates with parabolic equation methods quite accurately [74].

This is done by starting behind the zone plate with a given source and by propagating the field through the zone plate, setting appropriate values of the refractive index inside the plate components. Beyond the zone plate the field propagates in free space and there is no particular difficulty. Errors are incurred at the boundaries of the optical elements because of the discontinuity in refractive index, but these remain small for typical materials. Table 13.1 gives the refractive index of various materials for a wavelength of 2.4 nm.

In what follows we consider axially symmetric problems only. The appropriate form of the parabolic equation is that given in eqn. 13.1. In the examples presented here, a point source is used. A large-angle filter is applied so that the initial field is negligible for r greater than a few wavelengths [74]. As a result the transparent boundary condition given by eqn. 13.13 can be used to truncate the integration domain without incoming energy correction.

We first look at the case of an ideal thin Fresnel zone plate consisting of 300 concentric zones, alternatingly perfectly opaque and perfectly transparent,

Table 13.1 Refractive index for X-ray wavelength of 2.4 nm

Material	Refractive index
Germanium	$1 - 0.00260 + 0.00094i$
Nickel	$1 - 0.00442 + 0.00101i$

with outer radius $a = 18000$ nm. A 2.4 nm point source is located at a distance $d = 0.9$ mm from the lens. The focal length f is given by

$$f = \frac{R^2}{N\lambda} \tag{13.21}$$

where N is the number of Fresnel zones [110]. The corresponding image is at a distance d' given by

$$\frac{1}{f} = \frac{1}{d} + \frac{1}{d'} \tag{13.22}$$

For this case, $f = 0.45$ mm, so we expect to see the image at $d' = 0.9$ mm. Figure 13.5 shows field intensity contours, and indeed a sharp focus appears at 0.9 mm. Now thin Fresnel zone plate theory also predicts the existence of higher odd order foci [110], with the focal length f_k of order k given by

$$f_k = \frac{f}{k} \tag{13.23}$$

The corresponding image focal point is at a distance d'_k given by

$$\frac{k}{f} = \frac{1}{d} + \frac{1}{d'_k} \tag{13.24}$$

In Figure 13.5, the third order focus is clearly visible at a distance of 0.18 mm as predicted by the theory. The fifth order focus is weaker and cannot be distinguished on this contour plot, although it can be seen in Figure 13.9 below. Higher order foci, which come from diffraction at large angles, are not accurately represented with the narrow-angle PE code used here. We note that negative diffraction orders producing diverging energy are also present.

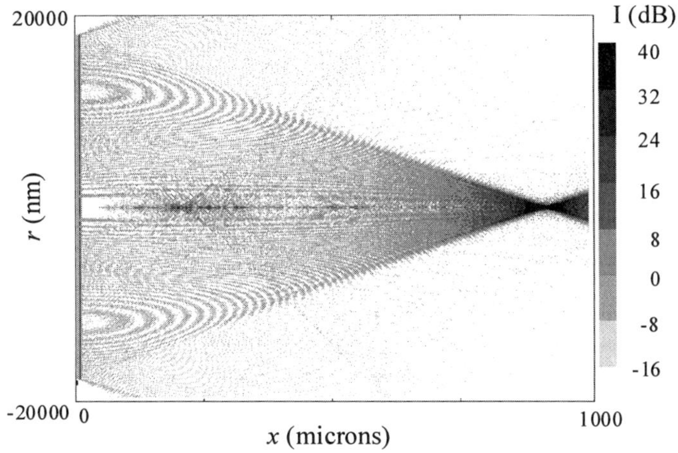

Figure 13.5 Diffraction by a thin zone plate [74]

It is quite instructive to look at an equivalent ideal thin lens with the same focal length f and outer radius a. This equivalent lens is modelled by the transmission function $T(r)$ given by

$$
\begin{aligned}
T(r) &= \exp\left(-ik\frac{r^2}{2f}\right), \quad r \leq a \\
T(r) &= 0, \quad r > a
\end{aligned}
\tag{13.25}
$$

This truncated Gaussian phase screen has support inside the circle of radius a, as required, and produces the appropriate focal effect. To see this, consider Figure 13.6.

The distance between points F and F' is $d + d'$. In the absence of the phase screen, the phase retardation ϕ_1 between F and F' is simply given by the optical path length of the direct path FF'

$$\phi_1 = k(d + d') \tag{13.26}$$

which corresponds to the geometrical path length f. When the phase screen is present, we have to add contributions from all paths FPF'. If we denote by r the radial distance of point P, the geometrical path length $l(r)$ is given by

$$l(r) = \sqrt{r^2 + d^2} + \sqrt{r^2 + d'^2} \tag{13.27}$$

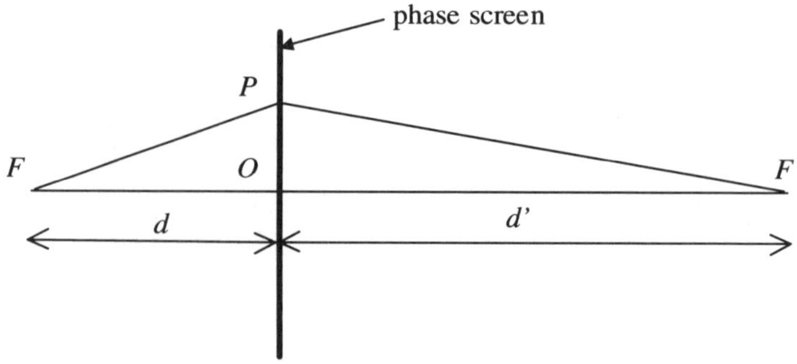

Figure 13.6 Equivalent perfect lens geometry

If d and d' are both large compared to the outer radius a, corresponding the to narrow-angle PE approximation, we have

$$l(r) \sim d + d' + \frac{r^2}{2}\left(\frac{1}{d} + \frac{1}{d'}\right) = d + d' + \frac{r^2}{2f} \tag{13.28}$$

corresponding to the phase factor

$$\phi(r) = k\left(d + d' + \frac{r^2}{2f}\right) \tag{13.29}$$

The transmission factor given by eqn. 13.25 provides exactly the right phase offset to make all the FPF' paths of equal optical length ϕ_1. To get the numerical PE solution at all points in the computational domain, the incident field is multiplied by the transmission coefficient on the transverse plane containing the equivalent lens, and then propagated in a vacuum in the usual way. Figure 13.7 shows field intensity contours for this equivalent ideal thin lens. There are no higher order foci, and the focus is sharper than for the zone plate: for the latter, zero order diffraction and out-of-focus higher order diffraction produce a background field which reduces the efficiency of the zone plate. The efficiency factor predicted by theory is $1/\pi^2$, which is close to the value of around 10% obtained with this numerical example (see Table 13.2 below).

We now look at the more realistic case of a thick Fresnel zone plate constructed from nickel elements of thickness b. We assume here that the refractive index in the transparent zone is 1, and that the nickel elements have a rectangular cross-section. More complicated situations involving for example grooved elements could be considered without difficulty. In order to optimize zone plate efficiency, the thickness b should be chosen such that the opaque

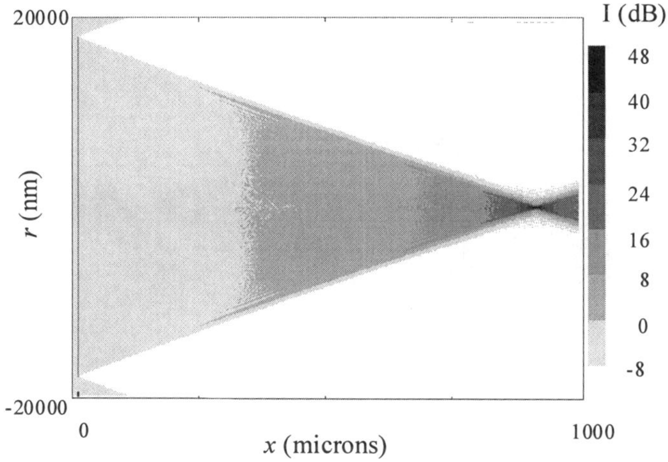

Figure 13.7 Diffraction by an ideal lens [74]

zones provide a phase shift of π. An estimate of the phase shift ϕ can be obtained by assuming plane wave propagation through the dielectric medium. Then we have

$$\phi \sim \frac{2\pi\delta b}{\lambda} \tag{13.30}$$

where λ is the wavelength in vacuum and δ is the real part of $1 - n$. A phase shift of π is obtained for

$$b \sim \frac{(l + 1/2)\lambda}{\delta} \tag{13.31}$$

where l is an integer. Perfect cancellation does not occur since the zone plate elements always have some degree of absorption at X-ray frequencies. Minimal absorption occurs for $l = 0$, or

$$b \sim \frac{\lambda}{2\delta} \tag{13.32}$$

For the parameters used here, this gives $b \sim 270$ nm. Figure 13.8 shows field intensity contours for $b = 300$ nm. We notice the first and third order foci, but also the second order focus at 0.3 mm corresponding to the second order focal length.

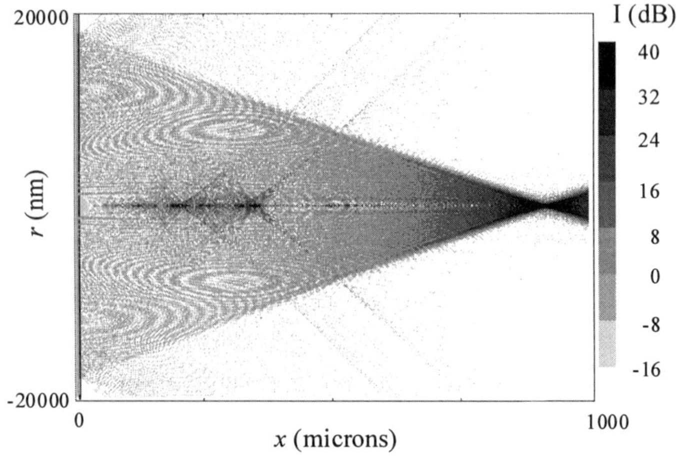

Figure 13.8 Diffraction by a thick Fresnel zone plate [74]

The efficiency of a lens is defined as the part of incoming energy flux directed into the main focus [110]. It is also the ratio of maximum intensity in the focal plane to that of a perfect lens [74]. Table 13.2 gives efficiency values for different zone plate thicknesses, the other parameters being as before. It is interesting to note that due to constructive interference of the radiation penetrating through the nickel elements, the diffraction efficiency of the thick zone plates is actually better than that of the thin zone plate. The best efficiency is obtained for the case $b = 300$ nm, for which it is close to the theoretical estimate [74]. Figure 13.9 shows the variations of normalized field amplitude along the axis of symmetry for the three cases considered. The third and fifth order foci are clearly visible for both zone plates, and the second order focus for the thick zone plate. We see that the peak amplitude is indeed higher for the thick zone plate.

13.4 General 3D case

We work in Cartesian coordinates (x, y, z) and start with the three-dimensional wave equation

$$\frac{\partial^2 \psi}{\partial x^2} + \frac{\partial^2 \psi}{\partial y^2} + \frac{\partial^2 \psi}{\partial z^2} + k^2 n^2 u = 0 \tag{13.33}$$

The three-dimensional parabolic equation is derived in a similar way to the 2D case. Choosing the positive x-direction as the paraxial direction and defining

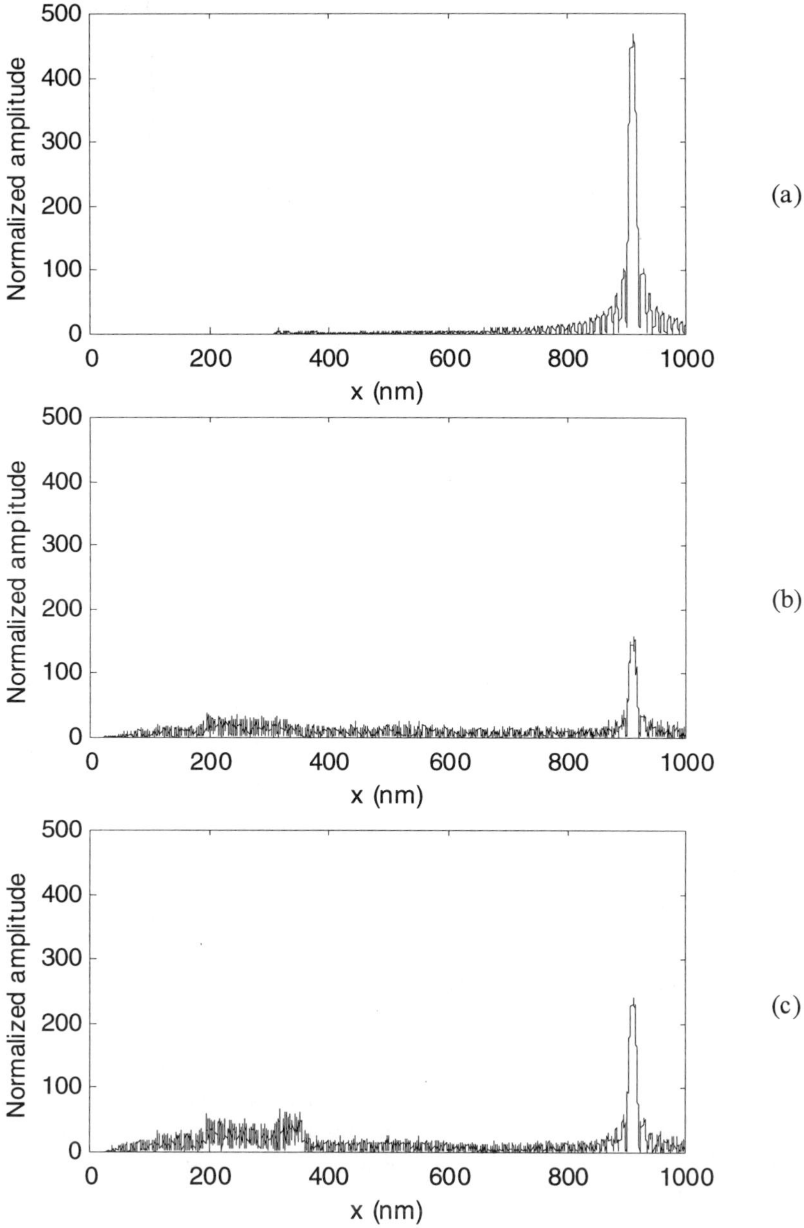

Figure 13.9 *Normalized field amplitude along axis for (a) ideal lens, (b) thin Fresnel zone plate and (c) thick Fresnel zone plate with b = 300 nm [74]*

Table 13.2 Diffraction efficiency for varying zone plane thickness

Zone plate type	Efficiency
thin zone plate	10.0%
thick zone plate, $b = 200$ nm	22.7%
thick zone plate, $b = 300$ nm	25.8%
thick zone plate, $b = 400$ nm	18.8%

the reduced field u by

$$u(x, y, z) = \mathrm{e}^{-ikx}\psi(x, y, z) \tag{13.34}$$

the 3D parabolic equation is given by

$$\frac{\partial u}{\partial x} + ik(1 - Q)u = 0 \tag{13.35}$$

where the pseudo-differential operator Q is defined by

$$Q = \sqrt{\frac{1}{k^2}\frac{\partial^2}{\partial y^2} + \frac{1}{k^2}\frac{\partial^2}{\partial z^2} + n^2} \tag{13.36}$$

As in the 2D case, the solution of eqn. 13.35 is given formally by

$$u(x + \Delta x, y, z) = \exp\left(ik\Delta x(Q - 1)\right) \cdot u(x, y, z) \tag{13.37}$$

For numerical implementation, we need a suitable approximation of the exponential operator. The situation is more complex than in two dimensions, since the operator Q now involves the two transverse variables y and z. In order to make the calculations tractable, it is desirable to decouple y and z. Following [172], we use the notations

$$Y = \frac{1}{k^2}\frac{\partial^2}{\partial y^2} \tag{13.38}$$

$$Z = \frac{1}{k^2}\frac{\partial^2}{\partial z^2} + n^2 - 1 \tag{13.39}$$

In what follows, we assume that the refractive index depends only on height z, which implies that Y and Z commute. First we separate terms in Y and Z, writing

$$e^{ik\Delta x(\sqrt{1+Y+Z}-1)} \sim e^{ik\Delta x(\sqrt{1+Y}-1)}\, e^{ik\Delta x(\sqrt{1+Z}-1)} \tag{13.40}$$

This neglects coupling between Y and Z. We can now find separate approximations for the factors in Y and Z. A further simplification is to use a first order Taylor expansion of the square-root for both operators, yielding

$$u(x+\Delta x, y, z) = e^{\frac{ik\Delta x}{2}Y}\, e^{\frac{ik\Delta x}{2}Z} \cdot u(x, y, z) \tag{13.41}$$

corresponding to the 3D narrow-angle PE

$$\frac{\partial u}{\partial x} = \frac{ik}{2}(Y+Z)u \tag{13.42}$$

Alternatively, one may keep the square-root expressions in Y and Z on the RHS of eqn. 13.40 to have wider angle modelling, or a combination of narrow and wide-angle expressions. For example if wide-angle modelling in y is required, but narrow-angle modelling in z is sufficient, one can use

$$u(x+\Delta x, y, z) = e^{ik\Delta x(\sqrt{1+Y}-1)}\, e^{\frac{ik\Delta x}{2}Z} \cdot u(x, y, z) \tag{13.43}$$

Either of eqns. 13.42 and 13.43 requires further treatment for numerical solution. For scattering applications, we shall be dealing with complicated boundaries for which Fourier transform solutions are not adequate. Instead we use finite-difference algorithms with various Padé schemes. One possibility is to use a combination of Padé-(1,1) and Padé-(2,1) schemes, as in the two-dimensional case [170, 172]. This is suitable when boundary conditions on scattering objects can be split into the y and z directions. This occurs for the homogeneous Dirichlet condition (the field is zero on the object) or the Neumann boundary condition (the normal derivative of the field is zero on the object) when the outer normals to scattering objects do not have non-zero components in y and z simultaneously. The solution is then obtained by inverting tridiagonal or pentadiagonal matrices.

To illustrate the use of Padé-(1,1)/Padé-(2,1) combinations, we apply them to acoustic scattering by a soft sphere, using the same parameters as in the axial symmetry simulation of Section 13.2.1 [95]. Figure 13.10 shows the field computed in a transverse plane located 2 m behind the sphere. For this calculation, the code was wide angle in y and narrow angle in z. This explains the flattening in the z-direction, where less interference fringes are visible than

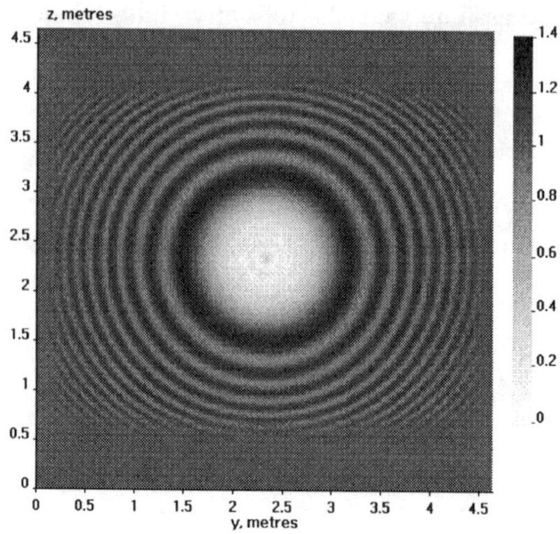

Figure 13.10 *Near-field computed with combination of Padé schemes, 2 m behind acoustically soft sphere of radius* $5\lambda = 0.5$ *m*

in the y-direction. Perfectly matched layers (see Section 8.3) were used to truncate the domain in the transverse plane.

Figure 13.11 shows a comparison of near-field results computed with the 3D Padé method and with the axial symmetry PE method, on the central vertical of the transverse plane located 2 m behind the sphere. Excellent agreement is obtained up to 1 m, which is the radius at which the non-local boundary condition was applied to truncate the axial symmetry PE. This confirms the efficiency of the axial symmetry NLBC.

Since the Padé-(1,1) and Padé-(2,1) schemes are second-order accurate (see Section 12.2), the range step can be large compared to the wavelength and quite large problems can be treated in reasonable time on a desktop computer. This type of implementation is particularly useful for building scatter applications where the scatterers have simple geometries [170, 172]. For more general boundary conditions however, the lack of L_0-stability can become a problem with Padé-(1,1) schemes, particularly for "collapsing" objects like spheres [94]. Then it is more appropriate to use a Padé-(1,0) scheme, i.e. the classical implicit solver. The main drawback is of course that this is only first-order accurate, and hence very small range steps must be used. This is not a serious problem for radar-cross-section applications with object dimensions less than a few tens of wavelengths, but computation times can become prohibitively large for other applications, for example urban propagation at millimetric wavelengths.

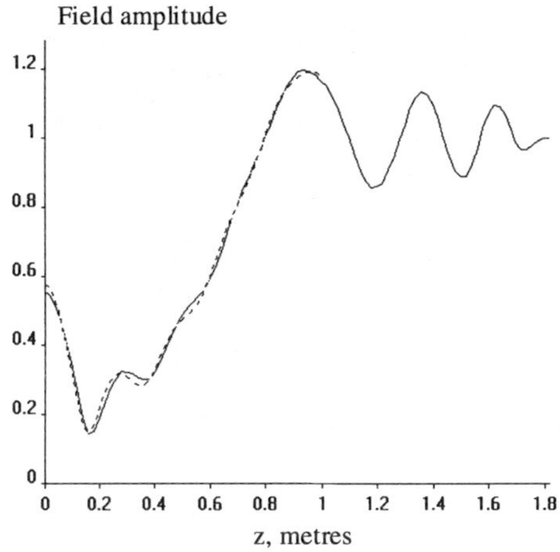

Figure 13.11 *Comparison of 3D Padé (continuous line) and axial symmetry PE (dotted line) results 2 m behind acoustically soft sphere of radius* $5\lambda = 0.5$ *m*

13.4.1 Implementation of the 3D Padé-(1,0) scheme

For the 3D narrow-angle PE of eqn. 13.42, the Padé-(1,0) scheme is given by

$$\frac{u_j^{m+1} - u_j^m}{\Delta x} = \frac{i}{2k\Delta z^2}\left(u_{i-1,j}^{m+1} + u_{i,j-1}^{m+1} - 4u_{i,j}^{m+1} + u_{i+1,j}^{m+1} + u_{i,j+1}^{m+1}\right) \qquad (13.44)$$

where Δx is the range step size and Δz is the grid spacing in the transverse plane. For simplicity, the same grid spacing has been used the y and z coordinates. The superscript index m refers to points along the x-axis, while the subscript indices i and j refer to grid points along the y and z axes respectively. Figure 13.12 shows the position of the grid points necessary to advance the scheme from range m to range $m + 1$.

Assume there are N grid points in the y direction. If the points in the transverse plane are numbered row by row, point (i, j) corresponds to index $i + N(j - 1)$. For Dirichlet boundary conditions, the resulting matrix has the structure shown in Figure 13.13, with three diagonals in the centre and two diagonals further away. Inversion of this pentadiagonal matrix can be efficiently carried out by factoring the 3D operator into 2D operators for the y and z coordinates [80].

Other types of boundary conditions on the scattering objects introduce additional non-zero entries in the matrix. In general the partial derivatives along

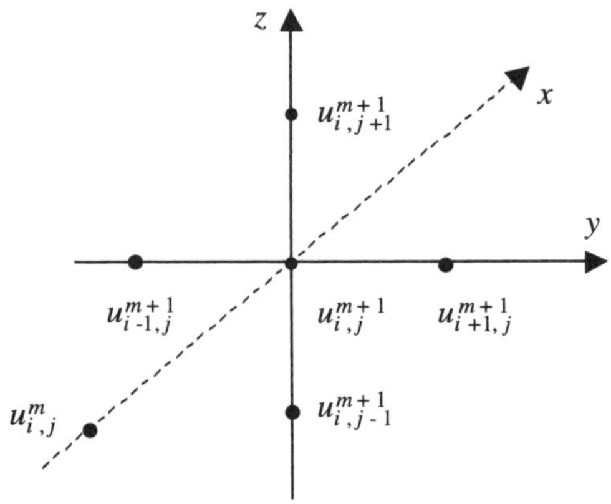

Figure 13.12 Position of grid points used for classical implicit solver

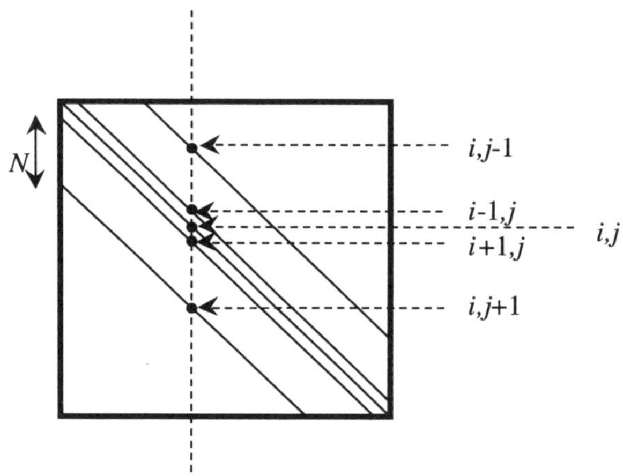

Figure 13.13 Structure of sparse matrix for Dirichlet case

the transverse coordinates are coupled and there is no obvious factorization of the matrix. A sparse matrix solver must then be used to invert the matrix. Efficient solvers are available, for example the bi-conjugate gradient method [6, 130], which works by iteration. We wish to find a solution x of the linear system

$$Ax = b \tag{13.45}$$

where A is a square matrix with complex coefficients, and x and b are complex vectors. Starting with a first guess x_0, we define the initial residuals r_0, s_0 and search directions p_0, q_0 by

$$r_0 = s_0 = p_0 = q_0 = b - Ax_0 \tag{13.46}$$

We now carry out the following inductive construction of estimates x_n

$$\begin{aligned}
\alpha_n &= \frac{< r_n, s_n >}{< Ap_n, q_n >} \\
r_{n+1} &= r_n - \alpha_n A p_n \\
s_{n+1} &= s_n - \alpha_n \overline{A}^T q_n \\
\beta_n &= \frac{< r_{n+1}, s_{n+1} >}{< r_n, s_n >} \\
p_{n+1} &= r_n + \beta_n p_n \\
q_{n+1} &= s_n + \beta_n q_n \\
x_{n+1} &= x_n + \alpha_n p_n
\end{aligned} \tag{13.47}$$

where $\overline{A}^T = (\overline{a_{ji}})$ denotes the conjugate transpose matrix of $A = (a_{ij})$ and $<,>$ denotes the complex inner product. By construction if $i \neq j$, we have the orthogonality relations

$$\begin{aligned}
< r_i, s_j > &= 0 \\
< r_i, q_j > &= 0 \\
< s_i, p_j > &= 0
\end{aligned} \tag{13.48}$$

and the biconjugacy relations

$$\begin{aligned}
< Ap_j, q_j > &= 0 \\
< p_j, Aq_j > &= 0
\end{aligned} \tag{13.49}$$

The orthogonality relations imply that the construction must stop after at most N steps, where N is the order of the matrix A. In practice, a convergence test must be applied, stopping the algorithm when the residual r_n is sufficiently small. The number of iterations depends on the accuracy of the first guess. For the PE algorithm, the use of the solution at the previous PE step as the first guess for the next step works well. The number of iterations is typically

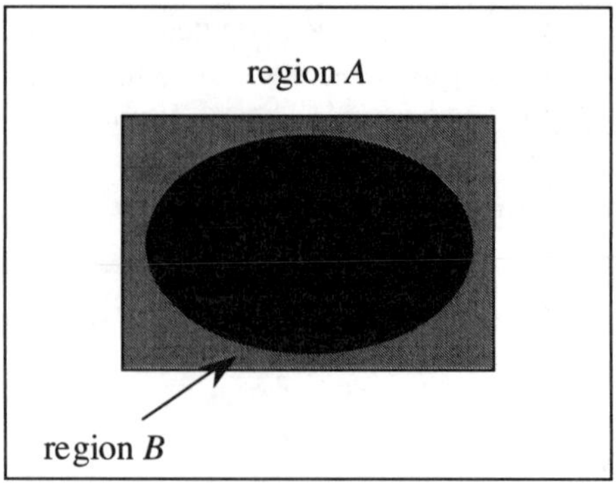

Figure 13.14 Domain decomposition into two regions: transverse plane view

between 50 and 100, depending on the complexity of the boundary conditions on the scatterers.

For many applications, the integration domain must extend in the transverse plane quite far away from the scatterer and consequently the number of operations required can become prohibitively large for 3D scattering calculations. Computation times can be substantially reduced by the double pass method described below.

13.4.2 Double pass method

We decompose the transverse domain into two regions, a small region B around the object and an outer region A, as shown in Figure 13.14. Accurate boundary conditions will be applied in region B, while an approximate method is used in region A. The first pass takes the field at range x and propagates it to range $x+\Delta x$ assuming that the object is not present at that range. Since there are no boundary conditions to satisfy inside the domain, the 3D propagation operator splits into 2-dimensional operators and matrix inversion for this pass is fast. To truncate the domain at the outer boundaries of region A, any domain truncation method can be used. Perfectly matched layers are particularly suitable for this application.

The resulting field is taken as the solution at range $x + \Delta x$ in region A. We now recompute the field in region B, this time accounting for the object. However we limit the integration domain to region B only, by using the first pass values as boundary values at the border of region B, as shown in Figure 13.15. Of course this method introduces an additional source of error since the values at the outer boundaries of B are approximate. Bearing in mind the paraxial nature of the PE, this additional error is negligible for an appropriate choice of region A.

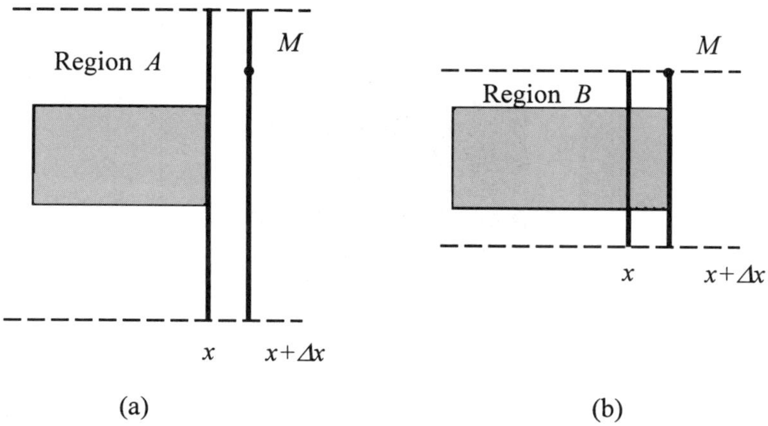

Figure 13.15 *Double pass method for 3D simulations, (a) first pass in region A ignoring object and (b) second pass in region B with object*

For far-field applications, the size of region A is governed by the necessity to obtain meaningful scattering patterns in the angular sector of interest. Typically region A is taken to be two to three times the maximum diameter of the scattering object in the transverse plane.

13.5 Rough surface modelling

Certain types of rough surfaces can be modelled explicitly with 3D scalar PE codes by using facet representation [95, 170, 171]. We consider surfaces constructed from rectangular blocks as shown in Figure 13.16. Boundary conditions are applied to each facet, neglecting the side areas. The roughness of the surface is parametrized by the position of outer facets along the normal. Random surfaces are constructed by taking a realization of the required probability distribution for this parameter.

Figure 13.17 shows the field scattered by a smooth perfectly conducting cube of side 1 m, under incidence from a vertically polarized plane wave of wavelength 10 cm and unit amplitude, propagating in the positive x-direction. This simulation represents the forward scattered field, as it just leaves the cube. Figure 13.18 shows the scattered field for a rough cube, modelled by square plates of side 10 cm. The plate offsets have Gaussian distribution with variance 1.5 cm. The perturbations due to the rough surface are clearly visible, causing ripples on the field contours. However these perturbations quickly die out as the field propagates forward, and they have no effect on the forward bistatic radar-cross-section.

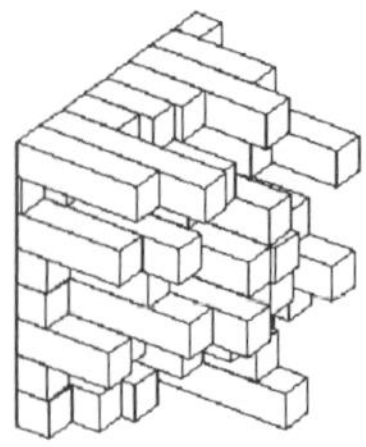

Figure 13.16 Explicit rough surface modelling using rectangular blocks

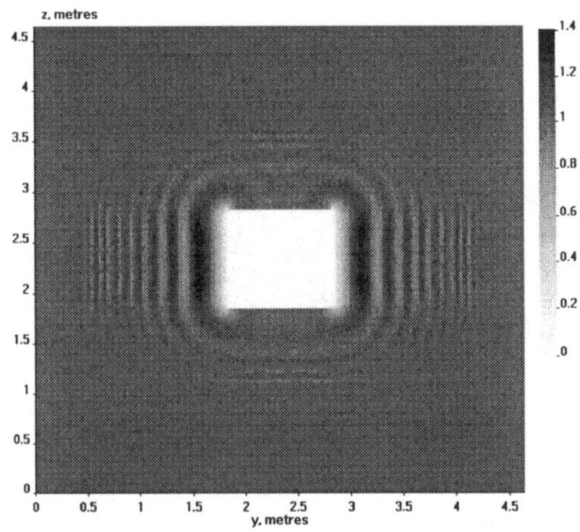

Figure 13.17 Forward scatter for smooth cube of side 1 m, 2 m behind cube; the incident wavelength is 10 cm

The situation is very different for backscatter, as illustrated in Figure 13.19, which shows the backscattered field 10 m from the same cube, along the central vertical. Backscattered energy in the specular direction is greatly diminished by surface roughness and the symmetry of the scattering pattern is lost in the rough case.

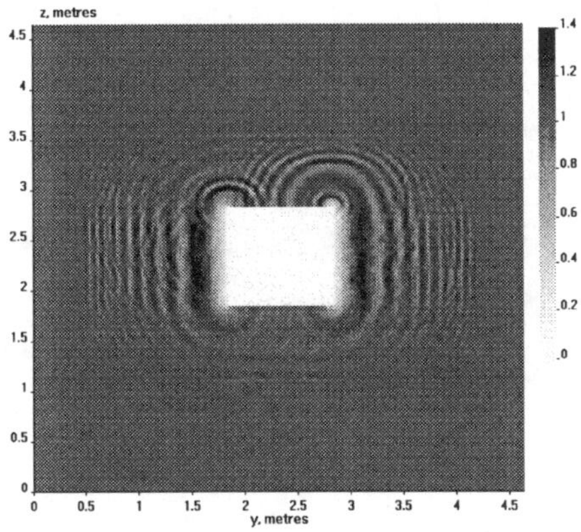

Figure 13.18 Forward scatter by rough cube of side 1 m, 2 m behind cube; the rms roughness is 1.5 cm

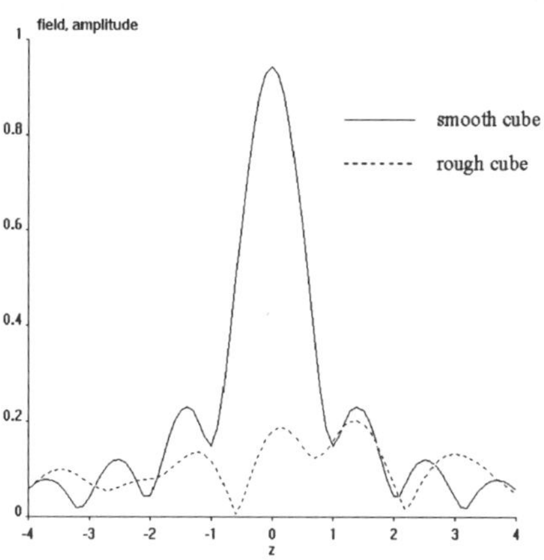

Figure 13.19 Backscatter from smooth and rough cubes, showing values on the central vertical 10 m from the cube

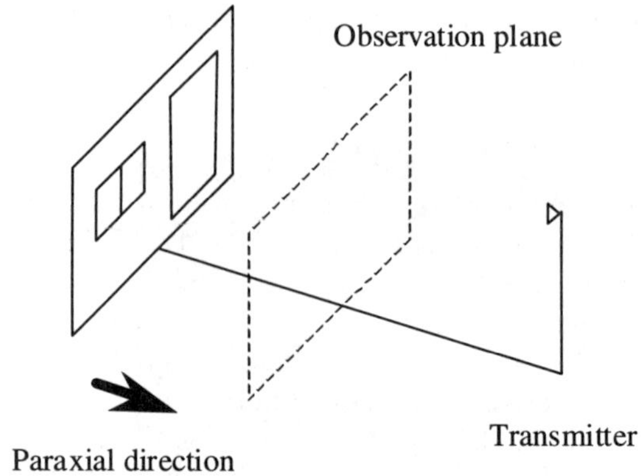

Figure 13.20 *Geometry for building scatter simulations under normal incidence*

13.6 Building scatter

13.6.1 Normal incidence

We first consider the case of normal incidence on a rectangular building, as shown in Figure 13.20. The presence of the ground is ignored at this stage. The building scatter problem is particularly simple in this case: the calculation starts with initial values given by the reflected field on the building surface, and this initial scattered field is just propagated in a vacuum, using as the paraxial direction the normal to the building wall, which is also the specular direction. The reflected field is obtained from the incident field by applying appropriate boundary conditions on the building surface. Detailed building features can be modelled in this way, using variable reflection coefficients. The source for this example is a vertically polarized omnidirectional 3 GHz antenna, located 50 m from the building and 3 m above the ground. The rectangular wall has width 4 m and height 3 m.

Figure 13.21 shows the effects of simple building features on the scattered field. Contour plots of the field amplitude on the building surface and on a transverse plane located 10 metres away from the wall are given. The values are normalized by the incident field at the centre of the wall. Three cases are considered: a brick wall with no features, a brick wall with window and door represented by areas of zero reflection coefficient, and a brick wall with framed window and door. The reflection coefficient for brick is taken to be 0.3, while the frames are perfectly conducting. As expected the unframed door and window introduce large gaps in the scattered field and reduce significantly the overall scattered energy. By contrast the presence of metallic frames modifies the scattering pattern, but causes strong reflections which compensate for the gaps of zero reflection coefficient.

13.6.2 Oblique incidence

The case of oblique incidence on a building wall is more challenging, and a combination of Padé schemes is required [170, 172]. The geometry for this case is shown in Figure 13.22. The natural way to solve the problem is again to choose the specular direction as the paraxial direction, as shown in Figure 13.23. However in this case the building wall acts as a boundary to the integration domain, whereas in the case of normal incidence the wall was taken into account simply by modifying the initial field.

For this geometry, the building is represented through boundary conditions as the solution is marched along the direction of specular reflection, using the appropriate reflection coefficients to compute the value of the reflected field on the wall. It is easy to implement this type of boundary condition with a Padé-(1,1) scheme. However we have seen that boundary discontinuities tend to cause propagating ripples with this scheme. Fortunately once the field has left the building, there is no difficulty using the Padé-(2,1) scheme, which quickly eliminates the unwanted ripples. The boundaries of the integration domain are shown as a dotted line in Figure 13.23, which displays a plane view of the oblique incidence simulation set-up: integration starts at the bottom left-hand corner A of the building, using the Padé-(1,1) scheme until the top right-hand corner B of the building is reached. During this phase, the upper boundary of the domain is the building wall AB. After B, the integration domain is increased upwards, and the Padé-(2,1) scheme is used.

Figure 13.24 shows the scattered electric field computed along the line with coordinates (in metres): $x = 0$, $z = 0.5$, $y = \pm 4$. Results are shown for the Padé-(1,1) scheme, the combination Padé-(1,1)/Padé-(2,1) scheme and a Physical Optics code. For clarity the Padé-(1,1) solution is offset by 5 dB. The results are normalized by the free-space path loss 60 m from the transmitter. There is excellent agreement between the Physical Optics and hybrid Padé-(1,1)/Padé-(2,1) solutions, while the pure Padé-(1,1) solution is extremely noisy.

13.6.3 Millimetre-wave building scatter

Measurements were conducted to assess the magnitude of building scatter effects at 38 GHz, using an isolated building of width 16 m and height 6.2 m [142]. The experimental set-up is shown in Figure 13.25. The transmitter was located 30 m from the building, at a height of 1.7 m above the ground. The transmitter was at the same height and was moved along a circular arc of radius 30 m. The measurements were compared to PE simulations [170]. The measured radiation pattern of the transmitter was used to calibrate the

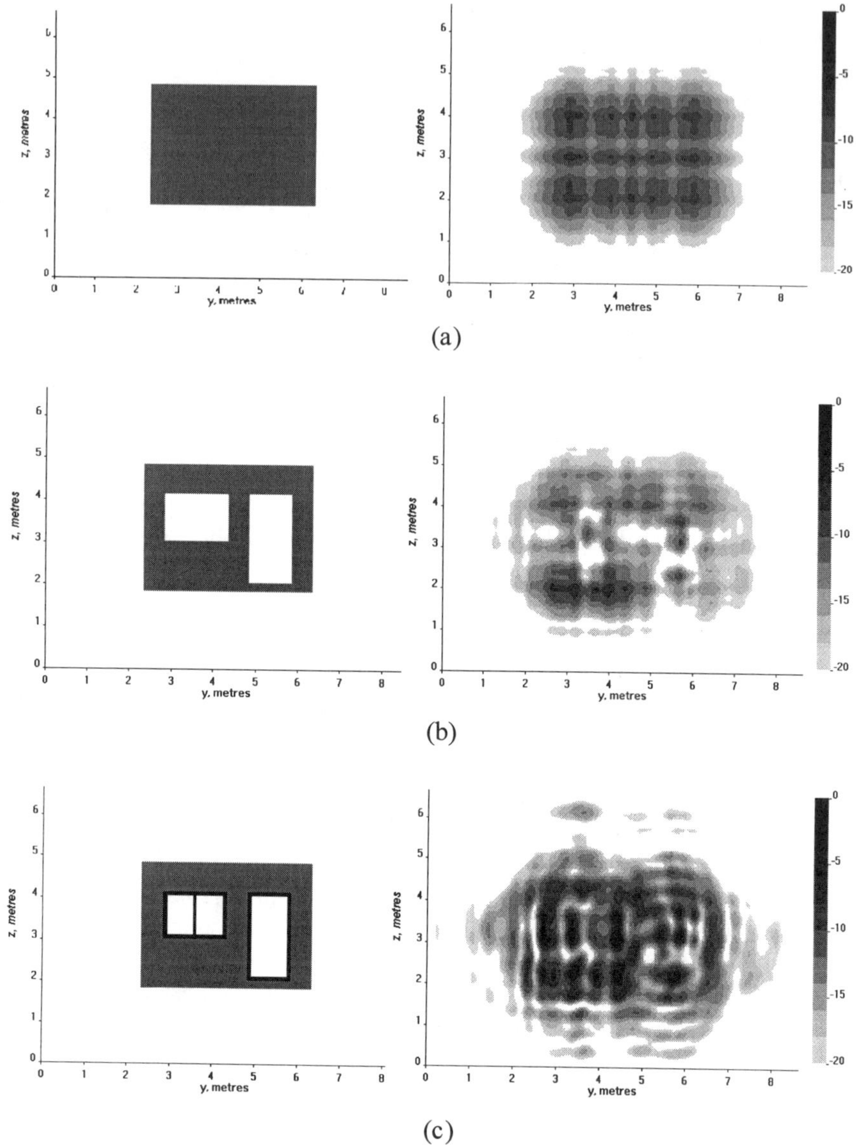

Figure 13.21 *Effect of building features under 3 GHz normal incidence, showing scattered field amplitude on the building surface (left column) and 10 m from the wall (right column) for (a) plain brick wall (b) unframed window and door and (c) framed window and door*

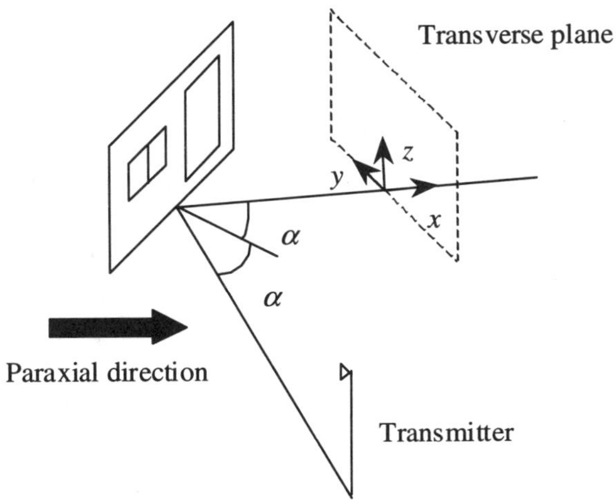

Figure 13.22 *Geometry for building scatter under oblique incidence*

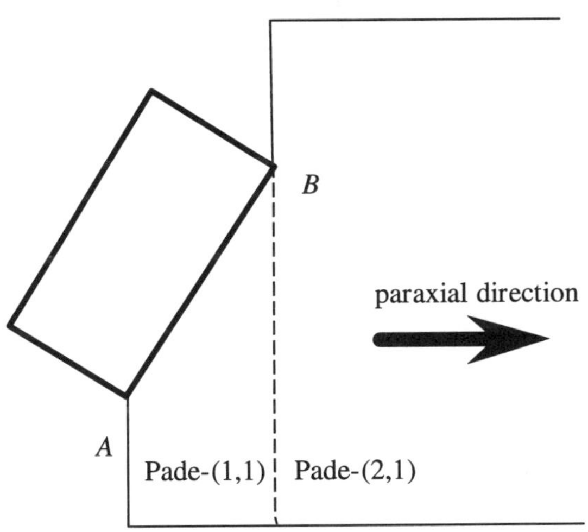

Figure 13.23 *Plan view showing use of Padé schemes for simulation of building scatter under oblique incidence*

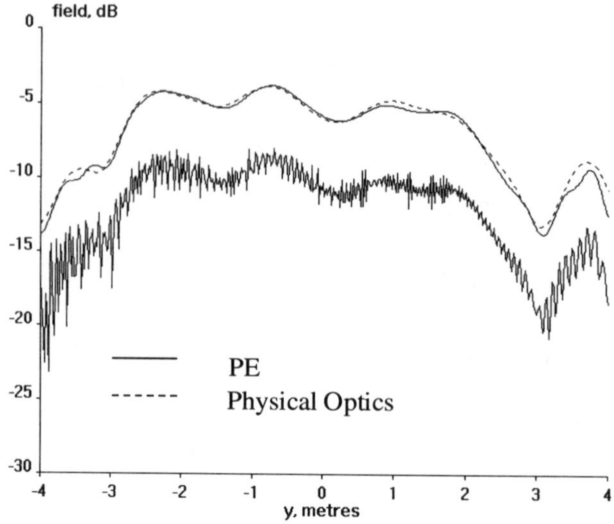

Figure 13.24 *Comparison of PE results and Physical Optics for oblique incidence*

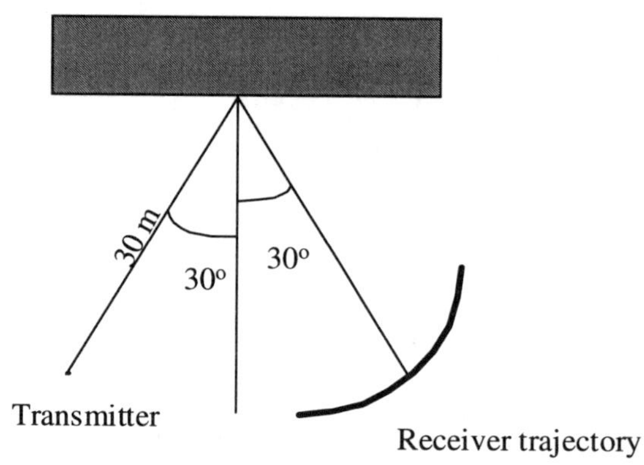

Figure 13.25 *Plan view of experimental set-up for 38 GHz building scatter measurements*

incident field. Figure 13.26 shows the reflected field on the building surface, assuming a smooth surface with reflection coefficients 0.3 for the bricks, 0 for the glass and 1 for the window frames.

Figure 13.27(a) shows the calculated and measured scattered fields for the smooth surface case. The PE results overestimate the specular peak, and do

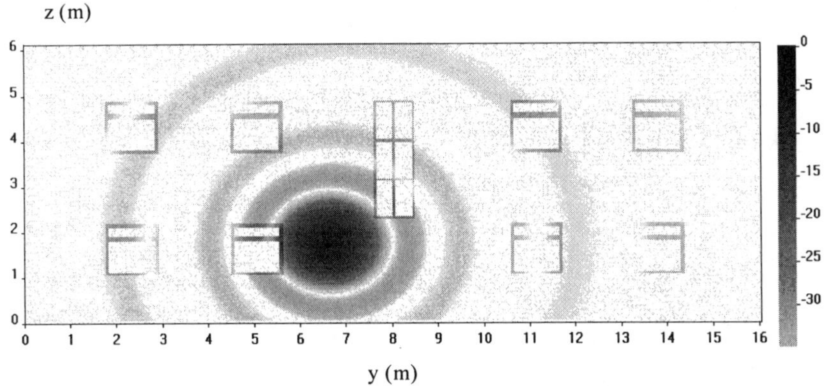

Figure 13.26 *PE simulation of reflected field on building wall, 38 GHz incidence*

not display the multiple lobing pattern of the measured data inside the specular peak. The reason for this is that the brick wall is significantly rough under 38 GHz incidence: the offsets of individual bricks are of the order of the 4 mm wavelength.

A more realistic simulation was carried out by modelling each individual brick as a rectangle of 17 cm by 7 cm, with random offsets having uniform distribution and a standard deviation of 1.5 mm. The results are shown in Figure 13.27(b). The magnitude of the specular peak is now in good agreement with the measured value, and the field exhibits multiple lobing, although this is not as marked as the measured variations. We conclude that effective reflection coefficients for large samples of building materials can be computed quite accurately using 3D scalar PE methods.

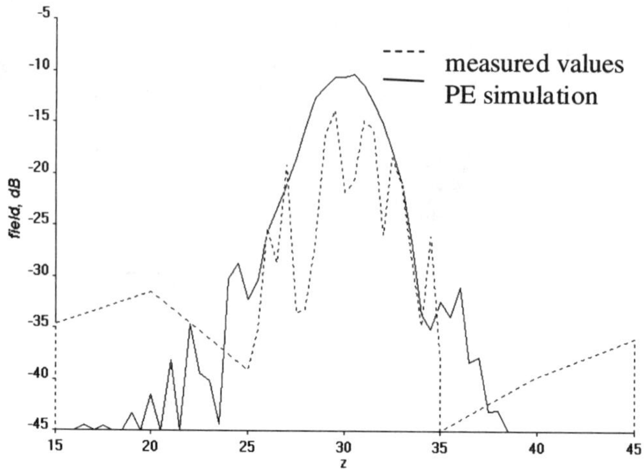

(a)

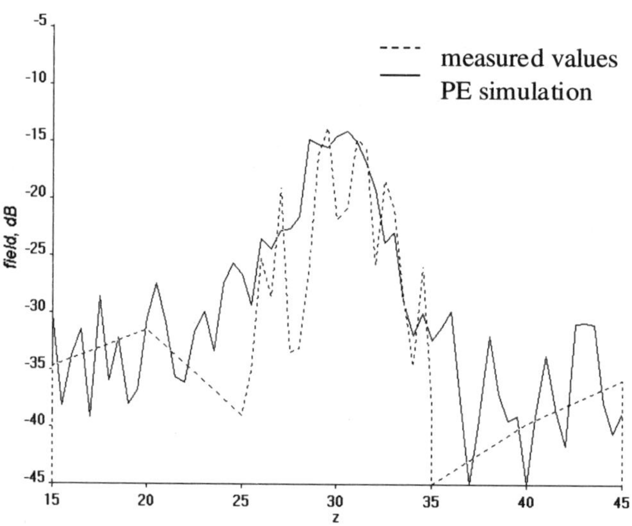

(b)

Figure 13.27 *Comparison of 38 GHz measurements and PE simulations for (a) smooth and (b) rough wall*

Chapter 14

Vector PE

14.1 Introduction

A vector version of the parabolic equation method is required to treat general three-dimensional electromagnetic problems. The vector PE is obtained by coupling component scalar parabolic equations via suitable boundary conditions on the scatterers [173, 174]. This allows accurate treatment of polarization effects within the paraxial constraints. The vector PE derivation is given in Section 14.2. Implementation aspects are briefly discussed in Section 14.3, and examples are given in Section 14.4.

In what follows, we consider scattering objects embedded in a vacuum, a situation which is the most common for bistatic RCS applications.

14.2 Vector PE framework

We assume harmonic time dependence in $\exp(-i\omega t)$ where ω is the angular frequency. Outside the scatterers, the electric and magnetic fields $\mathbf{E}$ and $\mathbf{H}$ satisfy the Maxwell curl and divergence equations

$$\nabla \times \mathbf{E} = i\omega\mu_0\mathbf{H} \tag{14.1}$$

$$\nabla \times \mathbf{H} = -i\omega\epsilon_0\mathbf{E} \tag{14.2}$$

$$\nabla \cdot \mathbf{E} = 0 \tag{14.3}$$

$$\nabla \cdot \mathbf{H} = 0 \tag{14.4}$$

where ϵ_0, μ_0 are the electric and magnetic permittivities of a vacuum. Denoting by C the velocity of light in a vacuum and k the wavenumber in a vacuum, we recall that these quantities are related by

$$\omega = kC = \frac{k}{\sqrt{\mu_0 \epsilon_0}} \tag{14.5}$$

Taking the curls of the first two equations, we obtain vector wave equations for $\mathbf{E}$ and $\mathbf{H}$

$$\nabla^2 \mathbf{E} + k^2 \mathbf{E} = 0 \tag{14.6}$$

$$\nabla^2 \mathbf{H} + k^2 \mathbf{H} = 0 \tag{14.7}$$

We solve in terms of the electric field $\mathbf{E}$, bearing in mind that the magnetic field can be obtained from the curl equation, eqn. 14.2 if required. We denote by $\mathbf{E}^i$ and $\mathbf{H}^i$ the fields that would exist in the absence of the scattering objects, and by $\mathbf{E}^s$ and $\mathbf{H}^s$ the scattered fields, which correspond to the perturbation caused by the scatterers:

$$\mathbf{E}^s = \mathbf{E} - \mathbf{E}^i \tag{14.8}$$

$$\mathbf{H}^s = \mathbf{H} - \mathbf{H}^i \tag{14.9}$$

We note that since $\mathbf{E}^i$ and $\mathbf{H}^i$ satisfy Maxwell's equations everywhere, the scattered fields satisfy them in the domain outside the scatterers. Working in Cartesian coordinates (x, y, z), we have three scalar wave equations for the components E^s_x, E^s_y, E^s_z of $\mathbf{E}^s$. As in the scalar case, we now choose x as the paraxial direction and define the reduced scattered field $\mathbf{u}$ in the x-direction by

$$\mathbf{u}^s = e^{-ikx} \mathbf{E}^s \tag{14.10}$$

We now carry out the usual PE derivation for each component wave equation, getting three scalar parabolic equations for the components (u^s_x, u^s_y, u^s_z) of $\mathbf{u}^s$. With the standard PE approximation, we obtain

$$\begin{cases} \frac{\partial^2 u_x^s}{\partial y^2}(x,y,z) + \frac{\partial^2 u_x^s}{\partial z^2}(x,y,z) + 2ik\frac{\partial u_x^s}{\partial x}(x,y,z) &= 0 \\ \frac{\partial^2 u_y^s}{\partial y^2}(x,y,z) + \frac{\partial^2 u_y^s}{\partial z^2}(x,y,z) + 2ik\frac{\partial u_y^s}{\partial x}(x,y,z) &= 0 \\ \frac{\partial^2 u_z^s}{\partial y^2}(x,y,z) + \frac{\partial^2 u_z^s}{\partial z^2}(x,y,z) + 2ik\frac{\partial u_z^s}{\partial x}(x,y,z) &= 0 \end{cases} \quad (14.11)$$

These must now be coupled in order to enforce boundary conditions on the scattering objects, and also the divergence free condition of eqn. 14.4. We first consider the case of perfectly conducting scatterers. Then continuity conditions state that the tangential electric field must be zero on the surface of the scattering objects, or equivalently that the electric field must be parallel to the normal to the surface. In terms of $\mathbf{E}^s$, this yields the non-homogeneous vector equation

$$\mathbf{n}(P) \times \mathbf{E}^s(P) = -\mathbf{n}(P) \times \mathbf{E}^i(P) \quad (14.12)$$

where P is a point on the surface of one of the scattering objects and $\mathbf{n}(P) = (n_x, n_y, n_z)$ is the outer normal to the surface at P. In terms of the reduced scattered field $\mathbf{u}^s$, we obtain the system of equations

$$\begin{cases} n_z u_y^s(P) - n_y u_z^s(P) &= -e^{-ikx}\left(n_z E_y^i(P) - n_y E_z^i(P)\right) \\ n_x u_z^s(P) - n_z u_x^s(P) &= -e^{-ikx}\left(n_x E_z^i(P) - n_z E_x^i(P)\right) \\ n_y u_x^s(P) - n_x u_y^s(P) &= -e^{-ikx}\left(n_y E_x^i(P) - n_x E_y^i(P)\right) \end{cases} \quad (14.13)$$

In the two-dimensional case, fields are automatically divergence-free since the problem is solved for E_y^s or H_y^s, which do not depend on the transverse coordinate y. This is no longer true for three-dimensional problems, and we must make sure that our solution is indeed divergence-free. Using the standard parabolic equation to estimate the range derivative, we obtain the divergence-free condition for the reduced scattered field in terms of transverse coordinates only in the form

$$\frac{i}{2k}\left(\frac{\partial^2 u_y^s}{\partial z^2} + \frac{\partial^2 u_z^s}{\partial z^2}\right) + iku_x^s + \frac{\partial u_y^s}{\partial z} + \frac{\partial u_z^s}{\partial z} = 0 \quad (14.14)$$

Using energy-conservation arguments, one can show that if the PE divergence-free condition of eqn. 14.14 is enforced on the boundary, then it is satisfied everywhere. The proof is given in Appendix D.

Hence for points P on the boundary of the scattering objects, we add to system 14.13 the divergence equation

$$\frac{i}{2k}\left(\frac{\partial^2 u_y^s}{\partial z^2}(P) + \frac{\partial^2 u_z^s}{\partial z^2}(P)\right) + iku_x^s(P) + \frac{\partial u_y^s}{\partial z}(P) + \frac{\partial u_z^s}{\partial z}(P) = 0 \qquad (14.15)$$

In fact this last equation is necessary to obtain a well-determined system of equations, since the linear system given by eqn. 14.13 is of rank 2 only.

More general boundary conditions on the scattering objects can be treated in a similar way. Assume that the Leontovich boundary conditions hold [82, 141]. Then the boundary conditions are given locally in terms of surface impedance, in the form

$$\mathbf{n} \times \mathbf{E}(P) = Z(P)\,\mathbf{n} \times \{\mathbf{n} \times \mathbf{H}(P)\} \qquad (14.16)$$

where $Z(P)$ is the impedance of the object at point P. The impedance Z is defined as

$$Z = \sqrt{\frac{\mu}{\eta}} \qquad (14.17)$$

where μ is the magnetic permittivity and η the complex electric permittivity of the object. Writing

$$\eta = \epsilon + i\frac{\sigma}{\omega} \qquad (14.18)$$

where ϵ is the electric permittivity and σ the conductivity, we have

$$\frac{Z}{i\omega\mu} = \frac{1}{ik}\sqrt{\frac{\mu}{\mu_0}}\sqrt{\frac{\epsilon_0}{\eta}} \qquad (14.19)$$

Assuming the object has the magnetic permittivity of a vacuum, we get

$$\frac{Z}{i\omega\mu} = \frac{1}{ik\sqrt{\eta_r}} \qquad (14.20)$$

where

$$\eta_r = \frac{\eta}{\epsilon_0} = \frac{\epsilon}{\epsilon_0} + i\frac{\sigma}{\epsilon_0 \omega} \tag{14.21}$$

is the relative complex electric permittivity. Using the curl equation, eqn. 14.2, we obtain an equation in terms of **E** only in the form

$$\mathbf{n} \times \mathbf{E} = \frac{1}{ik\sqrt{\eta_r}} \mathbf{n} \times \{\mathbf{n} \times (\nabla \times \mathbf{E})\} \tag{14.22}$$

Simplifying the double outer product on the RHS, we get

$$\mathbf{n} \times \mathbf{E} = \frac{1}{ik\sqrt{\eta_r}} \{\mathbf{n} \cdot (\nabla \times \mathbf{E})\, \mathbf{n} - \nabla \times \mathbf{E}\} \tag{14.23}$$

It remains to eliminate the range-derivatives in the curl using the SPE formulation to obtain a system of equations in terms of transverse coordinates only. As before, we need the divergence-free condition to obtain a system of rank 3. The resulting algorithm is quite flexible since the surface impedance may vary along the scattering objects.

14.3 Implementation aspects

Numerical implementation is a straightforward generalization of the scalar case. As the scalar components are coupled through non-separable boundary conditions, each step of the algorithm requires the inversion of a sparse matrix. A biconjugate gradient method like that presented in Section 13.4.1 may be used. Computational requirements for the vector case are naturally even more demanding than for the scalar case. The double pass method of Section 13.4.2 addresses this problem, by limiting the sparse matrix solver to a small transverse region enclosing the scatterer.

As in the scalar case, we may rotate the paraxial direction to solve for the scattered field in the desired angular sector. Bistatic radar-cross-section results are obtained via the near-field/far-field relationships of Appendix B. The field in a transverse plane beyond the object is used to retrieve the horizontal and vertical plane patterns. Complete patterns are obtained by rotating the object around the appropriate axis.

14.4 Examples

14.4.1 Bistatic scattering geometry

In the examples that follow, we assume that the incident electric field is parallel to the z-axis. Horizontal and vertical plane patterns then correspond to dif-

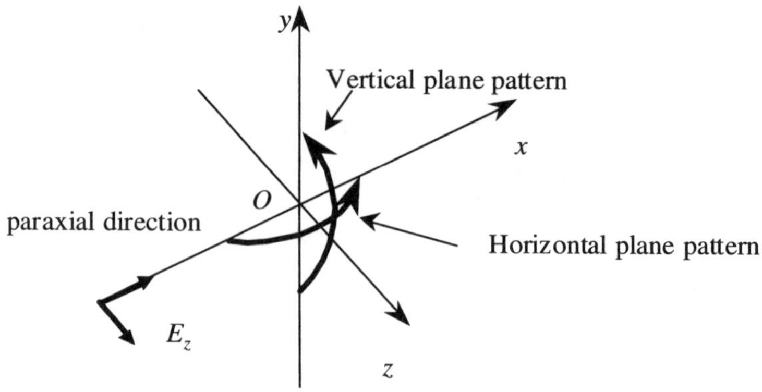

Figure 14.1 Scattering geometry for horizontally polarized incident wave

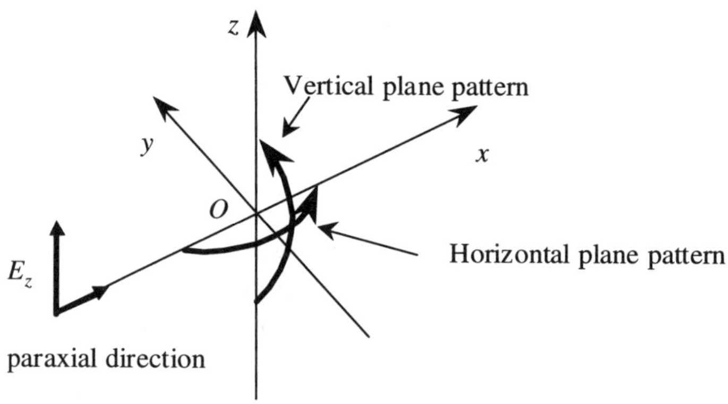

Figure 14.2 Scattering geometry for vertically polarized incident wave

ferent coordinate axes depending on the polarization of the incident wave. For a horizontally polarized incident wave, the vertical plane pattern is retrieved from results in the y-direction of the transverse plane, and the horizontal plane pattern is retrieved from results in the z-direction, as shown in Figure 14.1. This is reversed for vertical polarization, as shown in Figure 14.2.

14.4.2 Circular cylinder

A comparison with scalar PE results for a two-dimensional case provides an interesting validation of the vector PE. We consider a vertically polarized plane wave of wavelength λ incident on a perfectly conducting cylinder of radius 5λ. The geometry is shown in Figure 14.3. We can treat this as a scalar problem

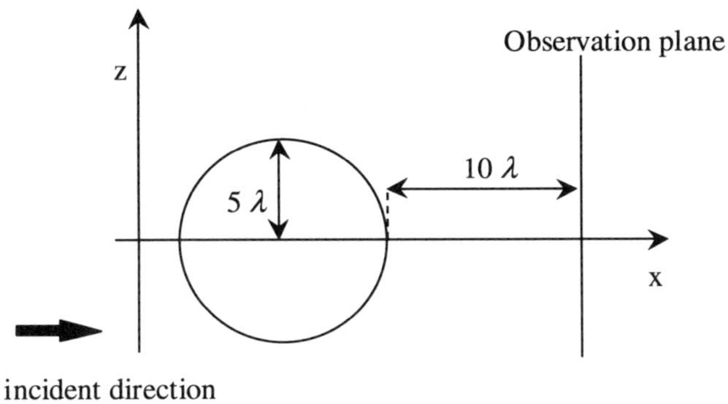

Figure 14.3 Plane wave incident on perfectly conducting cylinder

by solving for the magnetic field component H_y, or as a vector problem by solving for the electric field components E_x, E_z.

Figure 14.4 presents near-field forward scatter results in the transverse plane located 10 wavelengths behind the cylinder. Scalar PE results for H_y were obtained with a standard PE code, and then differentiated numerically to calculate E_x and E_z from the curl equation, eqn. 14.2. The vector PE results for this case do not require a full 3D simulation since the fields do not depend on y. Sparse matrix inversion in the whole domain is then tractable, and a single pass method was used. Scalar and vector PE results are almost identical. However the PE results differ substantially from the theoretical solution based on a Hankel and cosine function expansion [24].

This stems from the fact that a narrow-angle PE code has been used. The near-field results do not represent wide-angle diffraction correctly. For example interference between the fields diffracted by the top and bottom of the cylinder is inaccurately modelled. This however does not affect far-field results. Figure 14.5 shows bistatic RCS results obtained from theory, and forward scatter scalar and vector PE calculations for a single forward scatter run. The normalization assumes a wavelength of 1 m. Results are shown for angles up to 45° from the paraxial direction. Agreement is good up to 25° or so, which confirms that the near-field results contain the correct narrow-angle components.

14.4.3 Spheres

Perfectly conducting spheres provide a good test of scattering algorithms, as the theoretical solution given by Mie expansions [24] is available for comparison. We assume the incident field is a vertically polarized plane wave propagating in the positive x direction. The incident field is then parallel to the

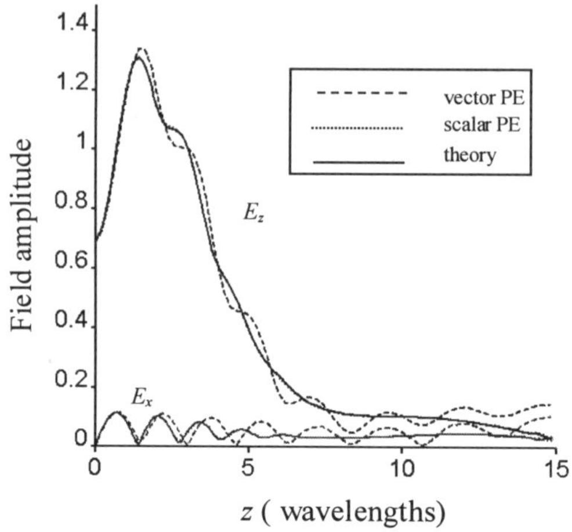

Figure 14.4 Near-field results for perfectly conducting cylinder of radius 5λ

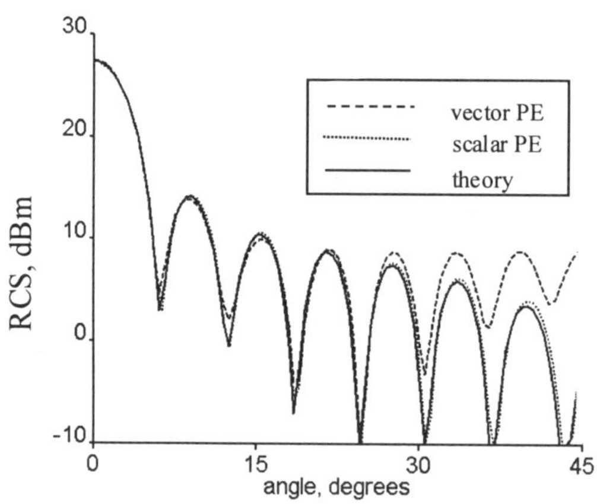

Figure 14.5 Bistatic RCS results for perfectly conducting cylinder of radius 5λ

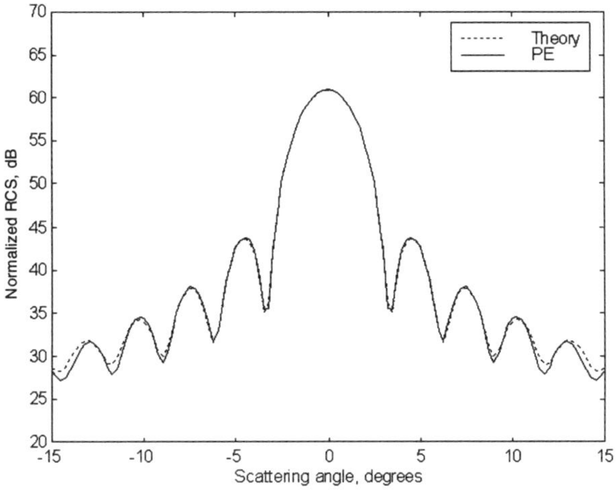

Figure 14.6 *Normalized total bistatic RCS for single forward scatter run, perfectly conducting sphere of radius* 10λ*, horizontal plane pattern*

z-axis, with

$$E_z = e^{ikx} \tag{14.24}$$

We look at vector PE results for perfectly conducting spheres of varying radius. Figures 14.6 and 14.7 show horizontal and vertical plane patterns for a sphere of radius 10λ obtained with a single forward scatter run. Agreement with the theory is very good within the accuracy range of the narrow-angle PE. Figures 14.8, 14.9 and 14.10 show the complete horizontal and vertical plane patterns for perfectly conducting spheres of radius 5λ, λ and $\lambda/4$ respectively. Each pattern has been obtained with seven rotating PE runs covering angular sectors of 30° each.

Results are in excellent agreement with the theory for the sphere of radius 5λ. When the radius becomes comparable to the wavelength, the PE approximation loses accuracy, as expected since it cannot represent surface waves propagating all around the object. This is not a narrow-angle effect, but a fundamental limitation of the paraxial approximation. Results are still acceptable for a sphere of radius λ, but the method breaks down quite dramatically for a sphere of radius $\lambda/4$.

The vector PE method has been applied to finitely conducting objects [171]. A Mie series solution is available for the case of a sphere with a surface

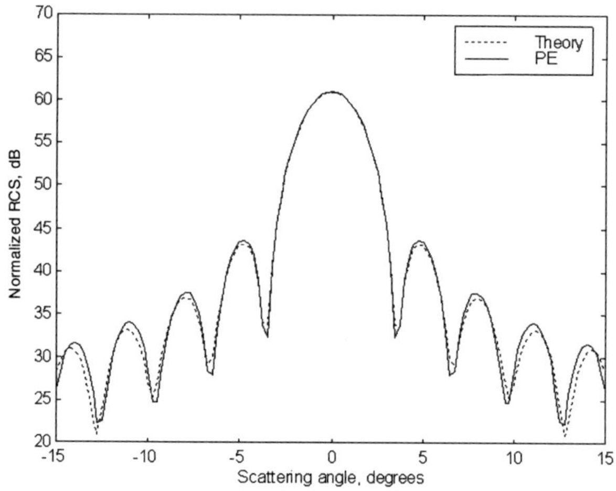

Figure 14.7 *Normalized total bistatic RCS for single forward scatter run, perfectly conducting sphere of radius* 10λ*, vertical plane pattern*

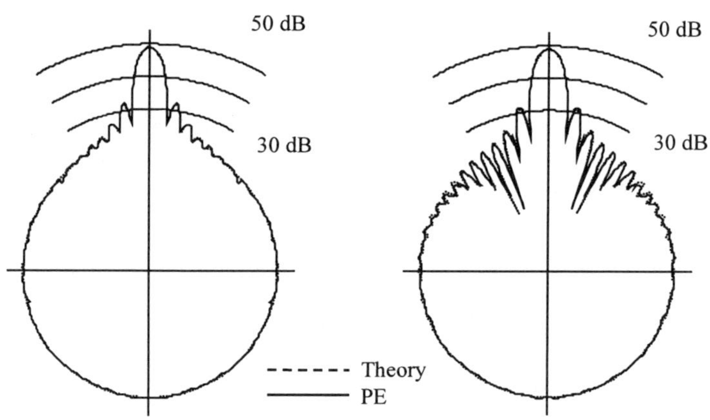

(a) (b)

Figure 14.8 *Normalized bistatic RCS, perfectly conducting sphere of radius* 5λ*, (a) horizontal and (b) vertical plane patterns*

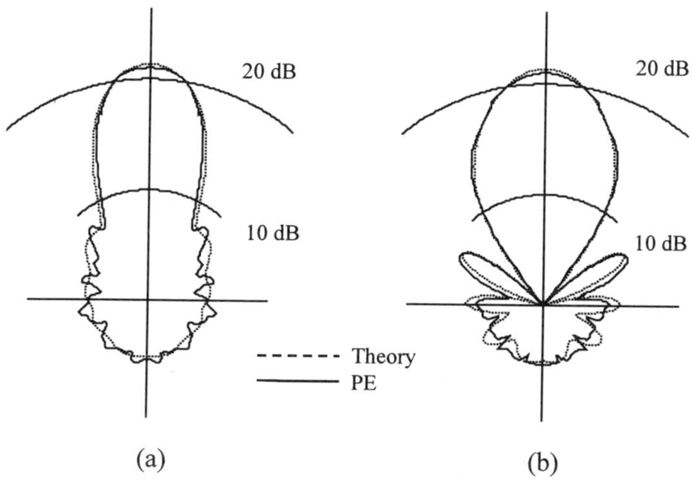

Figure 14.9 *Normalized bistatic RCS, perfectly conducting sphere of radius λ, (a) horizontal and (b) vertical plane patterns*

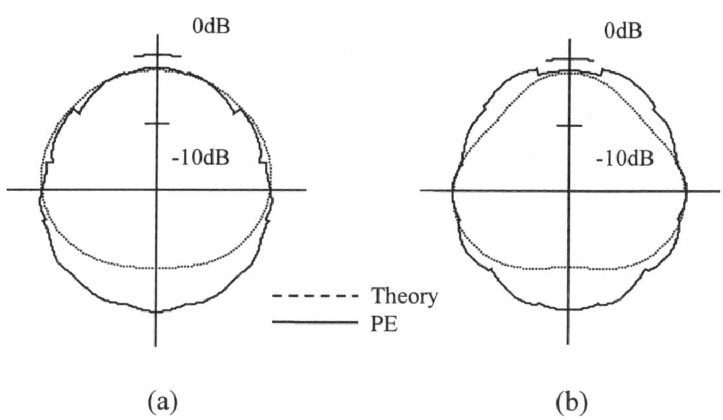

Figure 14.10 *Normalized bistatic RCS, perfectly conducting sphere of radius $\lambda/4$, horizontal and vertical plane patterns*

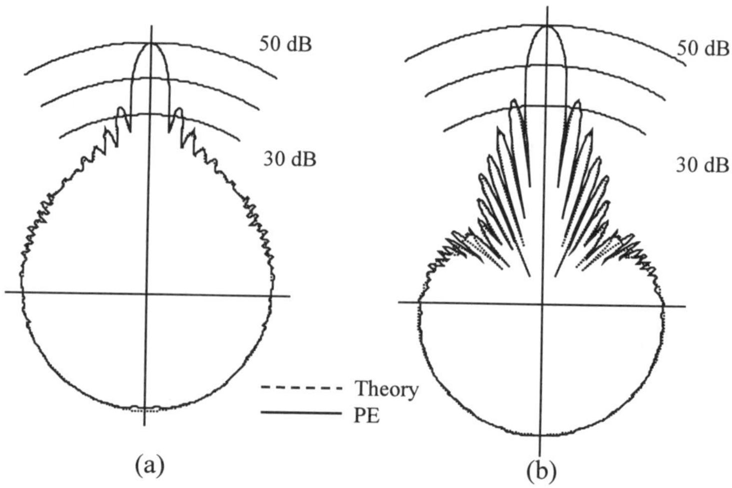

Figure 14.11 *Normalized bistatic RCS, finitely conducting sphere of radius 5λ ($n = 2+0.04i$), (a) horizontal and (b) vertical plane patterns*

impedance boundary condition [141]. We note that this solution differs from that obtained with the exact boundary conditions, involving continuity of the tangential fields at the surface. This is the appropriate reference to use, since our purpose is to test the accuracy of the vector PE method rather than that of the Leontovich boundary conditions.

We consider a sphere of radius 5λ, with relative permittivity $\epsilon_r = 3.84$ and conductivity $\sigma = 0.027$ S/m. The incident wave has a frequency of 0.3 GHz, corresponding to a wavelength of 1 metre. At that frequency, the refractive index inside the sphere is equal to $2 + 0.04i$. Figure 14.11 shows the horizontal and vertical plane patterns for this case. They are substantially different from those obtained for a perfectly conducting sphere of the same radius (see Figure 14.8). In particular the backscattering cross-section is reduced to 13 dB/m^2, when it was 19 dB/m^2 for the perfectly conducting case. There is good agreement between the vector PE solution and the surface impedance analytical results, except in the deep null around 60° for the vertical plane pattern. At that angle the scattered field is so weak that parasitic reflections from the edges of the domain become visible.

14.4.4 NASA almond

The NASA almond, shown in Figure 14.12, is a well-known test case for scattering algorithms [166]. The maximum dimension of the object is 25.2 cm. The very flat shape of the almond together with its electrically small tip make this object difficult to model with parabolic equation techniques. Moreover

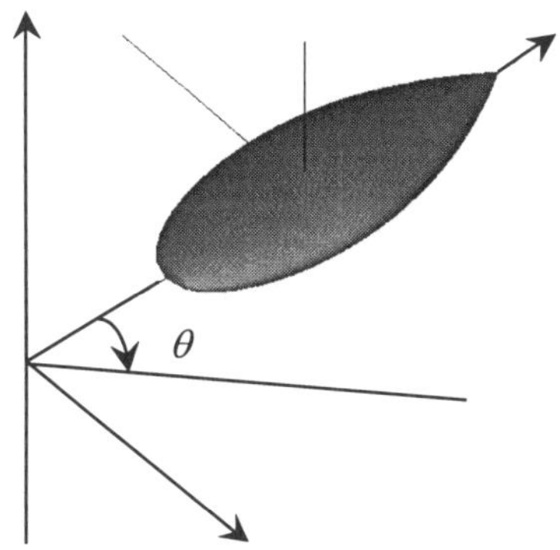

Figure 14.12 NASA almond

published results are for backscatter only, which is the most stressful situation for PE methods. In several respects, this example clearly shows the limitations of the vector PE method.

Measurements of the monostatic RCS of the NASA almond were made under 9.92 GHz incidence, corresponding to a wavelength of 3.022 cm. The almond was positioned with its broad side flat. The RCS was measured as a function of the azimuthal angle θ as shown in Figure 14.12, keeping a fixed elevation of 0°. Azimuth angle 0° corresponds to incidence on the tip.

A backscatter Physical Optics [24] code was used for comparison. For the backscatter case, the PO formulation simplifies to a scalar integral which is very fast to compute. Figures 14.13 and 14.14 show the monostatic RCS of the NASA almond computed as a function of azimuthal angle respectively for VV and HH polarization. The PE and PO results differ substantially from the measurements. The models behave reasonably well when the almond is viewed from the side (90° azimuth), but become increasing inaccurate as the view moves towards the tip. Of course this case is also stressful for MOM codes since the object is large relative to the wavelength [166]. However in principle a powerful enough computer could produce an accurate MOM solution, whereas fundamental limitations preclude this for the PO and PE methods.

14.4.5 F117 aircraft

One of the main strengths of the vector PE is that it can treat fairly large objects on a desktop computer. We illustrate this with an idealized representation of the F117 stealth aircraft, as shown in Figure 14.15 It should be noted

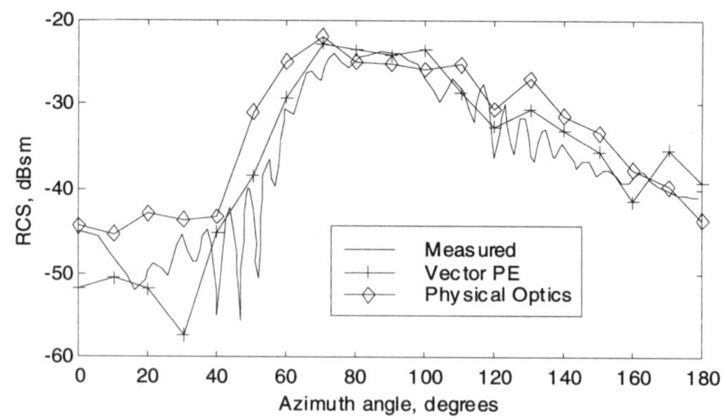

Figure 14.13 Monostatic VV RCS of NASA almond at 9.92 GHz

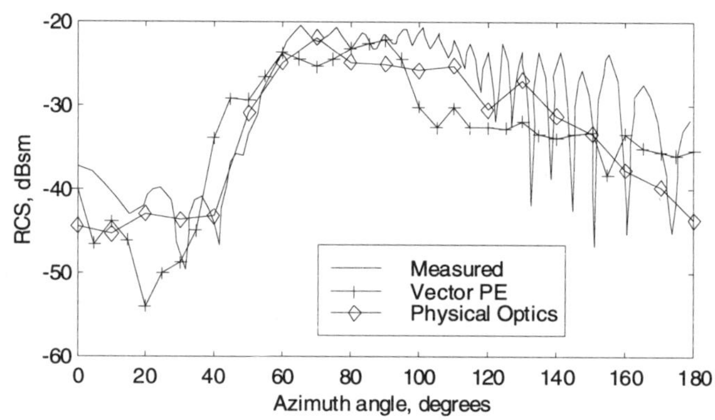

Figure 14.14 Monostatic HH RCS of NASA almond at 9.92 GHz

here that the PE method does not require a decomposition of the target into canonically shaped elements and can deal directly with an arbitrarily shaped target, given for example in one of the standard CAD formats. The length of the aircraft is 20.08 m.

We consider sideways incidence, where the incident direction is orthogonal to the main target axis, as shown in Figure 14.15. The incident wave is vertically polarized. Figure 14.16 shows forward bistatic RCS results in the horizontal plane for an incident frequency of 400 MHz. Angles are measured from the direction of the incident wave. Both the VV and HH vector PE results are shown. For comparison purposes, the forward bistatic RCS was also computed with a "shadow" Physical Optics method implemented along the

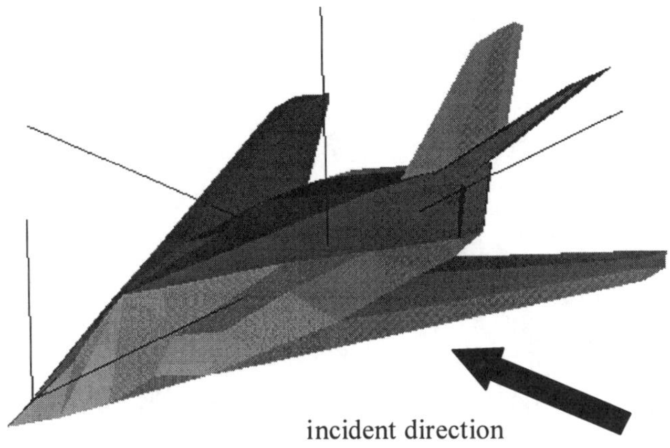

Figure 14.15 Idealized F117 aircraft shape

lines of [52]. This technique treats the object as a two-dimensional silhouette, thus losing three-dimensional properties and polarization effects. As expected, the polarization-independent "shadow" PO results are between the two PE curves. Figure 14.17 shows similar results for an incident frequency of 800 MHz. Computation times for these runs were 40 minutes for the 400 MHz case and 100 minutes for the 800 MHz case on a 200 MHz Pentium machine.

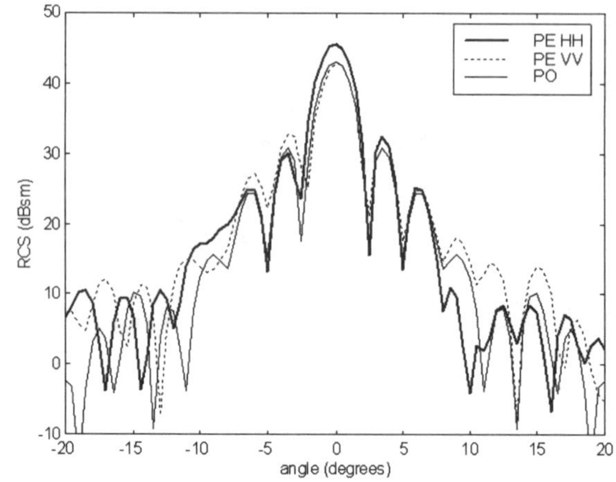

Figure 14.16 Forward bistatic RCS of F117 in horizontal plane for incident frequency of 400 MHz, sideways incidence

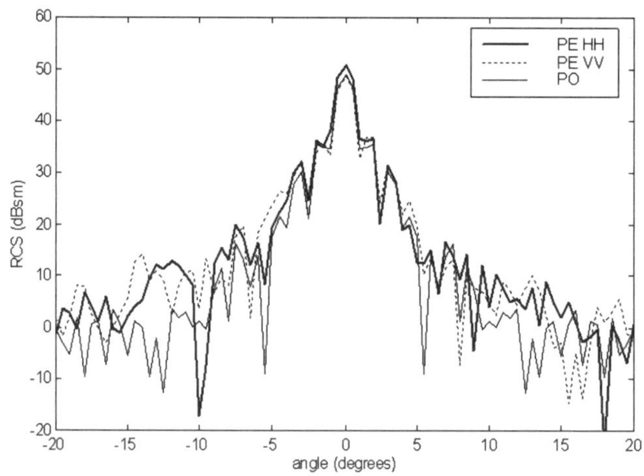

Figure 14.17 Forward bistatic RCS of F117 in horizontal plane for incident frequency of 800 MHz, sideways incidence

Appendix A

Airy functions

A.1 The Airy differential equation

The Airy differential equation is

$$f''(z) = zf(z) \tag{A.1}$$

The solutions are entire functions of the complex variable z. Any solution is a linear combination of the two Airy functions Ai and Bi, which can be defined by their Taylor expansions (valid for all z)

$$\mathrm{Ai}(z) = \mathrm{Ai}(0) \sum_{j=0}^{\infty} \frac{1 \cdot 4 \cdots (3j-2)}{3j!} z^{3j} + \mathrm{Ai}'(0) \sum_{j=0}^{\infty} \frac{2 \cdot 5 \cdots (3j-1)}{(3j+1)!} z^{3j+1} \tag{A.2}$$

where

$$\mathrm{Ai}(0) = \frac{\Gamma\left(\frac{1}{3}\right)}{2\pi\, 3^{1/6}}, \quad \mathrm{Ai}'(0) = -\frac{1}{3^{1/3}\Gamma\left(\frac{1}{3}\right)} \tag{A.3}$$

and

$$\mathrm{Bi}(z) = \mathrm{Bi}(0) \sum_{j=0}^{\infty} \frac{1 \cdot 4 \cdots (3j-2)}{3j!} + \mathrm{Bi}'(0) \sum_{j=0}^{\infty} \frac{1 \cdot 4 \cdots (3j-2)}{3j!} \tag{A.4}$$

where

$$\mathrm{Bi}(0) = \frac{3^{1/3}\Gamma\left(\frac{1}{3}\right)}{2\pi}, \quad \mathrm{Bi}'(0) = \frac{3^{1/6}}{\Gamma\left(\frac{1}{3}\right)} \tag{A.5}$$

Here Γ denotes the Gamma function [117]. It follows from eqns. A.3 and A.5 that the Wronskian of $(\mathrm{Ai}, \mathrm{Bi})$ is equal to $1/\pi$.

Extensive information on the solutions of the Airy equation can be found in Chapters 2, 4 and 11 of [117]. The numerical behaviour of the solutions at infinity depends on the angular sector under consideration (Stokes phenomenon). The first term of the asymptotic expansion of the Airy function Ai for large $|z|$ is given by

$$\mathrm{Ai}(z) \sim_{z\to\infty} \frac{1}{2\sqrt{\pi} z^{\frac{1}{4}}} \exp\left(-z^{\frac{3}{2}}\right) \tag{A.6}$$

provided z is in an angular sector of the form $-\pi + \epsilon < \mathrm{ph}(z) < \pi - \epsilon$, where $\mathrm{ph}(z)$ is the phase of z and ϵ is an arbitrary constant in the interval $]0, \pi[$. Similarly, we have

$$\mathrm{Ai}\left(z\mathrm{e}^{-\frac{i\pi}{3}}\right) \sim_{z\to\infty} \frac{\mathrm{e}^{7i\pi/12}}{2\sqrt{\pi} z^{\frac{1}{4}}} \exp\left(\frac{2i}{3} z^{\frac{3}{2}}\right) \tag{A.7}$$

provided z is in an angular sector of the form $-\frac{4\pi}{3} + \epsilon < \mathrm{ph}(z) < \frac{2\pi}{3} - \epsilon$, and

$$\mathrm{Ai}\left(z\mathrm{e}^{\frac{i\pi}{3}}\right) \sim_{z\to\infty} \frac{\mathrm{e}^{-i\pi/12}}{2\sqrt{\pi} z^{\frac{1}{4}}} \exp\left(-\frac{2i}{3} z^{\frac{3}{2}}\right) \tag{A.8}$$

provided z is in an angular sector of the form $-\frac{2\pi}{3} + \epsilon < \mathrm{ph}(z) < \frac{4\pi}{3} - \epsilon$. For both eqns. A.7 and A.8, ϵ is an arbitrary constant in the interval $]0, 2\pi/3[$. The functions

$$w_1(z) = \mathrm{Ai}(z\mathrm{e}^{2i\pi/3}), \quad w_2(z) = \mathrm{Ai}(z\mathrm{e}^{-2i\pi/3}) \tag{A.9}$$

are also independent solutions of the Airy equation. Eqns. A.7 and A.8 show that solutions of the Airy equation which are outgoing on the real positive axis are proportional to w_1, while incoming solutions are proportional to w_2.

A.2 Zeros of the Airy function

The zeros of the Airy function Ai are all real and negative (see [117], Chapter 11). We denote the sequence of zeros by $(a_j)_{j=1}^{\infty}$. The asymptotic behaviour of the zeros is given by [117]

$$a_j \sim -\left(\frac{3\pi}{8}(4j-1)\right)^{2/3}, \quad j \to +\infty \tag{A.10}$$

We now derive the convolution kernel for non-local boundary conditions in a linear medium, and the identity

$$\sum_{j=1}^{\infty} \frac{1}{a_j^2} = \left(\frac{\mathrm{Ai}'(0)}{\mathrm{Ai}(0)}\right)^2 \tag{A.11}$$

which are used in Chapter 8. It follows from Chapter 8 in [155] that if f is an entire function which does not vanish at zero, such that

$$|f(z)| \leq K\mathrm{e}^{|z|^\rho} \tag{A.12}$$

where K and ρ are constants, then f has the factorization

$$f(z) = \mathrm{e}^{Q(z)} \prod_{j=1}^{\infty} \left\{\left(1 - \frac{z}{z_j}\right) \mathrm{e}^{z/z_j}\right\} \tag{A.13}$$

where Q is a polynomial of degree $\leq \rho$ and the z_js are the zeros of f.

Now from the asymptotic expressions in eqns. A.7, A.8, we have

$$|\mathrm{Ai}(z)| \leq K \exp\left(|z|^{\frac{3}{2}}\right) \tag{A.14}$$

So eqn. A.12 applies to Ai with a polynomial Q of degree at most 1. Since the zeros of Ai are simple [117] and Q' is a constant, we have

$$\frac{\mathrm{Ai}'(z)}{\mathrm{Ai}(z)} = \frac{\mathrm{Ai}'(0)}{\mathrm{Ai}(0)} + \sum_{j=1}^{\infty} \frac{z}{a_j(z - a_j)} \tag{A.15}$$

It follows from eqn. A.13 that the series must be convergent. We note that this is indeed the case in view of the asymptotic behaviour of the zeros of the Ai function, given by eqn. A.10. Dividing eqn. A.15 by z, we get the inverse Laplace transform h of $H(z) = \mathrm{Ai}'(z)/z\mathrm{Ai}(z)$,

$$h(x) = \frac{\mathrm{Ai}'(0)}{\mathrm{Ai}(0)} + \sum_{j=1}^{\infty} \frac{\mathrm{e}^{a_j x}}{a_j} \tag{A.16}$$

Eqn. 8.27 defining the convolution kernel g_α for the non-local boundary condition in a linear medium follows readily from that expression. Now since the Airy function satisfies the Airy differential equation, differentiation of eqn. A.15 yields

$$z - \left(\frac{\mathrm{Ai}'(z)}{\mathrm{Ai}(z)}\right)^2 = \sum_{j=1}^{\infty}\left(\frac{1}{a_j(z-a_j)} - \frac{z}{a_j(z-a_j)^2}\right) \tag{A.17}$$

Setting $z = 0$ in this last equation, we obtain eqn. A.11.

Appendix B

Far-field expressions

B.1 Two-dimensional case

We have seen in Chapter 2 that for $x_0 \leq x$ the solution of the two-dimensional parabolic equation in a vacuum can be written as a convolution

$$u(x, z) = \int_{-\infty}^{+\infty} u(x_0, z')g(x - x_0, z - z')dz' \tag{B.1}$$

where the function $g(x, z)$ is the inverse Fourier transform of the paraxial propagator $\mathrm{e}^{-ikx(1-S)}$. The outgoing square-root S is given by

$$S(p) = \begin{cases} \sqrt{1 - \dfrac{4\pi^2 p^2}{k^2}} & \text{for} \quad |p| \leq \dfrac{k}{2\pi} \\ \\ i\sqrt{\dfrac{4\pi^2 p^2}{k^2} - 1} & \text{for} \quad |p| > \dfrac{k}{2\pi} \end{cases} \tag{B.2}$$

We now prove that g is expressed in terms of the Hankel function of the first kind of order one $H_1^{(1)}$ by

$$g(x, z) = \frac{ikx}{2\sqrt{x^2 + z^2}} \mathrm{e}^{-ikx} H_1^{(1)} \left(k\sqrt{x^2 + z^2}\right) \tag{B.3}$$

From [103], we know that the Hankel function of the first kind of order zero $H_0^{(1)}$ can be written as

$$H_0^{(1)} \left(k\sqrt{x^2 + z^2}\right) = 2 \int_{-\infty}^{+\infty} \frac{e^{ikxS(p)}}{kS(p)} \mathrm{e}^{2i\pi pz} dp \tag{B.4}$$

In other words the inverse Fourier transform of the function

$$F(x,p) = 2\frac{e^{ikxS(p)}}{kS(p)} \tag{B.5}$$

is $f(x,z) = H_0^{(1)}\left(k\sqrt{x^2+z^2}\right)$

This can be understood in terms of Green's functions: since $f(x,z)$ is an outgoing solution of the two-dimensional wave equation, it satisfies

$$\frac{\partial f}{\partial x} = i\sqrt{\frac{\partial^2}{\partial z^2} + k^2}\, f \tag{B.6}$$

We recall that

$$H_0^{(0)\prime}(\zeta) = -H_1^{(1)}(\zeta) \tag{B.7}$$

By taking the partial x-derivative of both sides of eqn. B.4, we obtain

$$\frac{i}{2}\frac{kx}{\sqrt{x^2+z^2}} H_1^{(1)}\left(k\sqrt{x^2+z^2}\right) = \int_{-\infty}^{+\infty} \mathrm{e}^{ikxS(p)}\mathrm{e}^{2i\pi pz}\,dp \tag{B.8}$$

Multiplying both sides by e^{-ikx}, we obtain eqn. B.3.

We can now write the PE solution in a vacuum at an arbitrary range x as a function of the solution at range $x_0 \leq x$ as

$$u(x,z) = \frac{ikx}{2}\mathrm{e}^{-ik(x-x_0)} \int_{-\infty}^{+\infty} u(x_0,z)\frac{H_1^{(1)}(k\rho(z'))}{\rho(z')}dz' \tag{B.9}$$

where

$$\rho(z') = \sqrt{(x-x_0)^2 + (z-z')^2} \tag{B.10}$$

which is the result announced in Chapter 2.

B.1.1 Far-field formulation

We now find a limiting expression for $u(x,z)$ as (x,z) tends to infinity along a given direction. We write

$$\begin{cases} x = r\cos\theta \\ z = r\sin\theta \end{cases} \tag{B.11}$$

Then for any z', we have

$$\rho(z') = r - x_0\cos\theta - z'\sin\theta + O\left(\frac{z'^2}{r}\right) \tag{B.12}$$

and using asymptotic expansions of Hankel functions (see for example [117])

$$H_1^{(1)}(k\rho(z')) = \sqrt{\frac{2}{\pi k\rho(z')}}\mathrm{e}^{ik\rho(z_p)-3i\pi/4} + O\left(\frac{z'^2}{r}\right) \tag{B.13}$$

Finally we have as r tends to infinity

$$\begin{aligned} u(r\cos\theta, r\sin\theta) &\sim \sqrt{\frac{k}{2\pi}}\mathrm{e}^{-i\pi/4}\frac{\cos\theta}{\sqrt{r}} \\ \times \quad &\mathrm{e}^{ikr(1-\cos\theta)+ikx_0}\int_{-\infty}^{+\infty} u(x_0, z')\mathrm{e}^{-ikz'\sin\theta}dz' \end{aligned} \tag{B.14}$$

Hence the far-field is expressed as a Fourier transform of the field on a fixed vertical.

B.1.2 Near-field/far-field transformation

In this section we assume a source with azimuthal symmetry and work in cylindrical coordinates. Eqn. B.14 provides the link between the aperture field of the source and the beam pattern. By definition, the beam pattern for a field component ψ in direction θ is

$$B(\theta) = \lim_{r\to\infty} r\mathrm{e}^{ikr}\psi(r\cos\theta, r\sin\theta) \tag{B.15}$$

Since in cylindrical coordinates, ψ is related to the reduced PE function u by

$$\psi(x,z) = \frac{\mathrm{e}^{ikx}}{\sqrt{kx}}u(x,z) \tag{B.16}$$

taking $x_0 = 0$, we can rewrite the far-field as

$$\psi(r\cos\theta, r\sin\theta) \sim \frac{1}{\sqrt{2\pi}} \mathrm{e}^{-i\pi/4} \frac{\sqrt{\cos\theta}}{r} \mathrm{e}^{ikr} \int_{-\infty}^{+\infty} u(0,z')\mathrm{e}^{-ikz'\sin\theta} dz' \quad \text{(B.17)}$$

It is reassuring to find that the far-field is indeed proportional to the spherical Green function. We can now rewrite the beam pattern as

$$B(\theta) = \frac{\sqrt{\cos\theta}}{\sqrt{2\pi}} \mathrm{e}^{-i\pi/4} \int_{-\infty}^{+\infty} u(0,z)\mathrm{e}^{-ikz\sin\theta} dz \quad \text{(B.18)}$$

Applying the inverse Fourier transform, we obtain the reduced aperture field as

$$u(0,z) = \sqrt{2\pi}\mathrm{e}^{i\pi/4} \int_{-\frac{1}{\lambda}}^{+\frac{1}{\lambda}} \frac{B(\theta(p))}{\sqrt{\cos(\theta(p))}} \mathrm{e}^{2i\pi pz} dp \quad \text{(B.19)}$$

where λ is the wavelength and the variable p is related to θ by

$$\sin\theta(p) = \lambda p \quad \text{(B.20)}$$

This is the correct wide-angle near-field/far-field formulation. For narrow-angle beams, we can write

$$\frac{B(\theta(p))}{\sqrt{\cos(\theta(p))}} \sim B(\lambda p) \quad \text{(B.21)}$$

which gives the better-known narrow-angle formulation

$$u(0,z) = \sqrt{2\pi} e^{i\pi/4} \int_{-\infty}^{+\infty} B(\lambda p) e^{2i\pi pz} dp \quad \text{(B.22)}$$

where we have assumed that B is negligible outside some interval $[-\alpha, \alpha]$ for a small angle α. It is clear that this last equation is not equivalent to eqn. B.19 for wide beams. Using it to generate aperture fields for wide-angle propagation would result in substantial errors.

The moral here is that the method used to derive the initial field for a PE computation must match the PE propagator used in that computation. With the narrow-angle SPE, eqn. B.22 is adequate, whether the more general eqn. B.19 must be used with wide-angle PE propagators. In fact eqn. B.19 is hardly more demanding computationally so this does not cause any particular difficulties.

B.1.3 Bistatic RCS

Recall that the 2-dimensional bistatic RCS of an object in direction θ is given by

$$\sigma(\theta) = \lim_{r\to\infty} 2\pi r \frac{|\psi^s(r\cos\theta, r\sin\theta)|^2}{|\psi^i(r\cos\theta, r\sin\theta)|^2} \tag{B.23}$$

where ψ^i and ψ^s are the incident and scattered fields respectively. If the background medium is a vacuum, we can use eqn. B.14 to relate the far-field to the field at a range x_0 just beyond the object. If the incident field is a plane wave of amplitude 1, the bistatic RCS in direction θ becomes

$$\sigma(\theta) = k\cos^2\theta \left| \int_{-\infty}^{+\infty} u(x_0, z) e^{-ikz\sin\theta} dz \right|^2 \tag{B.24}$$

This gives the bistatic RCS as a Fourier transform of the PE field on a given vertical. For small scattering angles, the normalization factor in $\cos^2\theta$ can be neglected, but this is no longer possible for larger angles.

B.2 Three-dimensional case

We work with the scalar three-dimensional case: for propagation in a vacuum, each component of the electromagnetic field propagates independently according to the wave equation, since no coupling takes place through the propagation medium or scattering objects. This enables us to reconstruct the electromagnetic far-field by superposition of its components.

Let u be the PE reduced field corresponding to an arbitratry electromagnetic field component. For propagation in a vacuum, the PE reduced field at range x can be written as a function of the reduced field at range $x_0 \leq x$ with the convolution expression

$$u(x, y, z) = \int_{-\infty}^{+\infty} \int_{-\infty}^{+\infty} u(x_0, y', z') g(x - x_0, y - y', z - z') dy' dz' \tag{B.25}$$

where the function $g(x, z)$ is the inverse Fourier transform of the 3D propagator $e^{-ikx(1-S)}$. The 3D outgoing square-root $S(p, q)$ is defined by

$$S(p, q) = \begin{cases} \sqrt{1 - \dfrac{4\pi^2(p^2 + q^2)}{k^2}} & \text{for} \quad \sqrt{p^2 + q^2} \leq \dfrac{k}{2\pi} \\ \\ i\sqrt{\dfrac{4\pi^2(p^2 + q^2)}{k^2} - 1} & \text{for} \quad \sqrt{p^2 + q^2} > \dfrac{k}{2\pi} \end{cases} \tag{B.26}$$

We have

$$\frac{\mathrm{e}^{ikr}}{r} = \frac{1}{2\pi}\int_{-\infty}^{+\infty}\int_{-\infty}^{+\infty}\frac{e^{ikxS(p,q)}}{kS(p,q)}\mathrm{e}^{2i\pi py+2i\pi qz}dpdq \tag{B.27}$$

where

$$r = \sqrt{x^2+y^2+z^2} \tag{B.28}$$

Differentiating eqn. B.27 with respect to x, we can express g as

$$g(x,y,z) = -\frac{1}{2\pi}\mathrm{e}^{-ikx}\frac{\partial\left(\frac{\mathrm{e}^{ikr}}{r}\right)}{\partial x} \tag{B.29}$$

Finally, the field at range $x \geq x_0$ can be written as

$$u(x,y,z) = -\frac{1}{2\pi}\int_{-\infty}^{+\infty}\int_{-\infty}^{+\infty} u(x_0,y,z)\frac{ik(x-x_0-1)}{\rho}\frac{\mathrm{e}^{ik\rho}}{\rho}dy'dz' \tag{B.30}$$

where

$$\rho\,(y',z') = \sqrt{(x_0-x)^2+(y-y')^2+(z-z')^2)} \tag{B.31}$$

B.2.1 Far-field formulation

The derivations in three dimensions are very similar to those for the two-dimensional case. We start with eqn. B.30 and work in spherical coordinates

$$\left\{\begin{array}{lcl} x & = & r\cos\theta \\ y & = & r\sin\theta\cos\phi \\ z & = & r\sin\theta\sin\phi \end{array}\right. \tag{B.32}$$

We now expand the integrand on the RHS of eqn. B.30 in $1/r$, getting

$$\begin{aligned} u(x,y,z) &= -\frac{ik\cos\theta\, e^{ikr}}{2\pi \quad r}\mathrm{e}^{-ik(1+\cos\theta)x_0} \\ &\iint u(x_0,y',z')\mathrm{e}^{-ik\cos\theta(y'\cos\phi+z'\sin\phi)}\,dy'dz' + O\left(\frac{1}{r^2}\right) \end{aligned} \tag{B.33}$$

This gives the solution directly on an arbitrary transverse plane as a function of the field at a given range, for propagation in a vacuum.

B.2.2 Near-field/far-field transformation

In three dimensions, the beam pattern for a field component ψ in direction (θ, ϕ) is defined by

$$B(\theta, \phi) = \lim_{r \to \infty} re^{-ikr} \psi(r \cos\theta, r \sin\theta \cos\phi, r \sin\theta \sin\phi) \tag{B.34}$$

Letting r tend to infinity in eqn. B.33 applied with $x_0 = 0$, we get

$$B(\theta, \phi) = \frac{-ik \cos\theta}{2\pi} \int_{-\infty}^{+\infty} \int_{-\infty}^{+\infty} u(0, y, z) e^{-ik \cos\theta (y \cos\phi + z \sin\phi)} dydz \tag{B.35}$$

Again we obtain the aperture field as a function of the far-field beam pattern by applying a suitable Fourier inversion to eqn. B.35.

B.2.3 Bistatic RCS

The total bistatic RCS in direction (θ, ϕ) is defined as

$$\sigma(\theta, \phi) = \lim_{r \to \infty} 4\pi r^2 \frac{\left|\mathbf{E}^s(x, y, z)\right|^2}{\left|\mathbf{E}^i(x, y, z)\right|^2} \tag{B.36}$$

where $\mathbf{E}^i$ and $\mathbf{E}^s$ are the incident and scattered electric fields respectively.

When receiver polarization has to be taken into account, the quantity of interest is the bistatic RCS in direction (θ, ϕ) assuming the receiver is polarized along vector $\mathbf{t}$. This is defined by

$$\sigma_{\mathbf{t}}(\theta, \phi) = \lim_{r \to \infty} 4\pi r^2 \frac{\left|\mathbf{E}^s(x, y, z) \cdot \mathbf{t}\right|^2}{\left|\mathbf{E}^i(x, y, z)\right|^2} \tag{B.37}$$

where the dot denotes the complex inner product. If the incident field is a plane wave of unit amplitude, we get from eqn. B.35

$$\sigma_{\mathbf{t}}(\theta, \phi) = \frac{k^2 \cos^2\theta}{\pi} \left| \iint \mathbf{E}^s(x_0, y, z) \cdot \mathbf{t} e^{-ik \sin\theta (y \cos\phi + z \sin\phi)} dydz \right|^2 \tag{B.38}$$

The cross-polarized and and co-polarized cross-sections are obtained by using eqn. B.38 with the appropriate transmitter and receiver polarizations. The total RCS is obtained by summing the x, y and z components of the RCS.

Appendix C

Theoretical derivation of mode series

C.1 Introduction

We consider solutions of the two-dimensional wave equation

$$\frac{\partial^2\psi}{\partial x^2}(x,z) + \frac{\partial^2\psi}{\partial z^2}(x,z) + k^2n^2(z)\psi(x,z) = 0 \tag{C.1}$$

in the half-plane $z \geq 0$, satisfying a surface impedance boundary condition of the form

$$\frac{\partial\psi}{\partial z}(x,0) = ik\delta\psi(x,0) \tag{C.2}$$

and the Sommerfeld radiation condition at infinity

$$\lim_{r\to\infty}\sqrt{r}\left\{\frac{\partial\psi(r\mathbf{e})}{\partial\mathbf{e}} - ik\psi(r\mathbf{e})\right\} = 0 \tag{C.3}$$

uniformly for all unit vectors $\mathbf{e}$ such that $e_z \geq 0$. We wish to expand these solutions in the form

$$\psi(x,z) = \sum_{j=1}^{\infty} A_j f_j(z)\exp(ikC_jx) \tag{C.4}$$

Recall that the modal expansion is related to the operator T defined by

$$Tf(z) = f''(z) + k^2(n^2(z) - A)f(z) \tag{C.5}$$

where $n(z)$ is the refractive index and A is a constant to be chosen later. Because of the radiation condition at infinity, the operator T is not self-adjoint and one cannot use the tools of Sturm-Liouville theory as in underwater acoustics problems [66].

The modal expansion of eqn. C.4 was derived by Watson [163] for the case of diffraction by a perfectly conducting Earth with a linear squared refractive index profile, using techniques of complex analysis. For more complicated cases involving finitely conducting ground or a layered atmosphere, the validity of the expansion is quite difficult to establish. We outline a derivation using the framework of functional analysis, which explains some of the features of mode theory. The use of the surface impedance boundary condition greatly simplifies the analysis.

In Section C.2, we show that for a linear profile, the modes always have a unique asymptote along the line of argument $\pi/3$. Section C.3 gives the function analysis framework for the derivation, looking at the modal problem in terms of operators on a suitable Hilbert space. Section C.4 proves the validity of modal expansions for linear profiles. Finally Section C.5 deals with perturbations of a linear profile.

C.2 Modes for linear profiles

For the moment, we assume that the squared refractive index profile is linear. We have for all $z \geq 0$

$$n^2(z) = n^2(0) + \alpha z \tag{C.6}$$

where α is positive. We take $A = n^2(0)$, and assume to simplify notations that $\alpha = 1$; this can be achieved without loss of generality by the change of variable

$$z' = (\alpha k^2)^{2/3} z \tag{C.7}$$

The eigenfunctions of T associated with an eigenvalue μ satisfy

$$f''(z) + (z - \mu) f(z) = 0 \tag{C.8}$$

The outgoing solutions are proportional to the Airy function

$$w_1(-z) = \mathrm{Ai}\left\{(z - \mu)\mathrm{e}^{-i\pi/3}\right\} \tag{C.9}$$

It remains to impose the boundary condition at the interface $z = 0$. Assume first an infinite surface impedance. Then μ must satisfy

$$\mathrm{Ai}\left(\mu e^{2i\pi/3}\right) = 0 \tag{C.10}$$

This means that the eigenvalues are proportional to the zeros of the Airy function Ai, so they form a discrete sequence of complex numbers located on the half-line of argument $\pi/3$ [117].

If the surface impedance δ is finite, the eigenvalues μ satisfy the modal equation

$$\mathrm{Ai}'\left(\mu e^{2i\pi/3}\right) - e^{-i\pi/6} ik\delta \mathrm{Ai}\left(\mu e^{2i\pi/3}\right) = 0 \tag{C.11}$$

which is equivalent to saying that μ must be a zero of a certain entire function which is not identically zero. This implies that the zeros form a discrete sequence [57]. In fact we can say more about the location of the zeros: when $\delta = 0$, the zeros are proportional to those of Ai′, and are again located on the ray Γ of argument $\pi/3$ [117]. Otherwise, we cannot pinpoint the location of the zeros, but we can state that the sequence of zeros has a unique asymptote along the half-line of argument $\pi/3$: indeed if we can find a sequence μ_j of zeros such that $|\mu_j|$ tends to infinity and $\mathrm{ph}(\mu_j \exp(2i\pi/3))$ tends to some constant $\theta \neq \pi$, then using asymptotic expansions of Ai and Ai′ away from the negative real line, we have

$$\frac{\mathrm{Ai}'(\nu_j)}{\mathrm{Ai}(\nu_j)} \sim -\nu_j^{1/2} \tag{C.12}$$

where

$$\nu_j = \mu_j \mathrm{e}^{2i\pi/3}(2\alpha k^2)^{1/3} \tag{C.13}$$

while on the other hand since the μ_js satisfy the modal equation, eqn. C.11, we have

$$\frac{\mathrm{Ai}'(\nu_j)}{\mathrm{Ai}(\nu_j)} = k\delta \mathrm{e}^{-i\pi/6} \tag{C.14}$$

Clearly eqns. C.12 and C.14 cannot hold simultaneously as $|\nu_j|$ tends to infinity, hence the only possible asymptote of eigenvalues is along the ray of argument $\pi/3$. There is indeed an asymptote of eigenvalues along this ray: putting $\nu = \mu \exp(2i\pi/3)$, we have

$$\frac{\mathrm{Ai}'(\nu)}{\mathrm{Ai}(\nu)} \sim (-\nu)^{1/2} \tan\left(\frac{2}{3}(-\nu)^{3/2} - \frac{\pi}{4}\right) \tag{C.15}$$

and we can find an infinite number of solutions in ν, which are asymptotically close to the zeros of Ai′. Finally, whatever the impedance δ, we have a unique asymptote of eigenvalues along the ray Γ of argument $\pi/3$. We denote by μ_j the sequence of eigenvalues, ordered in increasing modulus.

C.3 Functional analysis framework

C.3.1 Complex heights

For large values of z, we have for the eigenfunction f_j corresponding to eigenvalue μ_j,

$$f_j(z) \sim \frac{e^{7i\pi/12}}{2\sqrt{\pi}z^{1/4}} \exp\left(\tfrac{2}{3}iz^{3/2} - iz^{1/2}\mu_j\right) \tag{C.16}$$

If the imaginary part of μ_j is positive, we see that $|f_j(z)|$ tends to infinity as z tends to infinity on the real positive half-line. From the discussion in the previous section, we know that large eigenvalues are on or close to the ray Γ of argument $\pi/3$, so this will happen if j is large enough. We are going to exploit the behaviour of Airy functions of large argument in other parts of the complex plane to make our problem more tractable. Eqn. C.16 is valid for large z provided the argument of z remains in an interval of the form $[-2\pi/3+\epsilon, 4\pi/3-\epsilon]$, for some positive ϵ. In particular, if z tends to infinity on a ray of argument θ in the open interval $]0, 2\pi/3[$, we see that $f_j(z)$ tends to zero, the fastest decrease being on the ray Γ of argument $\pi/3$. This leads us to consider complex heights instead of solving the wave equation for real heights only. The extended operator T is defined on functions holomorphic in some region S of the complex plane by

$$Tf(z) = f''(z) + zf(z) \tag{C.17}$$

We look in particular at the angular sector shown in Figure C.1, with complex numbers of argument between 0 and $\pi/3$. We parametrize curve Γ by

$$z = te^{i\pi/3} \tag{C.18}$$

We define an inner product on Γ by putting

$$< f, g >= \int_0^{+\infty} f(z(t))\overline{g(z(t))}dt = \int_\Gamma f(z)\overline{g(z)}d|z| \tag{C.19}$$

We denote the associated Hilbert norm by $||\quad||_2$. The space $L_2(\Gamma)$ consists of functions f such that

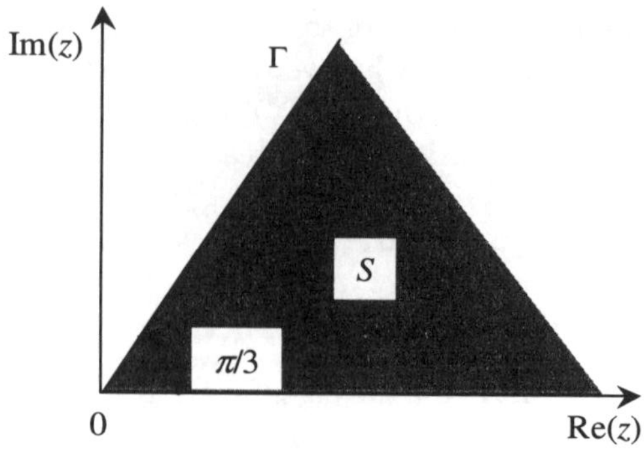

Figure C.1 Height coordinate in the complex plane

$$||f||_2 = \left\{ \int_0^{+\infty} |f(z)|^2 d|z| \right\}^{1/2} < +\infty \tag{C.20}$$

Eqn. C.16 shows that the eigenfunctions of T which are outgoing on the ray Γ_0 of argument zero are in $L_2(\Gamma)$. This leads us to consider the domain $D(T)$ of functions $f(z)$ on Γ which are are twice-differentiable along Γ, satisfy the impedance boundary condition at zero, are square-integrable, and are such that Tf is also square-integrable. We are going to show that the eigenfunctions of T form a basis of $L_2(\Gamma)$, in the sense that any function f in $L_2(\Gamma)$ can be expanded as

$$f = \sum a_j f_j \tag{C.21}$$

where the f_js are suitably normalized eigenfunctions of T corresponding to eigenvalues μ_j and the series converges in $L_2(\Gamma)$.

C.3.2 The resolvent of T

At this point we have to introduce some operator terminology [134]. An operator U in $L_2(\Gamma)$ is a Hilbert-Schmidt operator if it is of the form

$$Uf(z) = \int_\Gamma K(z, z')dz \tag{C.22}$$

where the kernel K is square-integrable in (z, z')

$$\int_\Gamma \int_\Gamma |K(z, z')|^2 d|z| d|z'| < \infty \tag{C.23}$$

A complex number μ is in the resolvent set of T if the operator $T - \mu I$, where I is the identity operator, has a bounded inverse. The inverse is the resolvent $R(T, \mu)$. It turns out that although the operator T is unbounded, its resolvent $R(T, \mu)$ exists and is a Hilbert-Schmidt operator for any complex μ which is not an eigenvalue of T. To see this, we construct a Green function for T on Γ. Suppose μ is not an eigenvalue of T. We know that solutions of the differential equation

$$Tf + (z - \mu)f = 0 \tag{C.24}$$

which are square-integrable on Γ are of the form

$$\chi(z) = a\mathrm{Ai}\left\{(z - \mu)\mathrm{e}^{-i\pi/3}\right\} \tag{C.25}$$

We take the coefficient a equal to 1. Now let ϕ be a solution of the differential equation C.24 which satisfies the impedance boundary condition at zero. Since μ is not an eigenvalue of T, χ and ϕ are independent, and their Wronskian $W(\chi, \phi)$ is a non-zero constant. We normalize ϕ in such a way that the Wronskian is equal to 1. Now we put

$$\begin{aligned} G(z, z') &= \phi(z)\xi(z') \qquad |z| \leq |z'| \\ G(z, z') &= \phi(z')\xi(z) \qquad |z| \geq |z'| \end{aligned} \tag{C.26}$$

It is easy to check that for any z, the function $G(z, z')$ is square-integrable in z'. Now let g be a square-integrable function on Γ. We put

$$f(z) = \int_\Gamma G(z, z')g(z')dz' \tag{C.27}$$

Applying the Cauchy-Schwarz inequality [134], we have

$$|f(z)| \leq \left[|\chi(z)|\left(\int_0^{|z|} |\phi|^2\right)^{\frac{1}{2}} + \phi(z)\left(\int_{|z|}^{\infty} |\chi|^2\right)^{\frac{1}{2}}\right] ||g||_2 \tag{C.28}$$

We are going to show that the function in square brackets is square-integrable. We know that ϕ is of the form

$$\phi(z) = a\mathrm{Ai}\left\{\zeta(z)\mathrm{e}^{i\pi/3}\right\} + b\mathrm{Ai}\left\{\zeta(z)\mathrm{e}^{-i\pi/3}\right\} \tag{C.29}$$

Now using asymptotic expansions [117], we can find a constant K_1 such that for all z in Γ

$$|\phi(z)| \le K_1|z|^{-1/4} \exp\left(\tfrac{2}{3}|z|^{3/2}\right) \tag{C.30}$$

and

$$\left\{\int_0^{|z|} |\phi(z)|^2 d|z|\right\}^{\frac{1}{2}} \le K_1|z|^{-\frac{1}{2}} \exp\left(\tfrac{2}{3}|z|^{\frac{3}{2}}\right) \tag{C.31}$$

Similarly we can find a constant K_2 such that for all z in Γ

$$|\chi(z)| \le K_2|z|^{-\frac{1}{4}} \exp\left(-\tfrac{2}{3}|z|^{\frac{3}{2}}\right) \tag{C.32}$$

and

$$\left\{\int_{|z|}^{\infty} |\chi(z)|^2 d|z|\right\}^{\frac{1}{2}} \le K_2|z|^{-\frac{1}{2}} \exp\left(-\tfrac{2}{3}|z|^{\frac{3}{2}}\right) \tag{C.33}$$

Hence we have

$$|\chi(z)|\left\{\int_0^{|z|} |\phi|^2\right\}^{\frac{1}{2}} + |\phi(z)|\left(\int_t^{\infty} |\chi|^2\right)^{\frac{1}{2}} \le 2K_1K_2|z|^{-\frac{3}{4}} \tag{C.34}$$

which is indeed square-integrable. This property implies that f is square-integrable and twice-differentiable. Using integration by parts and the fact that $W(\chi, \phi) = 1$, we have

$$Tf - \mu f = g \tag{C.35}$$

We note that f is indeed in the domain of T on Γ since g is square-integrable, so the resolvant $R(\mu)$ is given by

$$R(\mu)g = (T - \mu)^{-1} = f \tag{C.36}$$

Eqn. C.34 shows that $R(\mu)$ is a Hilbert-Schmidt operator.

C.4 Mode expansions for the linear case

C.4.1 The self-adjoint case

Assume first that the impedance δ is zero or infinite, corresponding to a perfectly conducting ground for vertical or horizontal polarization. Then the

operator $T_1 = \exp(-i\pi/3)T$ is self-adjoint: writing $f(z) = f_1(t)$, we have in terms of t,

$$\mathrm{e}^{-i\pi/3}Tf(z) = -f_1''(t) + tf_1(t) \tag{C.37}$$

Using two successive integrations by parts, we have

$$\begin{aligned} < \mathrm{e}^{-i\pi/3}Tf, g > &= \int_0^\infty (-f_1''(t) + tf_1(t))\overline{g_1(t)}dt \\ &= f_1(0)\overline{g_1'(0)} - f_1'(0)\overline{g_1(0)} \\ &\quad + \int_0^\infty f_1(t)\left(\overline{-g_1''(t) + tg_1(t)}\right)dt \\ &= < f, \mathrm{e}^{-i\pi/3}Tg > \end{aligned} \tag{C.38}$$

provided both f and g are in $D(T)$: if δ is infinite, we have $f_1(0) = g_1(0) = 0$, and if $\delta = 0$, we have $f_1'(0) = g_1'(0) = 0$, so in both cases the integrated part is zero.

The resolvent of T_1 is a Hilbert-Schmidt operator, and so it is compact [134]. It follows that the eigenvectors of T_1 (which are the same as those of T) form a complete orthonormal set [134]. Note that the eigenvalues of T_1 must be real, since T_1 is self-adjoint. This is indeed the case since the eigenvalues of T are all of argument $\pi/3$. Let (μ_j) be the sequence of eigenvalues of T, ordered in increasing modulus. We have

$$\sum_{j=1}^{\infty} |\mu_j|^2 < \infty \tag{C.39}$$

Let (f_j) be a corresponding sequence of orthonormal eigenfunctions on Γ. We can write

$$f_j(z) = \alpha_j \mathrm{Ai}\left\{(z - \mu_j)\mathrm{e}^{-i\pi/3}\right\} \tag{C.40}$$

where the normalization constant α_j has to be determined. We may constrain α_j to be real and positive. Then the f_js are real, and from eqn. 5.56, we have

$$\alpha_j^{-2} = |\mu_j|\mathrm{Ai}^2\left\{-\mu_j\mathrm{e}^{-i\pi/3}\right\} + \mathrm{Ai}''\left\{-\mu_j\mathrm{e}^{-i\pi/3}\right\} \tag{C.41}$$

For the case of infinite impedance (horizontal polarization), we have

$$\alpha_j = \mathrm{Ai}'\left\{-\mu_j\mathrm{e}^{-i\pi/3}\right\} \tag{C.42}$$

For the case of zero impedance (vertical polarization), we have

$$\alpha_j = |\mu_j|^{1/2}\mathrm{Ai}\left\{-\mu_j \mathrm{e}^{-i\pi/3}\right\} \tag{C.43}$$

Using information on the asymptotic behaviour of the zeros of the Airy function Ai and asymptotic expansions for Ai and Ai′, we find the same asymptotic behaviour of the normalization coefficients in both cases,

$$\alpha_j \sim \pi^{-1/2}|\mu_j|^{1/4} \sim \pi^{-1/2}\left\{\frac{3\pi}{2}j\right\}^{1/6}, \quad j \to \infty \tag{C.44}$$

Using asymptotic expansions again, this yields the asymptotic behaviour of the eigenfunctions: for fixed z, we have

$$|f_j(z)| \sim \frac{1}{2}|\mu_j|^{-1/2}\left|\sin\left(ze^{-i\pi/3}|\mu_j|^{1/2}\right)\right| \tag{C.45}$$

for horizontal polarization and

$$|f_j(z)| \sim \frac{1}{2}|\mu_j|^{-1/2}\left|\cos\left(ze^{-i\pi/3}|\mu_j|^{1/2}\right)\right| \tag{C.46}$$

for vertical polarization. We note that eqns. C.45 and C.46 are valid for any z in the complex plane.

The modes (C_j) corresponding to eigenvalues μ_j are given by

$$C_j = \sqrt{1 + \frac{\mu_j}{k^2}} \tag{C.47}$$

where we take the square-root with positive imaginary part. We have

$$ikC_j \sim \mathrm{e}^{2i\pi/3}|\mu_j|^{1/2} \quad j \to \infty \tag{C.48}$$

which shows that mode attenuation increases quite fast along the mode asymptote.

We know that any function f in $L_2(\Gamma)$ can be expanded as

$$f(z) = \sum_{j+1}^{\infty} < f, f_j > f_j \tag{C.49}$$

where the sum converges in the inner product norm. We are now going to connect expansions on Γ to expansions on the positive real half-line.

C.4.2 Connecting the solution

We make the following remark: if $\phi(z)$ is a reasonably well-behaved real function on the real half-line $z \geq 0$, we can approximate ϕ with a function f of the form

$$f(y) = P(z)\mathrm{e}^{-z} \tag{C.50}$$

where P is a polynomial. The function f is entire, and is square-integrable on the ray Γ of argument $\pi/3$. We apply this to the initial field for the transmitting antenna of interest. Since f is square-integrable on the ray Γ of argument $\pi/3$, we can write

$$f(z) = \sum_{j=1}^{\infty} A_j f_j(z) \tag{C.51}$$

where the series converges in $L_2(\Gamma)$, or equivalently

$$\sum_{j=1}^{\infty} |A_j|^2 < +\infty \tag{C.52}$$

Since the eigenfunctions f_j are real on Γ, the coefficients A_j are given by

$$A_j = e^{-i\pi/3} \int_{\Gamma} f(z) f_j(z) dz \tag{C.53}$$

We now look at the function ψ defined by

$$\psi(x, z) = \sum_{j=1}^{\infty} A_j f_j(z) \exp(ikC_j x) \tag{C.54}$$

When does this series converge? From the previous discussion of the asymptotic behaviour of the modes and height-gain functions, we have if z is not on the ray of argument $\pi/3$

$$|f_j(z)\exp(ikC_j x)| \sim \frac{1}{2}|\mu_j|^{-1/2} \exp\left(|\mu_j|^{1/2}\left\{\left|\mathrm{Im}(ze^{-i\pi/3})\right| - \frac{x}{2}\right)\right\} \tag{C.55}$$

as j tends to infinity. In view of eqn. C.52, the series converges provided the expression in square brackets is negative. On the real ray of argument 0, this is true for

$$z\sqrt{3} < x \tag{C.56}$$

corresponding to points of elevation less than $\pi/6$. Clearly the speed of convergence increases as elevation decreases. On the ray of argument $\pi/3$, the series converges for any z and any positive range x, since the sine functions remain bounded. In the (x, z)-region where the series converges, ψ is a solution of the wave equation which satisfies the boundary conditions at the ground and at infinity. Also, on the ray of argument $\pi/3$, the series at range zero converges in the L_2-sense to the correct initial values for the field.

We denote by S the complex numbers of argument in $[-\pi/3, +\pi/3[$. Consider the wave equation for an (x, z)-region including ranges $x \geq 0$ and complex heights z, with initial conditions given by $f(z)$ on the ray Γ of argument $\pi/3$, the radiation condition at infinity and the impedance boundary condition at $z = 0$. This has a unique solution $\phi(x, z)$ which at any fixed range x is holomorphic in z. Since the two holomorphic functions $\phi(0, z)$ and $f(z)$ are identical on Γ, they must be identical on any connected region of the complex plane containing Γ. We conclude that in any connected region where they are both defined, ψ must coincide with ϕ: if the modal series converges, the sum is indeed the solution of the wave equation which satisfies the initial conditions given by $f(z)$ and the boundary conditions at the ground and at infinity. This is the connection property, linking the solution for complex heights of argument $\pi/3$ to real heights. We note that the connection property rests on the crucial fact that the modal series converges in $L_2(\Gamma)$.

Once the connection principle has been established, it turns out that the excitation factors A_j can be computed without involving complex heights. Using the fact that f and f_j are holomorphic, we have

$$\int_\Gamma f(z) f_j(z) dz = \int_0^{+\infty} f(t) f_j(t) dt \tag{C.57}$$

This follows from integration on the closed contour C shown in Figure C.2, letting the radius of the circular arc tend to infinity. The integral on the RHS of eqn. C.57 is indeed convergent since f behaves as $\exp(-t)$ while f_j behaves as $\exp(a\sqrt{t})$ for some positive constant a.

C.4.3 Finite impedance case

In the finite impedance case, the operator T is not normal in general and we need another ingredient to prove convergence of the modal series. This is done by using contour integrals for the resolvent. The technique is similar to that used by Coddington and Levinson [30] for the equation $f'' = 0$ on a finite interval. We denote by T_0 the operator corresponding to zero impedance, and

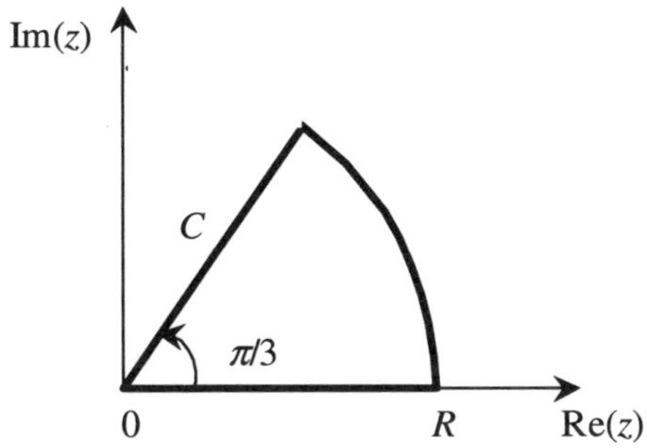

Figure C.2 Integration contour for excitation factors

by G_0 the corresponding Green function. We have seen in Section C.4.1 that the operator $\exp(i\pi/3)T_0$ is self-adjoint with Hilbert-Schmidt resolvent, and that there exists a complete sequence of orthonormal eigenfunctions h_j for T_0. A function f in $L_2(\Gamma)$ can be expanded as

$$f = \sum_{j=1}^{\infty} < f, h_j > h_j \tag{C.58}$$

This can be expressed in terms of contour integrals for the resolvent $R_0(\mu)$ of T_0. Let C_j be the circle of radius $|a_j|$, where $|a_j|$ is the jth zero of the function Ai. Since the eigenvalues of T_0 are of the form $\exp(i\pi/3)a'_j$, where $|a'_j|$ is the jth zero of the function Ai$'$, this circle does not go through any poles of the resolvent. Using residue theory, eqn. C.58 can be rewritten as

$$f = \lim_{j\to\infty} \int_{C_j} R_0(\mu) f d\mu \tag{C.59}$$

We shall show that for large j, C_j does not go through any poles of the resolvent $R(\mu)$ of T and that we can write

$$f = \lim_{j\to\infty} \int_{C_j} R(\mu) f d\mu \tag{C.60}$$

Then making use of the fact that the eigenvalues of T are all simple, we use residue theory to express this as

$$f = \sum_{j=1}^{\infty} < f, f_j > f_j \tag{C.61}$$

where the f_js are eigenfunctions of T, normalized to have

$$\int_\Gamma f_j^2(z)dz = 1 \tag{C.62}$$

The series of eqn. C.61 converges in $L_2(\Gamma)$. By contrast with the self-adjoint case, the eigenfunctions need not be orthogonal, but they satisfy the pseudo-orthogonality relation

$$\int_\Gamma f_j(z)f_l(z)dz = 0, \quad j \neq l \tag{C.63}$$

In other words f_j is orthogonal to $\overline{f_l}$ for $j \neq l$.

As in the self-adjoint case, we can connect modal expansions on the ray of argument $\pi/3$ and on the real ray of argument 0. Again this involves estimates of the eigenvalues μ_j and of the normalized eigenfunctions f_j as j tends to infinity.

In order to prove eqn. C.60, we consider the Green functions G_0 and G for T_0 and T. We know that the poles of the Green function G_0 are simple since T_0 is self-adjoint. This can also be seen by recalling that Ai and Ai$'$ have no common zeros. The poles of the Green function G are all simple except in the unusual case when γ^2 is an eigenvalue. If $|\gamma|$ is large, which is usually the case, this can only happen if the phase of γ is close to $\pi/6$. We shall assume that γ^2 is not an eigenvalue, so that the poles of G are all simple.

Assume μ is not an eigenvalue of T_0 or T. Putting $\gamma = ik\delta$, we define

$$g_1(\mu, z) = w_1(\mu - z) \tag{C.64}$$

$$g_2^0(\mu, z) = w_2(\mu - z) - \frac{w_2'(\mu)}{w_1'(\mu)} w_1(\mu - z) \tag{C.65}$$

$$g_2^0(\mu, z) = w_2(\mu - z) - \frac{w_2'(\mu) - \gamma w_2(0)}{w_1'(\mu) - \gamma w_1(\mu)} w_1(\mu - z) \tag{C.66}$$

The Green functions are given by

$$G_0(\mu, z, z') = \begin{cases} g_1(\mu, z)g_2^0(\mu, z'), & |z'| \leq |z| \\ g_2^0(\mu, z)g_1(\mu, z'), & |z'| \geq |z| \end{cases} \tag{C.67}$$

and

$$G(\mu, z, z') = \begin{cases} g_1(\mu, z)g_2(\mu, z'), & |z'| \leq |z| \\ g_2(\mu, z)g_1(\mu, z'), & |z'| \geq |z| \end{cases} \tag{C.68}$$

We have

$$(G - G_0)(\mu, z, z') = \frac{\gamma}{w_1'(\mu)\left(w_1'(\mu) - \gamma w_1(\mu)\right)} w_1(\mu - z)w_1(\mu - z_p) \tag{C.69}$$

We first show that the eigenfunctions of T form a complete set, using a theorem of Carleman. This is true if T has a Hilbert-Schmidt resolvent R_μ which satisfies the following property: $||R_\mu|| = O(|\mu|^{-1})$ as μ tends to infinity in at least five directions separated by angles of at most $\pi/2$ [134]. In fact for our operator T, the growth property of the resolvent holds along any direction except that of argument $\pi/3$. As μ tends to infinity along a ray of direction different from $\pi/3$, we have uniformly in z and z' in Γ,

$$(G - G_0)(\mu, z, z') \sim \frac{\gamma}{|\nu|^{1/2}} \mathrm{e}^{-\frac{2}{3}\left((\nu+t)^{3/2} - \nu^{3/2}\right)} \mathrm{e}^{-\frac{2}{3}\left((\nu+t')^{3/2} - \nu^{3/2}\right)} \tag{C.70}$$

where

$$\nu = -\mu \mathrm{e}^{-i\pi/3}, z = t\mathrm{e}^{-i\pi/3}, z' = t'\mathrm{e}^{-i\pi/3} \tag{C.71}$$

From the mean value theorem, we have

$$\frac{2}{3}\left\{(\nu + t)^{3/2} - \nu^{3/2}\right\} = t(\nu + \theta(t)t)^{1/2} \tag{C.72}$$

where $\theta(t)$ is a real number between 0 and 1. Now if μ tends to infinity on a ray of direction different from $\pi/3$, we have

$$\nu = r\mathrm{e}^{i\alpha}, -\pi < \alpha < \pi \tag{C.73}$$

If ν has positive real part, the real part $R(t)$ of the function in eqn. C.72 satisfies

$$R(t) \geq t|\nu|^{1/2} \cos\left(\frac{|\alpha|}{2}\right) \tag{C.74}$$

Otherwise, the phase of the RHS of eqn. C.72 satisfies

$$-\frac{\pi - |\alpha|}{2} < \mathrm{ph}\left\{(\nu + \theta(t)t)^{1/2}\right\} < \frac{\pi - |\alpha|}{2} \tag{C.75}$$

which implies that the real part $R(t)$ of the function in eqn. C.72 satisfies

$$R(t) \geq t|\mathrm{Im}(\nu)|^{1/2} \sin\left(\frac{|\alpha|}{2}\right) \tag{C.76}$$

Finally there is a positive constant K_1 such that

$$R(t) \geq K_1|\mu|^{1/2}t \tag{C.77}$$

From this, we get a bound on the modulus of the Green function: there is a positive constant K_2 such that we have, uniformly on z and z' in Γ and μ going to infinity in a direction different from $\pi/3$,

$$|(G - G_0)(\mu, z, z')| \leq \frac{K_2}{|\mu|} \exp(-K_1|\mu|^{1/2}|z|) \exp(-K_1|\mu|^{1/2}|z'|) \tag{C.78}$$

It follows that $G - G_0$ satisfies

$$\left\{\int_\Gamma \int_\Gamma |(G - G_0)(\mu, z, z')|^2 \, d|z| d|z'|\right\}^{1/2} \leq \frac{K_2}{2K_1|\mu|} \tag{C.79}$$

Since $\exp(-i\pi/3)T_0$ is self-adjoint, the resolvent R_0 of T_0 satisfies the growth property on any ray of direction different from $\pi/3$. Now eqn. C.79 implies that the resolvant R of T has the same growth property. We now know that the eigenfunctions of T form a complete system.

Let μ_j and μ_j^0 be the sequences of eigenvalues of T and T_0 respectively. The residue of $G - G_0$ at μ_j is

$$r_j(z, z') = \frac{w_1(\mu_j - z)w_1(\mu_j - z_p)}{(\mu_j - \gamma^2)w_1^2(\mu_j)} \tag{C.80}$$

and the residue of $G - G_0$ at μ_j^0 is

$$-r_j^0(z, z') = -\frac{w_1(\mu_j^0 - z)w_1(\mu_j^0 - z_p)}{\mu_j^0 w_1^2(\mu_j^0)} \tag{C.81}$$

We can write the integral of $G - G_0$ on contour C_J as

$$\int_{C_J} (G - G_0)(\mu, z, z')d\mu = \frac{1}{2\pi i} \sum_{j=1}^{J} \left(r_j(z, z') - r_j^0(z, z') \right) \tag{C.82}$$

Using asymptotic expansions [117], we find that

$$\left\{ \int \left| r_j(z, z') - r_j^0(z, z') \right|^2 d|z|d|z'| \right\}^{1/2} \sim \frac{|\gamma|^2}{|\lambda_j|^2} \sim Kj^{-4/3} \tag{C.83}$$

which proves that the series on the RHS of eqn. C.82 converges in $L_2(\Gamma \times \Gamma)$ to a function g. Now put

$$P_j f(z) = -\frac{1}{2\pi i} \int_\Gamma r_j(z, z') f(z') dz' \tag{C.84}$$

The operator P_j is the spectral projector corresponding to eigenvalue μ_j [134]. Since the eigenfunctions form a complete set, we have $f = g$. We now use the properties of T and its adjoint to describe the spectral projectors exactly: since the pole μ_j of G is simple, we have

$$\lim_{\mu \to \mu_j} (\mu - \mu_j)(T - \mu)^{-1} f = P_j f \tag{C.85}$$

Multiplying both sides by $T - \mu$, we obtain

$$(\mu - \mu_j) P_j f = 0 \tag{C.86}$$

which means that $P_j f$ is proportional to eigenfunction f_j. Hence we can write

$$f = \sum_{j=1}^{\infty} A_j f_j \tag{C.87}$$

Now since $\overline{f_j}$ is the eigenfunction of the adjoint T^* of T corresponding to eigenvalue $\overline{\mu_j}$, we have for distinct j and l

$$< Tf_j, \overline{f_l} >= \mu_j < f_j, \overline{f_l} >=< f_j, T^* \overline{f_l} >= \mu_l < f_j, \overline{f_l} > \tag{C.88}$$

which implies

$$< f_j, \overline{f_l} >= \int_\Gamma f_j f_l = 0 \tag{C.89}$$

We conclude that if f_j is suitably normalized, then

$$A_j = \int_\Gamma f f_j \tag{C.90}$$

so that the expansion given by eqn. C.61 indeed holds.

C.5 Perturbations of the linear case

We now treat the case of a square refractive index profile which is linear above a certain height. Assume that

$$n^2(z) = n^2(H) + z - H, \quad z \geq H \tag{C.91}$$

We put

$$q(z) = n^2(z) - n^2(H) - z + H \tag{C.92}$$

so that q vanishes for z greater than H. We approximate q with a function Q defined for complex numbers of argument strictly between $-\pi$ and $+\pi$, with the following properties: Q is holomorphic except at zero, and $Q - Q(0)$ is real on the positive real line and of argument $\pi/3$ on the ray Γ of argument $\pi/3$. More precisely, if we put $\gamma = \exp(i\pi/3)$, Q satisfies

$$Q(\gamma t) - Q(0) = \gamma\,(Q(t) - Q(0)) \tag{C.93}$$

for any t of argument 0. On the real positive half-line, we put

$$\psi(y) = \frac{q(y^{1/6}) - q(0)}{y^{1/6}} \tag{C.94}$$

Using the properties of Hermite polynomials [57], we can approximate ψ by a function g of the form

$$g(y) = P(y)\mathrm{e}^{-y^2} \tag{C.95}$$

where P is a polynomial. If we now put

$$Q(z) = q(0) + zg(z^6) \tag{C.96}$$

it is easy to see that Q has the desired properties. Now we can define the operator T_Q for complex heights or argument strictly between $-\pi$ and π by

$$T_Q f(z) = f''(z) + (z + Q(z))f(z) \tag{C.97}$$

Again we have taken the squared refractive index at zero as the reference, and modes are related to eigenvalues of T_Q by

$$C = \frac{1}{k^2}\sqrt{\mu + k^2 n^2(0)} \tag{C.98}$$

where we choose the square-root with positive imaginary part. Using WKB approximations [117] of the solutions of

$$T_Q f = \mu f \tag{C.99}$$

one can show that a solution is outgoing at infinity for real heights if and only if it is square-integrable on the ray Γ of argument $\pi/3$. As in the linear profile case, we show that functions on Γ can be expanded in eigenfunctions of T_Q, and use a connection property to derive the modal expansion for solutions of the wave equation.

At this stage we note that we may assume that $Q(0) = 0$. Adding a constant offset d to Q corresponds to adding a constant multiple of the identity operator to T_Q: the eigenfunctions remain the same, and the eigenvalues are shifted by d. The basis property of the eigenfunctions is unaffected. The modes are transformed according to eqn. C.98, which does not affect convergence of the modal series. It follows from the assumption $Q(0) = 0$ that γQ is real on the ray Γ.

We start with the case where the impedance is zero or infinite. Since γQ is real on Γ, the operator γT_Q is self-adjoint on Γ. Since q is bounded, we may assume that $|Q|$ is bounded by M for positive real heights, and hence also on Γ thanks to eqn. C.93. With a perturbation argument along the lines of that used in [30] for the differential equation $-f'' + qf = 0$ on a finite interval, one can prove that T_Q has a Hilbert-Schmidt resolvent. This is done by constructing the Green function for T_Q by induction. Let $G_0(\mu, z, z')$ be the Green function corresponding to the unperturbed linear profile operator T_0, and $R_{0,\mu}$ be the resolvent for T_0. We put $G_Q^0 = 0$, and for any positive integer p,

$$G_Q^{p+1}(\mu, z, z') = G_0(\mu, z, z') - \int_\Gamma G_0(\mu, z, \zeta)Q(\zeta)G_Q^p(\mu, \zeta, z')d\zeta \tag{C.100}$$

when μ is not an eigenvalue of T. We have

$$(G_Q^{p+1} - G_Q^p)(\mu, z, z') = R_{0,\mu} \left\{ Q(\zeta)(G_Q^p - G_Q^{p-1})(\mu, \zeta, z') \right\}(z) \qquad \text{(C.101)}$$

where the expression in curly brackets on the RHS is considered as a function of ζ with fixed z'. Taking the L_2-norms in z, we get for all z'

$$||G_Q^{p+1} - G_Q^p)(\mu, z, z')|| \leq M||R_{0,\mu}|| \, ||(G_Q^p - G_Q^{p-1})(\mu, z, z')|| \qquad \text{(C.102)}$$

Now if μ tends to infinity along a ray of direction different from $\pi/3$, we know that $||R_{0,\mu}|| = O(|\mu|^{-1})$. So if $|\mu|$ is large enough, we have

$$M||R_{0,\mu}|| \leq \frac{1}{2} \qquad \text{(C.103)}$$

By induction, we get

$$||(G_Q^{p+1} - G_Q^p)(\mu, z, z')|| \leq 2^{-p}||G_0(\mu, z, z'|| \qquad \text{(C.104)}$$

and for fixed z', the sequence $\left(G_Q^p(\mu, z, z')\right)$ converges in $L_2(\Gamma)$ to a function $G_Q(\mu, z, z')$ which satisfies

$$\left\{ \int_\Gamma |G_Q(\mu, z, z')|^2 \, d|z| \right\}^{1/2} \leq \left\{ \int_\Gamma |G_0(\mu, z, z')|^2 \, d|z| \right\}^{1/2} \qquad \text{(C.105)}$$

It is easy to check that G_Q is the Green function for T_Q. It follows from eqn. C.105 that

$$\int_{-\infty}^{+\infty} \int_{-\infty}^{+\infty} |G_Q(\mu, z, z')|^2 d|z| d|z'| \leq \int_{-\infty}^{+\infty} \int_{-\infty}^{+\infty} |G_0(\mu, z, z')|^2 d|z| d|z'| \quad \text{(C.106)}$$

so T_Q has a Hilbert-Schmidt resolvent. We note that the proof also shows that the only asymptote of eigenvalues is along the ray of argument $\pi/3$.

Since γT_Q is self-adjoint with compact resolvent, the expansion theorem holds: there exists a complete orthonormal sequence of eigenvectors h_j, corresponding to eigenvalues μ_j. Using WKB [117] approximations, we find that the eigenfunctions satisfy similar estimates to those obtained for the eigenfunctions associated with the linear profiles, namely there is a constant K such that

$$|h_j(z)| \leq K|\mu_j|^{-1/2} \exp\left(|\mu_j|^{1/2} |\mathrm{Im}(\gamma z)|\right) \tag{C.107}$$

From this, we can derive the same connection properties as for the linear case, finding that the modal expansion holds at range x and real height z provided the elevation at point (x, z) is less than $\pi/6$. As before, the finite impedance case follows by a perturbation argument on the resolvant. Provided the poles of the Green function for T_Q with the finite-impedance condition are simple, the modal expansion again holds for points of elevation less than $\pi/6$.

We have now shown that when the refractive index profile is a suitable holomorphic function, the modal expansion is valid for points of elevation less than $\pi/6$, and that in that case the only asymptote of eigenvalues is along the ray of direction $\pi/3$. Since we can approach any profile with the required type of holomorphic function, we find that the solution of the wave equation can be approximated to arbitrarily small error by a modal expansion.

Now it is well-known that for piecewise linear refractive index profiles, there is also an asymptote of eigenvalues along the ray of direction π [119]. There is no contradiction: as we approach the piecewise linear profile more and more closely with holomorphic profiles, the number of eigenvalues along direction π increases to infinity. In practice since these eigenvalues correspond to more and more heavily attenuated modes, they do not contribute significantly to the modal expansion.

Appendix D

Energy conservation

D.1 Two-dimensional problems

We first consider the following two-dimensional problem: find functions u which satisfy the narrow-angle PE

$$\frac{\partial^2 u}{\partial z^2} + 2ik\frac{\partial u}{\partial x} + k^2(n^2 - 1)u = 0 \tag{D.1}$$

in a domain bounded below by the curve $z = \zeta(x)$, as shown in Figure D.1, together with the Neumann boundary condition

$$u\left(x, \zeta(x)\right) = 0 \tag{D.2}$$

We make the further assumption that u is of finite energy, in the sense that

$$I(x) = \int_{-\infty}^{+\infty} |u(x,z)|^2 dz < +\infty \quad \text{for} \quad \text{all} \quad x \geq 0 \tag{D.3}$$

where we extend u downwards by setting $u(x, z) = 0$ below the boundary.

We are going to prove that the energy function $I(x)$ does not depend on x. At this point note that the refractive index $n(x, z)$ is allowed to vary with height and range, but must be real. This is a natural constraint since absorption in the propagation medium is incompatible with energy conservation.

To prove the energy conservation property, we compute the derivative of the energy function. Since we can also write

$$I(x) = \int_{-\zeta(x)}^{+\infty} |u(x,z)|^2 dz < +\infty \tag{D.4}$$

we can express the derivative as

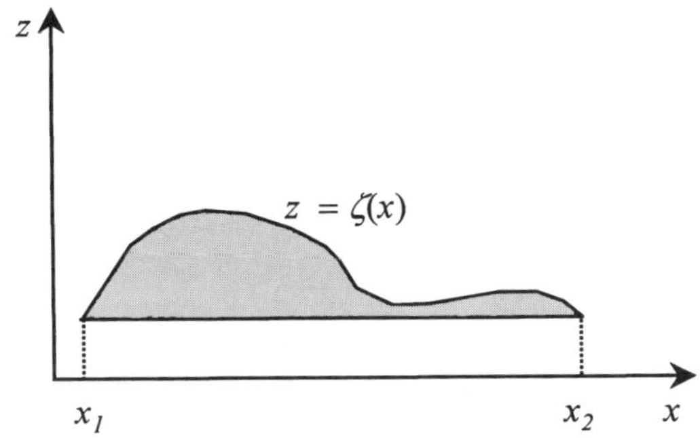

Figure D.1 *Scattering object delimited by one curve*

$$I'(x) = \int_{\zeta(x)}^{+\infty} \left(u\frac{\partial \bar{u}}{\partial x} + \bar{u}\frac{\partial u}{\partial x} \right) dz - \lim_{h \to 0} \frac{1}{h} \int_{\zeta(x)}^{\zeta(x+h)} |u(x,z)|^2 dz \tag{D.5}$$

If $\zeta(x)$ is continuous, we have

$$\lim_{h \to 0} \frac{1}{h} \int_{\zeta(x)}^{\zeta(x+h)} |u(x,z)|^2 dz = \zeta(x)|u(x,\zeta(x))|^2 \tag{D.6}$$

Since u vanishes on the boundary, the limit is zero. We now use the fact that u satisfies the SPE to substitute for the range derivatives in the integrand, getting

$$I'(x) = \frac{i}{2k} \int_{\zeta(x)}^{+\infty} \left(u\frac{\partial^2 \bar{u}}{\partial z^2} - \bar{u}\frac{\partial^2 u}{\partial z^2} \right) dz \tag{D.7}$$

The crucial point here is that the pure imaginary coefficient $2ik$ of the range derivative in the SPE combined with conjugation results in a change of sign in the integrand, which is now expressed as a difference of two terms. It is also important to note that the refractive index terms cancel out thanks to the fact that the refractive index remains real. To complete the proof, we integrate by parts, which yields

$$I'(x) = -\frac{i}{2k} \left(u\frac{\partial \bar{u}}{\partial z} - \bar{u}\frac{\partial u}{\partial z} \right) (x, \zeta(x)) \tag{D.8}$$

since the second integral obtained in the integration by parts procedure is

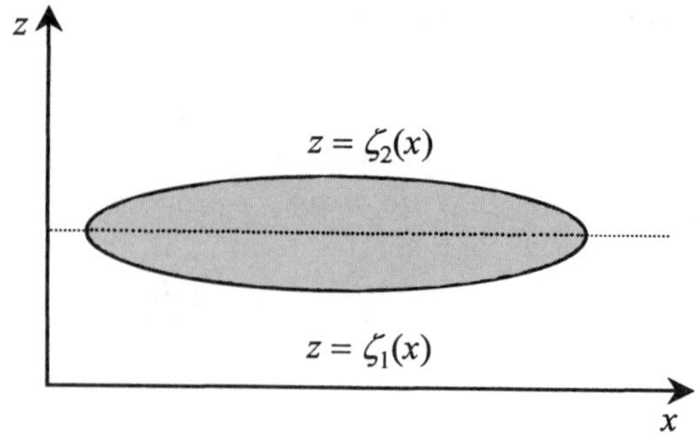

Figure D.2 Scattering object delimited by two curves

identically zero. We now use again the fact that u vanishes on the boundary to conclude that $I'(x) = 0$ for all $x \geq 0$, and hence $I(x)$ is constant.

Note that the proof only requires a continuous boundary curve. This allows us to consider for example piecewise linear boundaries. We can also allow the boundary to contain vertical segments, since their contribution to the integral appearing in the limit in eqn. D.5 is identically zero.

Finally the same reasoning can be applied to propagation outside a bounded scattering object delimited by two curves $z = \zeta_1(x), z = \zeta_2(x)$, as shown in Figure D.2, or more generally in a region Ω outside multiple scattering objects, still with the Neumann boundary condition

$$u(x,z) = 0 \quad \text{for} \quad \text{all} \quad (x,z) \quad \text{in} \quad \partial\Omega \tag{D.9}$$

where $\partial\Omega$ is the boundary of the region. The proof that energy is independent of range is a straightforward generalization of the previous derivation.

For a flat boundary located say at $z = 0$, the energy conservation result also holds for solutions satisfying the Dirichlet boundary condition

$$\frac{\partial u}{\partial z}(x,0) = 0 \tag{D.10}$$

Since the boundary is flat, the integral in the limit term in eqn. D.5 is zero, so eqn. D.8 still holds. The Dirichlet boundary condition ensures that the derivative of the energy function is zero, hence the result.

D.2 Three-dimensional problems

Generalization of the energy conservation principle to three-dimensional problems requires a slightly more technical derivation since we now have to deal with double integrals. We outline the proof for propagation outside a bounded surface Σ. Let u satisfy the SPE in the domain Ω outside Σ, together with the Neumann boundary condition on Σ. We assume that u has finite energy, where the energy function is defined by

$$I(x) = \int_{-\infty}^{+\infty}\int_{-\infty}^{+\infty} |u(x,z)|^2 dydz \tag{D.11}$$

If we denote by $\Omega(x)$ the intersection of the domain Ω with the transverse plane at range x, we can also express the energy function as

$$I(x) = \iint_{\Omega(x)} |u(x,z)|^2 dz \tag{D.12}$$

The derivative of the energy function is

$$I'(x) = \iint_{\Omega(x)} \left(f\frac{\partial \bar{f}}{\partial x} - \bar{f}\frac{\partial f}{\partial x}\right) dydz + \lim_{h\to 0}\frac{1}{h}\iint_{\partial\Omega(x)} |u(x,z)|^2 dydz \tag{D.13}$$

where $\partial\Omega(x,h)$ is the increase in the transverse section of the scatterer between ranges x and $x+h$. Note that this expression is valid at any range, including the minimum and maximum ranges where the transverse plane intersection with the scatterer is not empty. Since u vanishes on the boundary of the scatterer, the limit on the RHS is zero. We now substitute for the range derivatives using the SPE, which gives

$$I'(x) = \frac{i}{2k}\iint_{\Omega(x)} \left(f\left(\frac{\partial^2 \bar{f}}{\partial y^2} + \frac{\partial^2 \bar{f}}{\partial z^2}\right) - \bar{f}\left(\frac{\partial^2 f}{\partial y^2} + \frac{\partial^2 f}{\partial z^2}\right)\right) dydz \tag{D.14}$$

Applying Green's reciprocal theorem, we obtain

$$I'(x) = -\frac{i}{2k}\int_{\partial\Omega(x)} \left(f\frac{\partial \bar{f}}{\partial \mathbf{n}} - \bar{f}\frac{\partial f}{\partial \mathbf{n}}\right) ds \tag{D.15}$$

where $\mathbf{n}$ is the outer normal along the boundary. Again we use the fact that f vanishes on the boundary of the scatterer to conclude that $I'(x) = 0$ and hence energy is conserved. Generalization to the case of multiple scatterers is straightforward.

This energy conservation result is used in Chapter 14 to show that the vector PE solution automatically satisfies the divergence-free condition everywhere provided it satisfies it on the boundary of the scattering object.

D.3 Unicity of PE solutions

The energy conservation property is the basis for showing unicity of the PE solution in a variety of situations. For example take the case of two-dimensional propagation over perfectly conducting irregular terrain, for horizontal polarization. Assume we have two solutions u_1 and u_2 with the same initial field. Then $u = u_1 - u_2$ is also a solution. Since it vanishes at the ground interface, it satisfies the energy conservation property. Now the energy of u at zero range is zero, so it is zero at every range, which implies that u is identically zero and so $u_1 = u_2$.

Bibliography

[1] W.S. Ament, "Towards a theory of reflection by a rough surface", *Proc. IRE*, **vol. 41**, pp. 142–146, 1953.

[2] K.D. Anderson, "Inference of refractivity profiles by satellite-to-ground RF measurements", *Radio Sci.*, **vol. 17**, pp. 653–663, 1982.

[3] A. Arnold and M. Ehrhardt, "Discrete transparent boundary conditions for wide angle parabolic equations in underwater acoustics", *J. Comp. Phys.*, **vol. 145**, pp. 611–638, 1998.

[4] A. Bamberger, B. Engquist, L. Halpern and P. Joly, "Parabolic wave equation approximations in heterogeneous media", *SIAM J. Appl. Math.*, **vol. 48**, pp. 99–128, 1988.

[5] A. Bamberger, B. Engquist, L. Halpern and P. Joly, "Higher order paraxial wave equation approximations in heterogeneous media", *SIAM J. Appl. Math.*, **vol. 48**, pp. 129–154, 1988.

[6] R. Barrett, M. Berry, T.F. Chan, J. Demmel, J.M. Donato, J. Dongarra, V. Eijkhout, R. Pozo, C. Romine and H. Van der Vorst, *Templates for the Solution of Linear Systems: Building Blocks for Iterative Methods*, SIAM, Philadelphia, 1994.

[7] D.E. Barrick, "Theory of HF and VHF propagation across the rough sea, 1, The effective surface impedance for a slightly rough highly conducting medium at grazing incidence", *Radio Sci.*, **vol. 6**, pp. 517–526, 1971.

[8] D.E. Barrick, "Theory of HF and VHF propagation across the rough sea, 2, Application to HF and VHF propagation above the sea", *Radio Sci.*, **vol. 6**, pp. 527–533, 1971.

[9] D.E. Barrick, "Grazing behavior of scatter and propagation over any rough surface", *IEEE Trans. Antennas Propagat.*, **vol. 46**, pp. 73–83, 1998.

[10] A.E. Barrios, "Parabolic equation modelling in horizontally inhomogeneous environments", *IEEE Trans. Antennas Propagat.*, **vol. 40**, pp. 791–797, 1992.

[11] A.E. Barrios, "A terrain parabolic equation model for propagation in the troposphere", *IEEE Trans. Antennas Propagat.*, **vol. 42**, pp. 90–98, 1994.

[12] V.A. Baskakov and A.V. Popov, Implementation of transparent boundaries for numerical solutions of the Schrödinger equation, *Wave Motion*, **vol. 14**, pp. 123–128, 1991.

[13] F.G. Bass and I.M. Fuks, *Wave Scattering from Statistically Rough Surfaces*, Pergamon Press, Oxford, 1979.

[14] G.B. Baumgartner Jr., "XWVG: a waveguide program for trilinear tropospheric ducts", Technical Document No. 610, NCCOSC RDT&E Div. (formerly Naval Ocean Systems Center), 1983.

[15] G.B. Baumgartner Jr., H.V. Hitney and R.A. Pappert, "Duct propagation modelling for the integrated-refractive-effects prediction system (IREPS)", *IEE Proc.*, **vol. 130**, part F, pp. 630–642, 1983.

[16] B.R. Bean and E.J. Dutton, *Radio Meteorology*, Dover Publications Inc., New York, 1966.

[17] C.I. Beard, "Coherent and incoherent scattering of microwaves from the ocean", *IRE Trans.*, **vol. AP-9**, pp. 470–483, 1961.

[18] J.-P. Bérenger, "A perfectly matched layer for the absorption of electromagnetic waves", *J. Comp. Phys.*, **vol. 114**, pp. 185–200, 1994.

[19] L.A. Berry and M.E. Christman, "A FORTRAN program for calculation of ground wave propagation over homogeneous spherical earth for dipole antennas", Nat. Bureau Standards Rep., NBS 9178, 1978.

[20] A. Bespalov, "Application of fictitious domain method to the solution of Helmholtz equation in unbounded domain", INRIA Report No. 1797, 1992.

[21] M. Born and E. Wolf, *Principles of Optics*, 6th ed., Pergamon Press, 1986.

[22] V.A. Borovikov and B.Ye Kimber, *Geometrical Theory of Diffraction*, IEE Electromagnetic Wave Series 37, The Institution of Electrical Engineers, London, 1994.

[23] P.-P. Borsboom and A. Zebic-Le Hyaric, "RCS predictions using wide-angle PE codes, *IEE Conf. Pub.*, **no. 436**, pp. 2.191–2.194, 1997.

[24] JJ. Bowman, T.B.A. Senior and P.L.E. Uslenghi, *Electromagnetic Scattering by Simple Shapes*, revised ed., Hemisphere Publishing Corporation, New York, 1987.

[25] L.M. Brekhovskikh, *Waves in Layered Media*, 2nd ed., Academic Press, New York, 1980.

[26] H. Bremmer and S.W. Lee, "Propagation of a geometrical optics field in an isotropic inhomogeneous medium", *Radio Science*, **vol. 19**, pp. 243–257, 1984.

[27] K.G. Budden, *The Wave-Guide Mode Theory of Wave Propagation*, Logos Press, London, 1961.

[28] J.F. Claerbout , *Fundamentals of Geophysical Data Processing with Application to Petroleum Prospect*, McGraw-Hill, New York, 1976.

[29] R.H. Clarke and J. Brown, *Diffraction Theory and Antennas*, Ellis Horwood, 1980.

[30] E.A. Coddington and N. Levinson, *Theory of Ordinary Differential Equations*, McGraw-Hill, New York, 1964.

[31] F. Collino, "Perfectly matched absorbing layers for the paraxial equations", *J. Comp. Phys.*, **vol. 131**, pp. 164–180, 1997.

[32] F. Collino and P. Joly, "Splitting of operators, alternate directions, and paraxial approximations for the three-dimensional wave equation", *SIAM J. Sci. Comput.*, **vol. 16**, pp. 1019–1048, 1995.

[33] M.D. Collins, "A two-way parabolic equation for acoustic backscattering in the ocean", *J. Acoust. Soc. Am.*, **vol. 91**, pp. 1357–1358, 1992.

[34] M.D. Collins, "A split-step Padé solution for the parabolic equation method", *J. Acoust. Soc. Am.*, **vol. 94**, pp. 1736–1742, 1993.

[35] M.D. Collins, "Generalization of the split-step Padé solution", *J. Acoust. Soc. Am.*, **vol. 96**, pp. 382–385, 1994.

[36] M.D. Collins, W.A. Kuperman and W.L. Siegmann, "A parabolic equation for poro-elastic media", *J. Acoust. Soc. Am.*, **vol. 98**, pp. 1645-1656, 1995.

[37] D. Colton and R. Kress, *Integral Equation Methods in Scattering Theory*, John Wiley and Sons, New York, 1983.

[38] J. Cook and S. Burk, "Potential refractivity as a similarity variable", *Boundary-Layer Meteorology*, **vol. 58**, pp. 151–159, 1992.

[39] J.W. Cooley, P.A.W. Lewis and P.D. Welsh, "The fast Fourier transform algorithm: Programming considerations in the calculations of sine, cosine and Laplace transforms", *J. Sound. Vib.*, **vol. 12**, pp. 315–337, 1970.

[40] K.H. Craig, "Propagation modelling in the troposphere: parabolic equation method", *Electron. Lett.*, **vol. 24**, pp. 1136–1139, 1988.

[41] K.H. Craig and M.F. Levy, "Parabolic equation modelling of the effects of multipath and ducting on radar systems", *IEE Proc.*, part F, **vol. 138**, pp. 153–162, 1991.

[42] R.A. Dalrymple and P.A. Martin, "Perfect boundary conditions for parabolic water-wave models", *Proc. R. Soc. London A*, **vol. 437**, pp. 41–54, 1992.

[43] J. Deygout, "Multiple diffraction of microwaves", *IEEE Trans. Antennas Propagat.*, **vol. 14**, pp. 480–489, 1966.

[44] J. Deygout, "Correction factor for multiple knife-edge diffraction", *IEEE Trans. Antennas Propagat.*, **vol. 39**, pp. 1256–1258, 1991.

[45] G.D. Dockery, "Modeling electromagnetic wave propagation in the troposphere using the parabolic equation", *IEEE Trans. Antennas Propagat.*, **vol. 36**, pp. 1464–1470, 1988.

[46] G.D. Dockery and J.R. Kuttler, "An improved impedance boundary algorithm for Fourier split-step solutions of the parabolic wave equation", *IEEE Trans. Antennas Propagat.*, **vol. 44**, pp. 1592–1599, 1996.

[47] A. Erdelyi, *Tables of Integral Transforms*, vol. 1, McGraw Hill, 1954.

[48] M.D. Feit and J.A. Fleck, Jr., "Light propagation in graded-index fibers", *Appl. Opt.*, **vol. 17**, pp. 3990–3998, 1978.

[49] V.A. Fock, *Electromagnetic Diffraction and Propagation Problems*, Pergamon Press, 1965.

[50] M. Fournier, "Analysis of propagation in an inhomogeneous atmosphere in the horizontal and the vertical direction using the parabolic equation method", *AGARD CP*, **no. 453**, pp. 21.1–21.12, 1989.

[51] D. Givoli, "Non-reflecting boundary conditions", *J. Comp. Phys.*, **vol. 94**, pp. 1–29, 1991.

[52] J.L. Glaser, "Bistatic RCS of complex objects near forward scatter", *IEEE Trans. Aerospace and Electronic Systems*, **vol. 21**, pp. 70–78, 1985.

[53] I.S. Gradshteyn and I.M. Ryzhik, *Table of Integrals, Series and Products*, New York, Academic Press, 1980.

[54] K. Hacking, "U.H.F. propagation over rounded hills", *Proc. IEE*, **vol. 117**, pp. 499–511, 1970.

[55] M.P.M. Hall, L.W. Barclay and M.T. Hewitt, Eds., *Propagation of Radiowaves*, The Institution of Electrical Engineers, London, 1996.

[56] R.H. Hardin and F.D. Tappert, "Application of the split-step Fourier method to the numerical solution of nonlinear and variable coefficient wave equations", *SIAM Rev.*, **vol. 15**, p. 423, 1973.

[57] P. Henrici, *Applied and Computational Complex Analysis*, Wiley-Interscience, 1977.

[58] H.V. Hitney, "Hybrid ray optics and parabolic equation methods for radar propagation modeling", Proceedings of Radar 92, *IEE Conf. Pub.*, **no. 365**, pp. 58–61, 1992.

[59] H.V. Hitney, "Modelling tropospheric ducting effects on satellite to ground paths", *AGARD Conf. Proc.*, **no. 543**, pp. 16.1–16.5, 1994.

[60] J.T. Hviid, J. Bach Andersen, J. Toftgård and J. Bøjer, "Terrain-based propagation model for rural area – an integral equation approach", *IEEE Trans. Antennas Propagat.*, **vol. 43**, pp. 41–46, 1995.

[61] A. Ishimaru, *Wave Propagation and Scattering in Random Media*, vol. 2, Academic Press, San Diego, 1978.

[62] M. Israeli and S.A. Orszag, "Approximation of radiation boundary conditions", *J. Comp. Phys.*, **vol. 41**, pp. 115–131, 1981.

[63] International Telecommunication Union, "Propagation in non-ionized media", Reports of the CCIR, vol. 5, Geneva, 1990.

[64] G.L. James, *Geometrical Theory of Diffraction for Electromagnetic Waves*, IEE Electromagnetic Waves Series, Peter Peregrinus, Stevenage, 1980.

[65] R. Janaswamy, "A curvilinear coordinate-based split-step parabolic equation method for propagation predictions over terrain", *IEEE Trans. Antennas Propagat.*, **vol.46**, pp. 1089–1097, 1998.

[66] F.B. Jensen, W.A. Kuperman, M.B. Porter and H. Schmidt, *Computational Ocean Acoustics*, AIP series in Modern Acoustics and Signal Processing, 1994.

[67] D.S. Jones, *Methods in Electromagnetic Wave Propagation*, vol. 2, Oxford University Press, Oxford, 1986.

[68] D. Kannan, *An Introduction to Stochastic Processes*, North Holland, New York, 1979.

[69] J.B. Keller, "Geometrical theory of diffraction", *J. Opt. Soc. Am.*, **vol. 52**, pp. 116–130, 1952.

[70] J.B. Keller, "Diffraction by an aperture", *J. Appl. Phys.*, **vol. 28**, pp. 426–444 and 570–579, 1957.

[71] D.E. Kerr, ed., *Propagation of Short Radio Waves*, IEE Electromagnetic Wave Series, Peter Peregrinus, London, 1987.

[72] M. Kline, "A note on the expansion coefficient of geometrical optics", *Comms. Pure and Applied Math.*, vol. XIV, pp. 473–479, 1961.

[73] H.W. Ko, J.W. Sari and J.P. Skura, "Anomalous wave propagation through atmospheric ducts", *Johns Hopkins APL Tech. Dig.*, **vol. 4**, pp. 12–26, 1983.

[74] Y. V. Kopylov, A. V. Popov and A. V. Vinogradov, "Application of the parabolic wave equation to X-ray diffraction optics", *Optics Comms.*, **vol. 118**, pp. 619–636, 1995.

[75] G.A. Korn and T.M. Korn, *Mathematical Handbook for Scientists and Engineers*, McGraw-Hill Book Company, New York, 1968.

[76] J.R. Kuttler, "Differences between the narrow-angle and wide-angle propagators in the split-step Fourier solution of the parabolic wave equation", *IEEE Trans. Antennas Propagat.*, **vol. 47**, pp. 1131-1140, 1999.

[77] J.R. Kuttler and G.D. Dockery, "Theoretical description of the PE / Fourier split-step method of representing electromagnetic propagation in the troposphere", *Radio Sci.*, **vol. 26**, pp. 381–393, 1991.

[78] J. Larsen and H. Dancy, "Open boundaries in short-wave simulations: a new approach", *Coastal Engng.*, **vol. 7**, pp. 285–287, 1983.

[79] D. Lee and S.T. McDaniel, "Ocean acoustics propagation by finite difference methods", *Comput. Math. Applic.*, **vol. 14**, pp. 305–423, 1987.

[80] D. Lee, Y. Saad and M.H. Schultz, "An efficient method for solving the three-dimensional wide angle wave equation", in *Computational Acoustics, Wave Propagation*, edited by D. Lee, R.L. Sternberg and M.H. Schultz, pp. 75–88, Elsevier Science, New York, 1988.

[81] S.W. Lee, "Path integrals for solving some electromagnetic edge diffraction problems", *J. Math. Phys.*, **vol. 19**, pp. 1414–1422, 1978.

[82] M.A. Leontovich, *Investigations on Radiowave Propagation, Part II*, Printing House of the Academy of Sciences, Moscow, pp. 5–12, 1948.

[83] M.A. Leontovich and V.A. Fock, "Solution of propagation of electromagnetic waves along the Earth's surface by the method of parabolic equations", *J. Phys. USSR*, **vol. 10**, pp. 13–23, 1946.

[84] M.F. Levy. "Instability of the roots of the modal equation for the tropospheric waveguide", *Radio Sci.*, **vol. 22**, pp. 61–68, 1987.

[85] M.F. Levy, "Parabolic equation modelling of propagation over irregular terrain", *Electron. Lett.*, **vol. 26**, pp. 1153–1155, 1990.

[86] M.F. Levy, "Diffraction studies in urban environment with wide-angle parabolic equation method", *Electron. Lett.*, **vol. 28**, pp. 1491–1492, 1992.

[87] M.F. Levy, "Horizontal PE modelling of tropospheric effects on Earth-space paths", in *Proceedings of the 1994 Progress in Electromagnetics Research Symposium*, pp. 1489–1493, Eds. B. Arbesser *et al.*, Kluwer, Dordrecht, 1994.

[88] M.F. Levy, "Fast PE methods for mixed environments", AGARD SPP Symposium on propagation assessment in coastal environments, *AGARD Conf. Pub.*, **no. 567**, pp. 8.1–8.6, 1995.

[89] M.F. Levy, "Non-local boundary conditions for radiowave propagation", *Proceedings of Third International Conference on Mathematical and Numerical Aspects of Wave Propagation*, Ed. G. Cohen, SIAM, pp. 499–505, 1995.

[90] M.F. Levy, "Horizontal parabolic equation solution of radiowave propagation problems on large domains", *IEEE Trans. Antennas Propagat.*, **vol. 43**, pp. 137–144, 1995.

[91] M.F. Levy, "Transparent boundary conditions for parabolic equation solutions of radiowave propagation problems", *IEEE Trans. Antennas Propagat.*, **vol. 45**, pp. 66–72, 1997.

[92] M.F. Levy and P.-P Borsboom, "Radar cross-section computations using the parabolic equation method", *Electron. Lett.*, **vol. 32**, pp. 1234–1235, 1996.

[93] M.F. Levy and K.H. Craig, "TERPEM propagation package for operational forecasting with EEMS", Proceedings of the 1996 Battlespace Atmospherics Conference, Technical Document 2938, NCCOSC, RDT&E Division, San Diego, pp. 497–505, 1996.

[94] M.F. Levy and A.A. Zaporozhets, "Target scattering calculations with the parabolic equation method", *J. Acoust. Soc. Am.*, **vol. 103**, pp. 735–741, 1998.

[95] M.F. Levy, P.-P. Borsboom, A.A. Zaporozhets and A. Zebic-Le Hyaric, "RCS calculations with the parabolic wave equation", AGARD SPP symposium on radar signature analysis and imaging of military targets, *AGARD Conf. Proc.*, **no. 583**, pp. 5.1–5.9, 1996.

[96] M.F. Levy, K.H. Craig, R.J.B. Champion, J.D. Eastment and N.W.M. Whitehead, "Airborne measurements of anomalous propagation over the English Channel", *IEE Conf. Pub.*, **vol. 333**, Part 1, pp. 173–176, 1991.

[97] H. J. Liebe, "MPM–An atmospheric millimeter-wave propagation model", *Int. J. of Infrared and Millimeter Waves*, **vol. 10**, pp. 631–650, 1989.

[98] R.J. Luebbers, "Propagation prediction for hilly terrain using GTD wedge diffraction", *IEEE Trans. Antennas Propagat.*, **vol. 32**, pp. 951–955, 1984.

[99] Y.L. Luke, *Integrals of Bessel Functions*, McGraw-Hill Book Company, New York, 1962.

[100] T.S.M. Maclean and Z. Wu, *Radiowave Propagation over Ground*, Chapman & Hall, London, 1993.

[101] R.J. McArthur, "Propagation modelling over irregular terrain using the split-step parabolic equation method", Proceedings of Radar 92, *IEE Conf. Pub.*, **no. 365**, pp. 54–57, Oct. 1992.

[102] R.J. McArthur and D.H.O. Bebbington, "Diffraction over simple terrain obstacles by the method of parabolic equations", Proceedings of ICAP 1991, *IEE Conf. Pub.*, **no. 333**, Part 2, pp. 824–827, 1991.

[103] N.W. McLachlan, *Bessel Functions for Engineers*, Oxford University Press, Oxford, 1955.

[104] G.D. Malyuzhinets, "Progress in understanding diffraction phenomena", *Sov. Phys. Usp.*, **vol. 69**, pp. 321–334, 1959.

[105] S.W. Marcus, "A model to calculate em fields in tropospheric environments at frequencies through SHF", *Radio. Sci.*, **vol. 17**, pp. 895-901, 1982.

[106] S.H. Marcus, "A generalized impedance method for application of the parabolic approximation to underwater acoustics", *J. Acoust. Soc. Am.*, **vol. 90**, pp. 391–398, 1991.

[107] S.H. Marcus, "A hybrid (finite difference–surface Green's function) method for computing transmission losses in an inhomogeneous atmosphere over irregular terrain", *IEEE Trans. Antennas Propagat.*, **vol. 40**, pp. 1451–1458, 1992.

[108] P.A. Martin and R.A. Dalrymple, "On the propagation of water waves along a porous-walled channel", *Proc. R. Soc. London*, Part A, **vol. 444**, pp. 411–428, 1994.

[109] J. Maser and G. Schmahl, "Coupled wave description of the diffraction by zone plates with high aspect ratios", *Optics Comms.*, **vol. 89**, pp. 355–362, 1992.

[110] A.G. Michette, *Optical Systems for Soft X-Rays*, Plenum Press, New York, 1986.

[111] A.R. Miller, R.M. Brown and E. Vegh, "New derivation for the rough-surface reflection coefficient and for the distribution of sea-wave elevations", *IEE Proc.*, **vol. 131**, part H, pp. 114–116, 1984.

[112] G. Millington and G.A. Isted, "Ground-wave propagation over an inhomogeneous smooth earth: part 2, experimental evidence and practical implementation", *Proc. IEE*, **vol. 97**, Pt. III, pp. 209–221, 1950.

[113] G. Millington, R. Hewitt and F.S. Immirzi, "Double knife-edge diffraction in field-strength predictions", *Proc. IEE*, **vol. 109C**, pp. 419–429, 1962.

[114] G. Millington, R. Hewitt and F.S. Immirzi, "The Fresnel surface integral", *Proc. IEE*, **vol. 109C**, pp. 430–437, 1962.

[115] H.M. Nussenzveig, *Diffraction Effects in Semiclassical Scattering*, Cambridge University Press, 1992.

[116] J.A. Ogilvy, *Theory of Wave Scattering from Random Rough Surfaces*, Adam Hilger, Bristol, 1991.

[117] F.W.J. Olver, *Introduction to Asymptotics and Special Functions*, Academic Press, 1974.

[118] R.H. Ott, "An alternative integral equation for propagation over irregular terrain, 2", *Radio Sci.*, **vol. 6**, pp. 429–435, 1971.

[119] R.H. Ott, "Roots of the model equation for em wave propagation in a tropospheric duct", *J. Math. Phys.*, **vol. 21**, pp. 1256–1266, 1979.

[120] R.H. Ott and L.A. Berry, "An alternative integral equation for propagation over irregular terrain", *Radio Sci.*, **vol. 5**, pp. 767–771, 1970.

[121] J.S. Papadakis, "Exact nonreflecting boundary conditions for parabolic-type approximations in underwater acoustics", *J. Comput. Acoust.*, **vol. 2**, pp. 83–98, 1994.

[122] R.A. Pappert, "Conversion of radiation modes to trapped modes due to lateral inhomogeneity of a simple elevated layer", *Radio Sci.*, **vol. 17**, pp. 305–322, 1982.

[123] R.A. Pappert, "Field strength and path loss in a multilayer tropospheric waveguide environment", Technical Note 1366, Naval Ocean Systems Center, 1984.

[124] R.A. Pappert and C.L. Goodhart, "Case studies of beyond-the-horizon propagation in tropospheric ducting environments", *Radio Sci.*, **vol. 12**, pp. 75–97, 1977.

[125] W.L. Patterson, "Advanced Refractive Effects Prediction System (AREPS), Version 1.0 User's Manual", Technical Document 3028, Space and Naval Warfare Systems Center, San Diego, CA 92152-5001, 1998.

[126] R. Paulus, Ed., "Proceedings of the Electromagnetic Propagation Workshop", Technical Document 2891, Space and Naval Warfare Systems Center, San Diego, CA 92152-5001, 1995.

[127] C.L. Pekeris, "Accuracy of the earth-flattening approximation in the theory of microwave propagation", *Phys. Rev.*, **vol. 70**, pp. 518–522, 1946.

[128] O.M. Phillips, "Spectral and statistical properties of the equilibrium range in wind-generated gravity waves", *J. Flui. Mech.*, **vol. 156**, pp. 505–531, 1985.

[129] A.V. Popov, "Accurate modeling of transparent boundaries in quasioptics", *Radio Sci.*, **vol. 31**, pp. 1781–1790, 1996.

[130] W.H. Press, S.A. Teukolsky, W.T. Vetterling and B.P. Flannery, *Numerical Recipes*, Cambridge University Press, 1994.

[131] A.C. Radder, "On the parabolic equation method for water-wave propagation", *J. Fluid. Mech.*, **vol. 95**, part 1, pp. 159–176, 1979.

[132] C.M. Rappaport, "Perfectly matched absorbing boundary conditions based on anisotropic lossy mapping of space", *IEEE Microwave and Guided Wave Letters*, **vol. 3**, pp. 90–92, 1995.

[133] H.R. Reed and C.M. Russell, *Ultra High Frequency Propagation*, 2nd Ed., Boston Technical Publishers, Inc., Cambridge, USA, 1966.

[134] R.D. Richtmyer, *Principles of Advanced Mathematical Physics*, Volume I, Springer-Verlag, New York, 1978.

[135] R.D. Richtmyer and K.W. Morton, *Difference Methods for Initial Value Problems*, Second Ed., Interscience Publishers, John Wiley & Sons, New York, 1967.

[136] C. L. Rino and H.D. Ngo, "Forward propagation in a half-space with an irregular boundary", *IEEE Trans. Antennas Propagat.*, **vol. 45**, pp. 1340–1347, 1997.

[137] L.T. Rogers, "Effects of the variability of atmospheric refractivity on propagation estimates", *IEEE Trans. Antennas Propagat.*, **vol. 44**, pp. 460–465, 1996.

[138] A. Sammar abd J.-M. André, "Dynamical theory of stratified Fresnel linear zone plates", *J. Opt. Soc. Am.*, **vol. 10**, pp. 2324–2337, 1993.

[139] S.R. Saunders and F.R. Bonar, "Explicit multiple building diffraction attenuation function for mobile radiowave propagation", *Electron. Lett.*, **vol. 27**, pp. 1276–1277, 1991.

[140] R.O. Schmidt, "Multiple emitter location and signal parameter estimation", *IEEE Trans. Antennas Propagat.*, **vol. 34**, pp. 260–280, 1986.

[141] T.B.A. Senior and J.L. Volakis, *Approximate Boundary Conditions in Electromagnetics*, IEE Electromagnetic Wave Series, London, 1995.

[142] A. Seville, U. Yilmaz, P.R.V. Charriere, N. Powell and K.H. Craig, "Building scatter and vegetation attenuation measurements at 38 GHz", Proceedings of ICAP-95 conference, *IEE Conf. Pub.*, **no. 407**, pp. 2.46–2.50, 1995.

[143] P.A. Sharples and M.J. Mehler, "Cascaded cylinder model for predicting terrain diffraction loss at microwave frequencies", *Proc. IEE*, part H, **vol. 136**, pp. 331-337, 1989.

[144] W.T. Shaw and A.J. Dougan, "The failure of specular limit formulae for Kirchhoff integrals associated with Gaussian surfaces with ocean-like spectra", *Waves in Random Media*, **vol. 5**, pp. L1–L8, 1995.

[145] G.D. Smith, *Numerical Solution of Partial Differential Equations: Finite Difference Methods*, 3rd Edition, Clarendon Press, Oxford, 1985.

[146] I.N. Sneddon, *The Use of Integral Transforms*, McGraw Hill, New York, 1972.

[147] M. Spivack, "A numerical approach to rough-surface scattering by the parabolic equation method", *J. Acoust. Soc. Am.*, **vol. 87**, pp. 1999–2004, 1990.

[148] R.B. Stull, *An Introduction to Boundary Layer Meteorology*, Kluwer Academic Press, Dordrecht, 1989.

[149] F. Tappert, "The parabolic equation method", in *Wave Propagation in Underwater Acoustics*, eds. J.B. Keller and J.S. Papadakis, pp. 224–287, Springer-Verlag, New York, 1977.

[150] M.E. Taylor, *Pseudo-Differential Operators*, Princeton University Press, 1981.

[151] V.I. Tatarski, *Wave Propagation in a Turbulent Medium*, Constable and Co., London, 1967.

[152] D.J. Thomson and G.H. Brooke, "Non-local boundary conditions for 1-way wave propagation", in *Mathematical and Numerical Aspects of Wave Propagation*, Ed. J. A. DeSanto, pp. 348–352, SIAM, Philadelphia, 1998.

[153] D.J. Thomson and N.R. Chapman, "A wide-angle split-step algorithm for the parabolic equation", *J. Acoust. Soc. Am.*, **vol. 74**, pp. 1848–1854, 1983.

[154] D.J. Thomson and M.E. Mayfield, "An exact radiation condition for use with the a posteriori PE method", *J. Comp. Acoust.*, **vol. 2**, pp. 113–132, 1994.

[155] E.C. Titchmarsh, *The Theory of Functions*, Oxford University Press, 1939.

[156] B.J. Uscinski, "High-frequency propagation in shallow water. The rough waveguide problem", *J. Acoust. Soc. Am.*, **vol. 98**, pp. 2702–2707, 1995.

[157] B. van der Pol and H. Bremmer, "The propagation of radiowaves over a finitely conducting earth", *Philos. Mag.*, **vol. 27**, pp. 261–275, 1937.

[158] C. Vassalo and F. Collino, "Highly efficient absorbing boundary conditions for the beam-propagation method", *J. Lightwave Technology*, **vol. 14**, pp. 1570–1577, 1996.

[159] L.E. Vogler, "An attenuation function for multiple knife-edge diffraction", *Radio Sci.*, **vol. 17**, pp. 1541-1546, 1982.

[160] A.G. Voronovich, *Wave Scattering from Rough Surfaces*, Springer Series on Wave Phenomena, Springer-Verlag, Berlin, 1994.

[161] J.R. Wait, "Review of mode theory of radio propagation in terrestrial waveguides", *Rev. Geophys.*, **vol. 1**, pp. 481–505, 1963.

[162] J.R. Wait, "Coupled mode analysis for a nonuniform tropospheric waveguide", *Radio Sci.*, **vol. 15**, pp. 667–673, 1980.

[163] G.N. Watson, "The diffraction of radiowaves by the earth", *Proc. Roy. Soc. London, Ser. A*, **vol. 95**, pp. 83–99, 1918.

[164] J.H. Whitteker, "Diffraction over a flat-topped terrain obstacle", *IEE Proc.*, **vol. 137**, Pt H, pp. 113–116, 1990.

[165] J.H. Whitteker, "Fresnel-Kirchhoff theory applied to terrain diffraction problems", *Radio Sci.*, **vol. 25**, pp. 837–851, 1990.

[166] H.T.G. Woo, M.J. Schuh and M.L. Sanders, "Benchmark radar targets for the validation of computational electromagnetics programs", *IEEE Antennas and Propagat. Magazine*, **vol. 35**, pp. 84–89, 1993.

[167] Z. Wu, T.S.M. Maclean, N. Jayasundere, L.J. Carter and A.A. Williamson, "Recovery effect in radiowave propagation", *Electronics Letters*, **vol. 26**, pp. 162–163, 1990.

[168] D. Yevick, "A guide to electric field propagation techniques for guided-wave optics", *Opt. Quantum Electron.*, **vol. 26**, pp. S185–S197, 1994.

[169] D. Yevick, "Stability issues in vector electric field propagation", *IEEE Photonics Technology Letters*, **vol. 7**, pp. 658–660, 1995.

[170] A.A. Zaporozhets, "Modelling of radiowave propagation in urban environment", *IEE Conf. Proc.*, **no. 436**, pp. 2.83–2.89, 1997.

[171] A.A. Zaporozhets, "Application of the vector parabolic equation method to urban radiowave propagation problems", *IEE Proc. H*, **vol. 146**, pp. 253–256, 1999.

[172] A.A. Zaporozhets and M.F. Levy, "Modelling of radiowave propagation in urban environment with parabolic equation method", *Electron. Lett.*, **vol. 32**, pp. 1615–1616, 1996.

[173] A.A. Zaporozhets and M.F. Levy, "Radar cross section calculation with marching methods", *Electron. Lett.*, **vol. 34**, pp. 1971–1972, 1998.

[174] A.A. Zaporozhets and M.F. Levy, "Bistatic RCS calculations with the vector parabolic equation method", *IEEE Trans. Ant. Prop.*, **vol. 47**, pp. 1688–1696, 1999.

[175] A. Zebic-Le Hyaric, personal communication.

[176] R.W. Ziolkowski and G.A. Deschamps, "Asymptotic evaluation of high frequency fields near a caustic: an introduction to Maslov's method", *Radio Sci.*, **vol. 19**, pp. 1001–1025, 1984.

Index